AF517540

Eco-Degradation due to Air Pollution

Editors

Prof. Arun Arya

Prof. S.J. Bedi

Prof. V.S. Patel

Indian Association for Air Pollution Control
C/o Botany Department, Faculty of Science
The Maharaja Sayajirao University of Baroda
Vadodara 390 002

SCIENTIFIC PUBLISHERS (INDIA)

P.O. BOX 91 JODHPUR

Published by:
Tanay Sharma
Scientific Publishers (India)
5-A, New Pali Road, P.O. Box 91
Jodhpur - 342 001
E-mail: info@scientificpub.com
www.scientificpub.com

ISBN: 978-81-7233-598-4

Laser typeset: Rajesh Ojha
Printed in India

ॐ भद्रं कर्णेभिः शृणयाम देवा भद्रं पश्येमाक्षाभिर्यजत्राः।
स्थिरैरङ्गैस्तु एटुवा सस्तनूभिवर्यशेमहि देवहितं यदायुः।।

यजु.: 25-21

O Holy Gods! May we ever hear with our ears auspicious words. May we ever see with our eyes pleasing things. may we atain, simultaneously, good health and prosperous long life.

The task of the
enlightened is to
remove the poverty, the
inequality, the inlignity
and, the suffering which
weighs down countless
millions in this world. — *Dr. S. Radhakrishnan*

ॐ हिरण्यगर्भः समवर्त्तताग्रे भूतस्य जातः पतिरेक आसीत्
स दाधार पृथिवीं द्यामुतेमां, कस्मै देवाय हविषा विधेम।।

— यजु. 13-4

वर्तमान जो प्रलय काल में प्राणिमात्र का स्वामी है,
सूर्य, चन्द्र तारे अन्तर में, सबका अन्तर्यामी है,
द्यौ, पृथ्वी और अन्तरिक्ष का आश्रय वह कहलाता है
कौन नियन्ता? अर्चन किसका? जो सुखरूप विधाता है।

Foreword

It is with great pleasure that I write a Foreword for the proceedings volume of the seminar on Ecodegradation due to Air Pollution at the instance of the Editor Prof. Arun Arya of the M.S. University of Baroda.

The Blue Planet Earth is at cross roads with respect to the present and future scenario of degrading life-supporting environment. Most of the global problems like green-house gases, global warming, ice melts, sea level rise, stratospheric ozone hole, rise in tropospheric U.V.-B and ozone levels etc. are linked with air pollution. From the contents of the book it is evident that the diversity of air pollution aspects are very broad and there is something interesting for every one interested in the future of the Blue-Planet, teachers and students alike. Ecotechnology, ecodegradation, eco-auditing, air pollution controls, green belts, town planning, historic building problems, biodeterioration of cultural heritage, fungal biopollutants in fruit markets, air pollution due to stone crushing and dolomite grinding units, indoor pollution, environment awareness and air pollution terminologies are included. In view of the above this volume must be an essential addition to College and University libraries and the book shelfs of every ecologist.

Prof. R.S. Ambasht
B.H.U. Professor Emeritus
Department of Botany
Banars Hindu University,
Varanasi - 221 005, India

Preface

ओ३म् भूमे मार्तिन धेहि मा भद्रया सुप्रतिष्ठितम्।
संविदाना दिवा कवे, श्रियां मा धेहि भूत्याम्।।

Atharva Veda 12-1-63

O Mother like Earth! may you stabilize me with pious wealth. O omniscient Earth! you having your contacts with the heaven, make me wealthy and prosperous.

German author Max Muller has said, "If I were to like over the whole world to find out the country most richly endowed with all the wealth, power and beauty that nature can bestow-in-some parts of a very paradise on earth - I should point to India." Our earth and its ecosystems are every changing. It is said that next world war if any will be for water. A once abundant resource is increasingly being wasted, misused and polluted. India faces an impeding crisis most visible in many of its cities. Rivers are polluted due to industrial effluents and pesticide run off. It is advised to enforce strict regulation on ground water extraction and plan projects with a holistic understanding of water as an ecological concept rather than as a hydrological or engineering construct.

उपस्थास्ते अनमीवा अयक्ष्माः

Atharva Veda 12-1-62

We pray to the God that on this planet earth we may live healthy and happily without suffering from diseases. This is almost impossible when our soil is getting polluted with heavy metals and pesticides, air is toxic due to pollutants like particulate matter, VOCs, COx, NOx, SOx etc. Some of these primary pollutants undergo further changes to produce secondary pollutants like ozone, PANs, sulphuric and nitric acids. Our river water and ground water is getting more and more polluted. Most of Indian river water is not fit directly for human consumption.

Industrial disasters can create havoc. One of the worst industrial disaster occurred on December 2, 1984, when 60 tonnes of deadly Methyl isocyanate leaked out of the Union Carbide factory in Bhopal killing about

8,000 people. Ms. Sunita Narain points out about an advertisement in late 1990's which says "Roll down the window of your bullet-proof car, Mr. Prime Minister the security threat is not the gun it is the air of Delhi." Same is now true for other cities like Bangalore and Vadodara, each passing day, the capital city of Karnataka adds 1000 new vehicles. In Bangalore only 30 per cent of people travel by bus as against 60 per cent in Delhi. Kolkata is one of the most polluted cities of the world. During winter pollutants cannot disperse easily, mainly due to inversion, low wind speed and high congestion. The average benzene concentration in 1992 and 1993 were 1004 $\mu g/m^3$ and 7082 $\mu g/m^3$. More than 50% of the population of core Kolkata city live in slums. The pollutants like polynuclear aromatic hydrocarbon [PAH] and benzene soluble organic matter [BSOM] and CO are on higher side, poorly maintain cars, poor quality of fuel and bad condition of city streets are responsible for environmental pollution. During 1985, WHO in its monitoring programme on air claimed that SPM value of Kolkata is the highest in the world. To make the Beijing more cleaner for last Olympics, administration had removed the entire two stroke and wheelers. Can we do this in India to restore the pollution free air for the major cities. US institute selected top 10 severely polluted locations out of 400 across the globe. Linfen, La Oroya, Dzerzhinsk, Norilsk, Chernobyl and Kabwe were also part of the list last year as well. Vapi in Gujarat is new entrant. Planning should respect ecology of the area where urban centers are to be developed. Land use should be determined on the basis of social and particularly environmental considerations and not on the basis of value alone.

Global warming is a natural geological process that could begin to reverse itself. Many people blame that global warming is due to carbon dioxide relesed in the atmosphere from burning the fossil fuels. However, the greater amount of carbon dioxide that enters – and leaves – the atmosphere as the part of natural cycle of water exchange from, and back into the sea water and vegetation. Global warming is the part of the cycle of warming and cooling that happened every 1500 to 5000 years for the last million years.

Global warming and related climate changes are likely to significantly increase the weather related risks facing human settlements, including floods, water and power supply failures and associated economic collapse. Global emissions of green house gases are likely to raise the world temperature and the sea level at the rate of 0.3°C and 6 mm $10y^{-1}$ by the year 2100. This will result into increased water demand for crops and reduced crop yield. The Netherlands' Fortis Bank has won a bid to emit CO_2 in an auction that marks a new phase in the global carbon market. The bank paid more than 13 million Euros for the rights to emit 800,000 tonnes of CO_2. The credits were made available through Kyoto Protocol's clean development mechanism. It is interesting to note that the city of Sao Paulo had developed a 30 million tonnes mountain of rubbish. With this back drop, A National seminar was organized by **Indian Association for Air pollution Control (IAAPC)** on 16th October, 2004 in Vadodara, to discuss issues related to Eco-Degradation due to Air Pollution. Seventeen speakers presented their findings on varied subjects like Green belt technology, pollution management by eco-technology, eco-degradation around Dolomite grinding units, air pollution through use of appropriate technology, effect of pollution on kids lungs, metals as air pollutants, alternate fuels etc. the papers presented during seminar and some invited articles are compiled in this volume. We hope that articles will be accepted with great enthusiasm and will be read by researchers and common man with interest. We wish the methods suggested by speakers will be helpful in formulating national strategy to combat air pollution. Veda tells us, "**Om Shiva no vatah eva vantu Bhumo**." Can we make our planet free from pollution so that pleasant breeze could blow on this planet Earth.

This is our proud moment for publishing third book from IAAPC. We are thankful to all the contributors for submitting their contributions for the book. We are thankful to Profs. S.D. Sabnis, M. Daniel, H.V. Bhavnani, Y.T. Jasrai, Dr. S.S. Bala and Shri B.R. Naidu, Zonal officer CPCB (West Zone) for their suggestions and kind help in organizing the National Seminar. The encouragement given by late Professors C.H. Pathak and G.P. Senan are highly appreciated. Our sincere thanks are due to Shri Pawan Kumar Sharma of Scientific Publishers (India), Jodhpur for nice printing of the book.

Arun Arya
S.J. Bedi
V.S. Patel

Vadodara, Gujarat
3-4-2009

Contents

CHAPTER - 1

Green Belt Technology- Eco-technological Strategy for Pollution Abatement

P.S. Dubey

Chairman, Madhya Pradesh Pollution Control Board, Bhopal (M.P.), India

ABSTRACT

Change and development are essential to human progress, and exploiting nature in this process is not a new phenomenon. Increasing atmospheric pollution has resulted in poor air quality in different cities. In metropolitan cities, exhausts from automobiles cause pollution in high traffic density areas. The gases released from industries pollute the atmosphere. In India, over 20 million small scale industries are existing and about 2000 industrial estates are working. Though a single small scale industry may not look a severely polluting unit a large number of such industries can contribute a lot. Clean air is the most natural resource needed for the sustenance of all living activities.

Role of plants as pollutant sink is well recognized. Plants help to remove pollutants from the air in three ways: absorption by the leaves or the soil surface, deposition of particulate matters and aerosols on the leaf surface, and fall out of particulates on the downside of the vegetation because of the lowering of air velocity. Trees and green spaces help to keep cities cool, act as natural filters, noise absorbers, improve microclimate and protect and improves air quality of natural resources, including soil, water, vegetation and wildlife. They play a very important social role in easing tensions and improving psychological health. The paper describes application of Green belt in reducing the air pollution. Identification of pollution sensitive plant species is done and recommendation of different plants is made. Cultivation and conservation of certain resistant varieties of trees in urban areas and particularly around industries is essential. It improves the ambient air quality and reduces the air pollution.

Keywords: Greenbelt technology, Pollution abatement, Scavenging potential of plants, Designing of a Green Belt.

Environment is our article of faith in nature. Environmentalism also emerged as critically important function of people in general. The environment is not an issue to be discussed or disposed off but to be worshipped. While ecological considerations are shaping our design of development, the Ecocrisis occasionally got transformed to Eco-fever marring the basic objectives of "Gaya". The dilemma of development vs. environment still exists and perhaps now occupies a larger dimension much because of one more major issue, the pollution of the ecosystem components as air, water and soil.

1.1 AMBIENT AIR QUALITY

In fact, air pollution never appeared as our first priority while it has partially a global dimension as once a pollutant is released in air environment nothing can control it and no remedy can be prescribed. Again, the control technologies are specific and costlier and depend on many factors. But, even under the maximum permissible limits of SPM (500 μgm^3), SOx and NO_2 (120 μgm^3) adding to 740 μgm^3, we find that cumulative damages occur which are not insignificant (Dubey, ICAR Report, 1992). Hence, emerged the need for not only air pollution control but its abatement / mitigation which warrants the design and technique at field level, with an emphasis on low cost especially for developing economies.

And one such strategy is establishment and development of Green Belt which should not mere means a strip of land with some plants so that it looks green, absolutely not, true. In fact, green belt requires scientific knowledge of characters of tree morphology, metabolism and many other parameters; it requires atmospheric meteorological inputs and studies of chemical characteristics of pollutants. Also, the width of the Green Belt and the objectives with which it is desired as well as the edaphic character and its status. Obligatorily, it requires ecological inputs, mathematical design pattern and quantitative induction of possible mitigation impacts with manipulative scope to be modified to the purpose aimed for.

The data and information presented in this address are derived from almost five years study encompassing different locations like Nagda, Kota, Baroda, Nimbahera, Ujjain and Dewas and analyzing almost twenty thousand samples collected in all three major seasons. Since lots of different species exist at these locations, care has been taken to select same tree species and only thirty-two species could be worked out. The data explains and replies to all the relevant but commonly asked questions as:

1. What are the best parameters to assess the tolerance of a species?
2. How much is the total scavenging potential of species?

3. What is the comparative efficiency of total scavenging potential of species?
4. Whether the specificity of the specific pollutants vs species exists?
5. Can we exemplify the mitigation of air pollutant load through the existing Green Belts?

This address is further divided in two parts, one; plant parameters for tolerance and two; scavenging potential of species to identify the right species. The tables are arranged in order of higher to lower scavenging potential and tolerance.

1.2 IDENTIFICATION OF PARAMETERS FOR TOLERANCE

It may be relevant to know after all how we decide that one species is tolerant and other is susceptible to air pollutants. The normal criteria as proposed by BHU School or Bombay School or Kota School are hypothetical. But this study is different as the sample size is large, collected from fields, analyzing about ten parameters of about nine species and drawing statistical correlation with the level of significance with the number of parameters with which high level of significance occur for number of species, it becomes very clear that SOD, POD, leaf area dry weight ratio and stomatal conductance are the best parameters to assess the tolerance while most discussed parameters like chlorophyll and ascorbic acid fails to prove as best indicator of tolerance of species (Table 1a, b, c).

Table 1a. Number of parameters of certain plants with significant correlation coefficient value at different levels

Plant Species	Significance Level	
	95.0 ($P<0.05$)	99.0 ($P<0.1$)
Ficus benghalensis	3	2
Calotropis procera	3	2
Cassia siamea	3	2
Ipomea fistulosa	5	4
Mangifera indica	6	3
Syzigium cumini	6	3
Melia azadirachta	6	6
Azadirachta indica	10	5(6)
Butea monosperma	9	6(7)
Delbergia sissoo	9	6

The other factors which determine success of a species in selection are accommodative characters; leaf characters and canopy structure along with some other factors further helps in a decision making for plant species selection.

For a total strategy some basic criteria may be identified apart from the plant parameters i.e. pollutants chemistry and meteorological conditions.

Table 1b. Parameters wise distribution of number of plant species at different levels of Significance

Parameters	Level of Significance			
	99.9 (P<0.001)	99.0 (P<0.001)	95.0 (P<0.001)	90.0 (P<0.001)
LDR	6	10	10	10
SOD	5	8(9)	10	10
POD	4(5)	7	9	10
Sc	1(2)	6	7(8)	8
AmA	0	1	6(7)	10
Pt	2	2(3)	5	5
Chl	0	0	5	6
AsA	3	4	4	6(8)
NR	0(1)	1	3(4)	6
Sg	1	1	1	6

Table 1c. Correlation Coefficient (r) Values between Foliar Sulphate Sulphur and Different Parameters of Certain Plants

Parameters	Level of Significance		
	Tolerant *F. benghalensis*	Moderate *M. indica*	Susceptible *D. sissoo*
LDR	0.89	0.81	0.93
SOD	0.83	0.86	0.78
POD	0.79	0.86	0.84
Sc	0.53	0.78	0.81
AmA	0.74	0.75	0.78
Pt	0.55	0.76	0.89
Chl	0.68	0.77	0.77
AsA	0.48	0.73	0.86
NR	0.53	0.74	0.84
Sg	0.44	0.74	0.73

1.3. TOLERANCE Vs TOTAL SCAVENGING POTENTIAL

The other and final part is the scavenging efficiency of a species which depends on accumulation of a pollutant and its product when it enters through stomata. A major drawback of most of the Indian works is that they have neglected the stomatal conductance and diffusive resistance of species, its diurnal cycle and seasonal response. The contents of the sulphates, other metabolites of pollutants and their disappearance decides the scavenging potential. Again the entire Indian scenario of research on air pollution and plants never incorporated surface deposition which contributes to 30-40 and even 50 per cent of total scavenging. Hence, accounting only internal metabolism of pollutant does not provide the extent of scavenging potential. Obviously, total scavenging potential should incorporate surface deposition potential of leaf as well as its metabolic removal, degradation or accumulation in the leaf (Table 2a).

Table 2a. Total Scavenging Potential (TSP) of different Species

Sr. No.	Species	TSP	Sr. No.	Species	TSP
1	*B. variegata*	28.20	17	*A. indica*	15.48
2	*P. acerifolium*	24.39	18	*F. infectoria*	15.29
3	*Bambusa spp.*	24.26	19	*C. sebestina*	15.23
4	*D. sissoo*	22.13	20	*B. spectabilis*	14.60
5	*Eucalyptus spp.*	22.11	21	*M. indica*	13.94
6	*Z. jujube*	20.27	22	*M. elengi*	13.29
7	*C. siamea*	20.00	23	*F. elastica*	13.19
8	*F. benghalensis*	18.61	24	*K. pinnata*	12.67
9	*C. lanceolatum*	18.12	25	*P. dulce*	12.80
10	*S. cumini*	17.82	26	*F. decora*	11.84
11	*A. cadamba*	17.70	27	*A. lebbek*	11.68
12	*A. auriculiformis*	17.30	28	*P. ferrugineum*	11.64
13	*F. nitida*	17.11	29	*C. pentandra*	11.60
14	*G. robusta*	16.67	30	*P. longifolia*	10.94
15	*S. campanulata*	15.95	31	*B. malabaricum*	9.99
16	*F. religiosa*	15.79	32	*C. fistula*	9.26

It is on this basis we have computed the total scavenging potential with all the pollutants deposited and internally accumulated. The total of all provide the TSP (Total Scavenging Potential) of a plant species. On this basis a list has been prepared while it is only on this basis that we can decide the comparative efficiency of one species with other species (Table 2b). Through this TSP itself one can identify the response of species against specific a pollutant by calculating individual metabolite like sulphate in case of sulfur dioxide or ammonical nitrogen in case of ammonia.

Table 2b. Comparative Efficiency between Best and Weak Scavenging species

Sr. No.	Species	TSP	Species
1	*B. variegata*	3.04 times to	*C. fistula*
2	*P. acerifolium*	2.43 times to	*B. malabaricum*
3	*Bambusa* spp.	2.21 times to	*P. longifolia*
4	*D. sissoo*	1.90 times to	*C. pentandra*
5	*Eucalyptus* spp.	1.89 times to	*P. ferrugineum*
6	*Z. jujube*	1.73 times to	*A. lebbek*
7	*C. siamea*	1.68 times to	*F. decora*
8	*F. benghalensis*	1.53 times to	*P. dulce*
9	*C. lanceolatum*	1.43 times to	*M. elengi*
10	*S. cumini*	1.35 times to	*K. pinnata*
11	*A. cadamba*	1.33 times to	*F. elastica*
12	*A. auriculiformis*	1.24 times to	*B. spectabilis*
13	*F. nitida*	1.16 times to	*C. sebestina*
14	*G. robusta*	1.04 times to	*F. infectoria*
15	*S. campanulata*	1.04 times to	*A. indica*
16	*F. religiosa*	1.01 times to	

It is quite interesting to note that how nature responds to all these situations and act, much more differently than expected. An analysis based on the first five or six parameters which are more significant in assessing the tolerance on these species are arranged on basis of tolerance while other list is prepared on basis of total scavenging potential. Arranging the species in a versus state (TSP Tolerance) indicate that some species have reverse relationship like *B. variegata* while species like *P. acerifolium, Bambusa*, have excellent tolerance and TSP, while a long list of moderate species appears which have equivalent potentials of both phenomenon. Obviously, we can identify best species out of the thirty two (Table 3, 4).

1.4 Screening of trees according to their TSP for Green Belts

Finally selection of a species will depend on TSP, tolerance, canopy structure, height, leaf characters and metabolism (Table 5a, b, c) (Table 6a, b). The green belt thus can be a combination of species to increase its efficiency to mitigate the potential load (Table 7a, b). Finally the plant species location can be decided on basis of fallout distance by exploring fallout models and meteorological conditions, which may decipher a range of 8x to 12x of the stack height and direction of wind. The wind rose pattern will further tell us the directions of the Green Belts around the industrial areas or stacks.

Table 3. Total Scavenging Potential Vs Tolerance

Sr. No.	TSP	Tolerance
1	*B. variegata*	*P. dulce*
2	*P. acerifolium**	*C. siamea**
3	*Bambusa* spp.*	*Z. jujube**
4	*D. sissoo*	*Bambusa* spp.*
5	*Eucalyptus* spp.*	*P. acerifolium**
6	*Z. jujube**	*F. nitida**
7	*C. siamea**	*F. elastica*
8	*F. benghalensis**	*F. decora*
9	*C. lanceolatum*	*S. cumini**
10	*S. cumini**	*Eucalyptus* spp.*
11	*A. cadamba*	*M. elengi*
12	*A. auriculiformis*	*B. spectabilis*
13	*F. nitida**	*A. indica*
14	*G. robusta*	*F. benghalensis**
15	*S. campanulata*	*M. indica*
16	*F. religiosa*	*C. sebestina*
17	*A. indica*	*F. infectoria**
18	*F. infectoria**	*D. sissoo*
19	*C. sebestina*	*G. robusta*
20	*B. spectabilis*	*A. cadamba*
21	*M. indica*	*S. campanulata*
22	*M. elengi*	*A. auriculiformis*
23	*F. elastica*	*K. pinnata**
24	*K. pinnata**	*F. religiosa*
25	*P. dulce*	*C. lanceolatum*
26	*F. decora*	*A. lebbek**
27	*A. lebbek**	*P. ferrugineum**
28	*P. ferrugineum**	*C. fistula**
29	*C. pentandra**	*P. longifolia**
30	*P. longifolia**	*C. pentandra**
31	*B. malabaricum**	*B. malabaricum**
32	*C. fistula**	*B. variegate*

Hence the Green Belt development is a definite design depending on various ecological and environmental considerations and these inputs will make ecotechnological strategy for a scientifically relevant Green Belt.

Table 4: Species arranged in decreasing order with reference to scavenging potential of different pollutants

S.N.	Ammonium Nitrogen	Chloride Chlorine	Sulphate Sulphur	Nitrate Nitrogen
1	*Bambusa* sp.	*F. religiosa*	*B. variegata*	*B. variegata*
2	*P. ferrugineum*	*Eucalyptus* sp.	*P. acerifolium*	*Z. jujuba*
3	*P. dulce*	*F. elastica*	*Bambusa* sp.	*S. cumini*
4	*Eucalyptus* sp.	*Z. jujuba*	*D. sissoo*	*M. elengi*
5	*M. indica*	*F. nitida*	*Eucalyptus* sp.	*Eucalyptus sp.*
6	*B. spectabilis*	*G. robusta*	*C. siamea*	*P. acerifolium*
7	*D. sissoo*	*B. spectabilis*	*Z. jujuba*	*M. indica*
8	*F. religiosa*	*F. decora*	*A. cadamba*	*S. campanulata*
9	*S. campanulata*	*C. pentandra*	*C. lanceolatum*	*D. sissoo*
10	*F. nitida*	*B. variegata*	*F. benghalensis*	*C. siamea*
11	*F. elastica*	*M. indica*	*S. cumini*	*F. decora*
12	*F. benghalensis*	*F. benghalensis*	*A. auriculiformis*	*P. ferrugineum*
13	*M. elengi*	*S. cumini*	*F. nitida*	*F. nitida*
14	*C. pentandra*	*C. fistula*	*G. robusta*	*F. religiosa*
15	*F. decora*	*S. campanulata*	*S. campanulata*	*F. benghalensis*
16	*A. indica*	*C. sebestina*	*F. religiosa*	*Bambusa* spp.
17	*S. cumini*	*D. sissoo*	*A. indica*	*A. auriculiformis*
18	*C. sebestina*	*A. indica*	*F. infectoria*	*A. indica*
19	*G. robusta*	*C. siamea*	*C. sebestina*	*F. elastica*
20	*Z. jujuba*	*F. infectoria*	*B. spectabilis*	*A. cadamba*
21	*C. siamea*	*P. acerifolium*	*M. indica*	*F. infectoria*
22	*K. pinnata*	*K. pinnata*	*K. pinnata*	*K. pinnata*
23	*P. acerifolium*	*Bambusa spp.*	*F. elastica*	*C. pentandra*
24	*B. variegate*	*P. ferrugineum*	*P. dulce*	*B. spectabilis*
25	*B. malabaricum*	*M. elengi*	*M. elengi*	*P. longifolia*
26	*A. auriculiformis*	*F. decora*	*A. lebbek*	*A. lebbek*
27	*A. cadamba*	*A. lebbek*	*F. decora*	*C. lanceolatum*
28	*F. infectoria*	*A. auriculiformis*	*C. pentandra*	*G. robusta*
29	*C. fistula*	*A. cadamba*	*P. ferrugineum*	*B. malabaricum*
30	*P. longifolia*	*B. malabaricum*	*P. longifolia*	*P. dulce*
31	*A. lebbek*	*P. dulce*	*B. malabaricum*	*C. sebestina*
32	*C. lanceolatum*	*C. lanceolatum*	*C. fistula*	*C. fistula*

Table 5. Basic Criteria for Green Belt Development

SET A: PLANT PARAMETERS

- Growth : Slow, Fast, Medium
- Average Life Span
- Height and Canopy Architecture
- Deciduous, Evergreen, Semi-evergreen

SET B: POLLUTANTS

- Nature and Class
- Chemical Characteristics
- Ambient Load and Interaction of Pollutants
- Mass Balance Computations

SET C: OTHER PARAMETERS

- Meteorological Conditions
- Stack Height
- Plume Characteristics and Settling Patterns
- Soil Characteristics
- Ground Water Levels and Additional Water
- Supply Facilities

Table 6a. Plant Surface Variables

Accommodations	**Canopy Structure**	**Leaf Vegetation**
Exudates	Areas Density	Boundary Layer
Trichomes	Bark	Change at High Winds
Pubescence	Bole	Flutter
Wase	Leaves	Stomatal resistance
	Porosity	
Biotic Surface	**Productive Structure**	**Non Biotic Surface**
	Soils	pH effects on:
		Reaction
Electrostatic Properties	Stem Type	Solubility

Table 6b Parameters for Optimization of Width of Green Belt

- Atmospheric Stability
- Height of Greenbelt (m)
- Dry Deposition Velocity (m/s) in Absence of Greenbelt
- Deposition Velocity onto Plant Canopy
- Separation Distance Between Source and Greenbelt (km)
- Foliage Surface Area of a Single Tree
- Pollutant Pathways via Greenbelt
- Species Composition

Table 7(a). Performance evoluation of green belt vs species composition

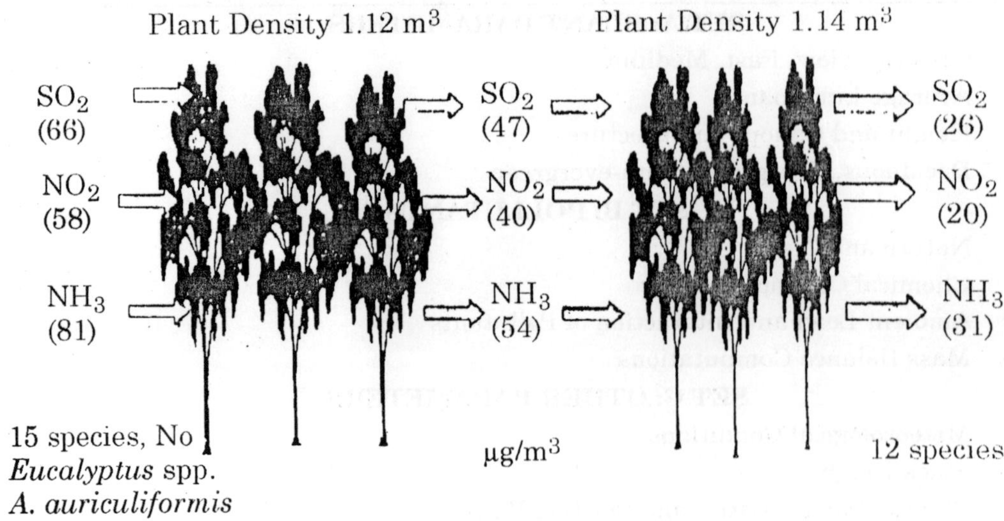

Table 7b. Performance Evaluation of Green Belt vs Species Composition

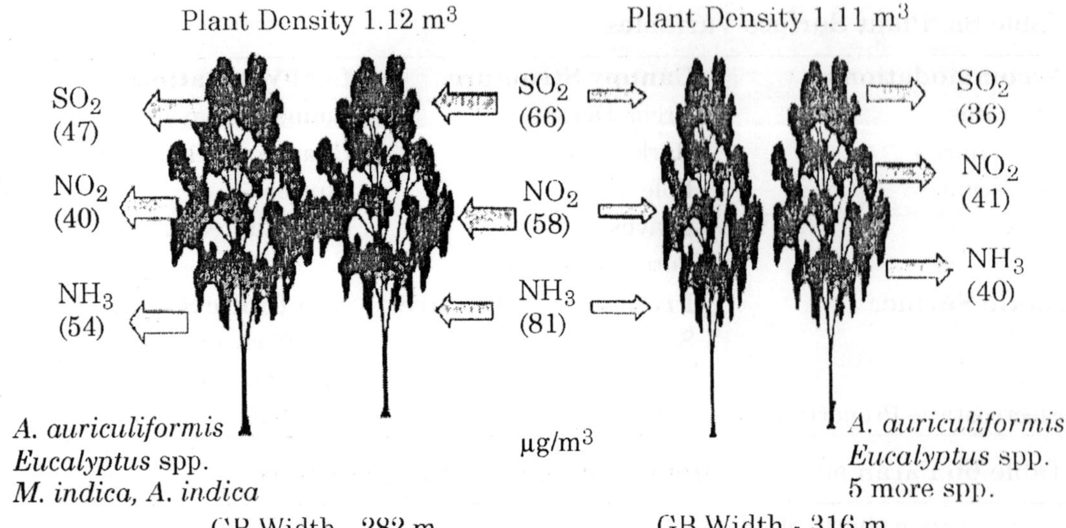

1.5 THE GREEN BELT WIDTH AND COMPOSITION

The Green Belt thus turns from a technique to technology with number of factors and inputs as well as the objectives. For example, your location does not permit Green Belt at 10x, or water scarcity is a limiting factor for species selection or even aesthetic sense as no one will like an exclusive Green Belt of bamboos at that point of time, the design will be a mixture of species of evergreen and deciduous ones and direction of the pollutants searching in to pass through a Green Belt (Tables 8a and 8b).

Table 8a.

Type *M. indica, C. myxa*

Type *D. regia, B. racemosa*

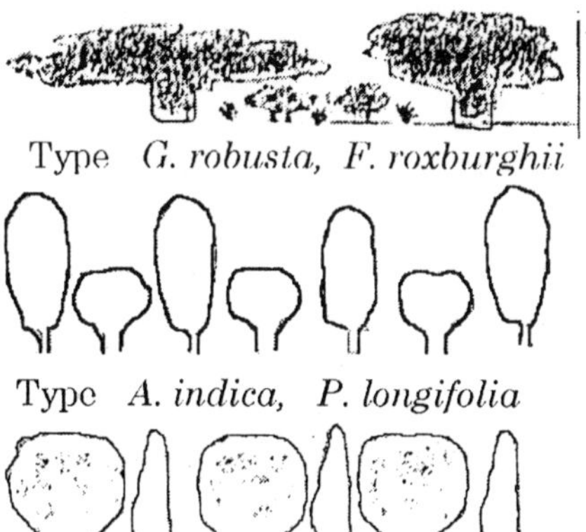

Table 8b. Stack Height and Green Belt Site

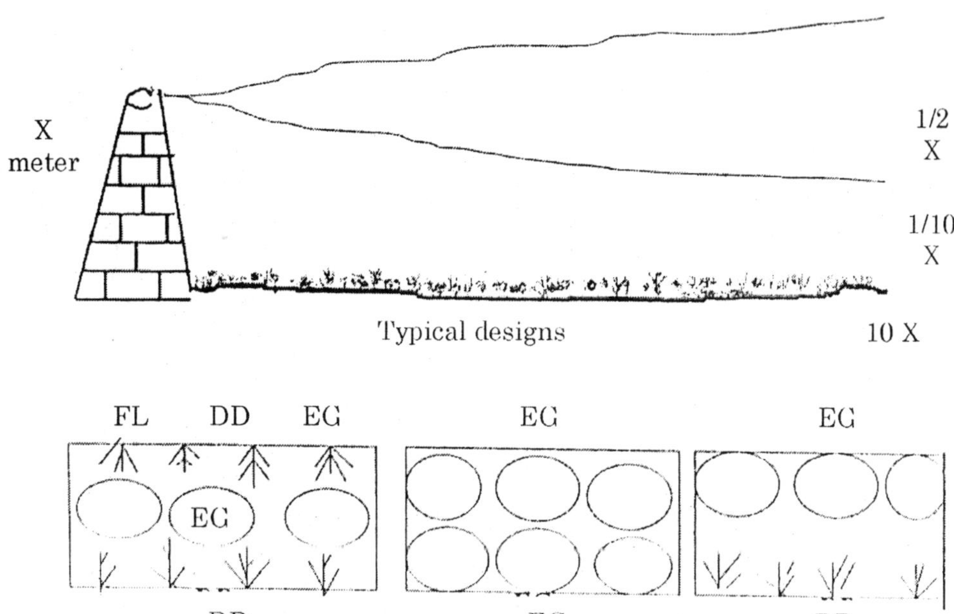

1.6 DESIGNING A GREEN BELT

It is very important to have into consideration all the above parameters and select appropriate species of plant/trees for designing a green belt for –

(a) Industrial Areas

(b) Highways, Road Sides

(c) Urban Areas.

CHAPTER - 2

Green Belts along roads and highways

S.B. Chaphekar*, Ramesh Madav and Anuradha Mancharkar

Department of Botany, University of Pune, Pune - 411 007
**14 Dushwadi, Thakar Dwar, Mumbai - 400 002*
e mail: sharad.chaphekar@gmail.com

ABSTRACT

Within the next twenty years, it is said the number of vehicles in the world is likely to double from its present total of 700 million vehicles. As cities expand into mega-cities, more people will drive more vehicle, greater distance and for longer times. In the absence of controls, the automotive emissions will likewise increase. In cities where a substantial proportion of the motor vehicle fleet is diesel-powered there are additional problems of black smoke and greater particulate emissions. No doubt, the need for automotive emissions controls is paramount, but more important is the judicious transportation planning which will relieve the town and cities of the unwanted pollution.

Air pollution containment using Green Belts may be done in two ways – developing Green Belt around the source of pollution and the other by developing Green Belt encompassing the receptor. The former is recommended where the source, like an industry, is identified as a point source of pollution. The latter proves useful where the pollution source is diffused and the receptor is subjected to an area source.

Key words: Green belts, Roads, Highways, Pollution scavengers, SO_2, Vehicular pollution.

A large amount of information is available to explain air pollution-induced injury to plants. The injury is of two types – visible and subtle. Whereas the latter affects adversely biomass production by plants, the former is a setback to foliage, vitality and regeneration of plants and may also prove fatal to plants. Still, plants are always recommended for pollution abatement. A critical assessment of the paradox reveals that plants indeed provide the least expensive remedy from air pollution (Singh and Rao, 1980, Gupta and Kapoor, 1985, Chaphekar, 1982), but only within the range of their tolerance towards air pollution. Hence the concept of development of green belts to contain air pollution is being considered with due attention to the specifics, like sources and receptors of pollution, type and intensity of pollution, and composition, location and architecture of green belts.

Air pollution containment using Green Belts may be done in two ways – developing green belt around the source of pollution and the other by developing green belt encompassing the receptor. The former is recommended where the source, like an industry, is identified as a point source of pollution. The latter proves useful where the pollution source is diffused and the receptor is subjected to an area source. We shall consider here a situation that is combination of both the above.

2.1 ENVIRONMENT OF ROADS AND HIGHWAYS

Many of the cities were recognized as suffering mainly from pollution from automobiles (Treshow, 1970). Indian cities have also started feeling the same problem as per recent studies (NEERI Report, 2005). An automobile is a point source of pollution that is mobile when emissions occur. The ever increasing number of automobiles and their movement on the network of roads, convert the same into an area source. This type of pollution affects a large number of citizens residing in urban areas. It is estimated that almost 67% of air pollution in Mumbai city is caused by automobiles. On a national scale, the figure stands somewhere near 40% and is increasing rapidly (NEERI, 2005). It is logical hence, that the citizens are protected from air pollution on and along roads, by all available means, including with Green Belts.

Vehicular emissions include sulphur dioxide, oxides of nitrogen, carbon monoxide, lead, particulates and organics like benzene. Of these, lead has been eliminated from gasoline and is no more a serious hazard (Chaphekar, *et al.*, 2000). Efforts are being made to reduce sulphur content in the fuel. Oxides of nitrogen giving rise to ozone in the atmosphere, is so far unattended, even un-estimated to a large extent. Still, it is worth noting that under normal conditions, all these parameter values (except ambient ozone concentrations which are still to be extensively reported) are within permissible limits in most cities (NEERI, 2005). The major problem that seems to be not under control so far, is SPM. Road dust formed due to wear and tear of road surface and tyres, combined with emissions and frequent workings of road surfaces by different government agencies, remains airborne for long durations, due to heavy traffic (Rakesh Kumar, 2005, *Pers. Comm.)*.

Almost eight months of dry season in the country further aggravates the dust problem. It is this dust pollution problem on and along roads (up to a horizontal distance of 30 meters from edges of roads, and up to about 10 meters vertically above road surface, (Smith, 1981) that has been grossly neglected, both by the authorities as well as by researchers in this field. This, in spite of the fact that particulate pollution is always found to be crossing permissible limits in all the cities in India (NEERI, 2005). The latest observations (as reported in Indian Express dated Sept, 11, 2006) state that air pollution including SPM is high at 255 micrograms per cubic meter, as against the permissible 140 micrograms. Virtually no site monitored returned the values that were within the permissible levels. Moreover, it is known that when a vehicle slows down (as near signals) from 60 km p h to 30-40 km p h, its emission increases almost two fold. In most of our cities, and particularly in Mumbai, traffic is so slow due to so many obstructions including signals, that pollution levels are always high. It is also suspected that fine particulates in air function as nuclei for deposition of air-borne chemical pollutants and intensify pollution hazard (Chaphekar, 1972). This is also reported to be a major contributor to respiratory problems, especially of children and senior citizens in cities (IIEM, 2006).These facts should warrant more studies on dust pollution and utility of green belts for abatement of dust menace in urban environment.

Roads are of different types. At one end, there are Super-Express and Express highways connecting major cities and at the other end, there are small and big lanes separating houses in residential areas in cities. 'Paths', streets and roads of different lengths and widths form road network in cities. Vehicular pollution exists according to the vehicular population on these roads, i.e. all types of roads except 'gallies' and lanes. Trees in the housing premises form green belts in lanes and small roads, sometimes supplemented with a row of trees on either side, planted by local authorities, mainly for aesthetic purposes. Traffic here is sparse hence pollution is low, if any.

Traffic on city roads is heavy and slow moving, constituting mixed type of vehicles, mainly cars and mass transport vehicles. Citizens desire green trees along these roads for aesthetics, shade, as also for pollution abatement. Scope of plantation however, is limited, due to lack of space on footpaths which are needed for free movement of pedestrians.

Intercity highways carry heavy traffic of all types of vehicles, including heavy vehicles. Pollution potential here is high and though few citizens reside along these roads, pollution is likely to affect agricultural farms and forests through which these roads pass. It has been reported recently that agricultural production in the country reduced by 10 to 80% due to air pollution (Agrawal, 2005). Green Belt here is essential mainly for protecting farm produce and forest biota. Possibility for development of ideal Green Belts along these roads is better here, since most of these roads are in the process of development at present. The expressways and roads being redeveloped under 'Golden Quadrilateral' programme have adequately marked 'right-of-way' (ROW) areas reserved by highway authorities, as 'no development zones' (NDZ), where only green belts and related accessories can be established.

2.2 WHICH TREES FOR ROADSIDE GREEN BELT?

While looking for trees to be planted along a road for pollution abatement, following aspects need to be remembered –

(i) Pollution removal efficiency of foliage – efficient dust capturing leaves have to be broad, with stiff, short petioles holding laminas parallel to the ground and with epidermal features protective of stomatal apertures. Leaves with very different characters like laminas swinging freely on long, thin petioles, unprotected stomata can be very efficient for absorption of gaseous pollutants. It is hence, apt that a Green Belt has to consist of a mix of different types of plants with varying efficiencies for removal of different types of pollutants, prevailing in the area of concern.

(ii) Space required above and below ground, especially with reference to buttress roots,

(iii) Littering of road due to leaf fall (deciduous trees), flowers (*Peltophorum*), fruit (figs), etc.

(iv) Straightness of the main trunk growth, for at least first 3-4 meters (natural growth or manageable by manipulating by acceptable silvicultural methods),

(v) Water requirement in terms of quantity and frequency of watering,

(vi) Fruits and/or flowers that would attract undue attention of animals.

While this information is needed for trees to be planted along roadsides, plants to be used for dividers, have to be tall enough to protect vision of drivers from headlight glare from vehicles moving in opposite direction, but short enough to keep road visibility while driving, especially on curves and bends.

A number of tree and shrub species can be suggested for plantation along roads from the vast diversity available in the country. Some of these are planted regularly and traditionally (e.g. Mango, Rain-tree, etc.), some wild species also need to be tried with care, for their utility value and/or conservation needs and even special characters like salt tolerance (e.g. *Terminalias, Barringtonias, Mallotus, Grewia, Actinodaphne, Derris,* etc.).

Plate - I

Plate - II

Table 1. List of plant species recommended for plantation along highway in Pune district of Maharastra (Partial list only, consisting of generic names starting with A)

Name	(Partial list only, consisting of generic names starting with A)					Remark
	Approx. Ht. (m)	Gr. Rate	Reproduc.	Canopy	Leafliness	
Abutilon indicum	3	Fast	seeds	Oblong	Evergreen	
Acacia auriculiformis	15	Fast	Seeds	Oblong	Evergreen	Exotic
Acacia catechu	3	Fast	Seeds	Oblong	Evergreen	
Acacia leucophloea	3	Fast	Seeds	Oblong	Deciduous	
Achras sapota	10	Slow Fast	Grafting	Spreading	Evergreen	Exotic
Actinodaphne hohenakeri	10	Slow	Seeds	Round	Evergreen	
Aegle marmelos	12	Slow	Seeds/Rt	Oblong	Evergreen	Medicinal
Albizzia lebbeck	20	Fast	Seeds	Round/Spr	Deciduous	
Anona squamosa	10	Fast	Seed/Gra	Round	Evergreen	Fruits
Ailanthus excelsa	20	Fast	Seed?Cut	Round	Deciduous	
Alstonia scholaris	15	Fast	Seeds	Round	Evergreen	
Anogeissus latifolia	12	Slow	Budding	Round/Obl	Evergreen	
Anthocephalus chinensis	20	Fast	Seeds	Spreading	Evergreen	
Artocarpus heterophyllus	10	Slow	Seeds	Oblong	Evergreen	Fruits
Azadirachta indica	20	Fast	Seeds	Spreading	Evergreen	Medicinal

In a country of vast bioclimatic diversity, a single list of plants is obviously not suitable for all regions. At the same time, it must be appreciated that plantation cannot depend upon only natural availability of resources like water and nutrition. The Planning Commission has taken these facts into account and recommended fifteen agro-climatic regions, which are further divided into 68 sub-zones, based on detailed characteristic features, *viz.* natural resources, land productivity level, relative pressure on land and environmental factors like temperature, precipitation range, relative humidity, soil type, etc. (Alagh, 1990). This system is found to be more suitable than the systems based on only natural environmental assets, *viz.* Champion's Classification of Forest Types (1938) or its modified form. Carter (1954), or that of Krishnan based on Thorntwaite' system of hydrotherms (1988). The list of plants given in Table 1 refers to a part of recommendation for Zone No. IX – Western Plateau and Hills, Subzone No. 1 – Hilly districts that include revenue districts like Kolhapur, Nasik, Pune and Satara in Maharashtra and Jhabua in M.P. Similar lists are prepared for all the Zones and Sub-zones in the country (Chaphekar *et al.*, 1995). Such listing is extremely useful as a Ready Reckoner for making choice of plant species to be cultivated in green belts, in different parts of the country.

Table 2. List of plant species from which suitable combinations can be made for green belts in Pune region

Tree species	Tree Size	Shrub species
Anona squamosa	Medium	*Ravenia spectabilis*
Glyricidia sepium	Medium	*Duranta plumieri*
Grevillea robusta	Big	*Hamelia patens*
Guazuma ulmifolia	Medium	*Hibiscus rosa-sinensis*
Heterophragma roxburghii	Big	*Ixora coccinia*
Holoptelia integrifolia	Big	*Lantana camara*
Jacaranda mimosaefolia	Medium	*Lowsonia inermis*
Kigelia africana	Medium	*Nerium indicum*
Lagerstroemia parviflora	Medium	*Ricinus communis*
Lagerstroemia speciosa	Medium	*Sesbania sesban*
Madhuca longifolia	Big Medium	*Nyctanthus arbor-tristis*
Mangifera indica	Medium Big	*Melaleuca leucadendron*
Mimusops elingi	Medium Big	*Pithecellobium dulce*
Ouginia oojeniensis	Big	*Poinciana pulcherrima*
Peltophorum pterocarpum	Big	*Sesbania grandiflora*
Polyalthia longifolia	Medium	*Tecoma stans*
Prosopis cineraria	Medium	*Thevetia peruviana*
Sapindus emarginatus	Medium	*Tabernaemontana divaricata*
Spathodea campanulata	Big	*Holmskioldia sanguinea*
Syzygium cumini	Big	*Crotons*
Tamarindus indica	Big	*Dracaena* sp.

Plants listed in Table 1 also take into account another aspect of green belt. An efficient belt has to be like an absorbent wall of foliage running parallel to the road on either side. This can be achieved by planting big trees, small trees and shrubs, (as per availability of space) with their canopies forming an efficient pollution-absorbing wall along roadside. Position of plant species with reference to each other within the belt needs to be planned suitably. Ideally, a belt should be as broad as possible horizontally, with at least three rows of trees. This is possible along express highways between cities. Within urban areas, one has to do the best with available space enough for a single row of trees and bushes forming two vertical layers of canopies. Table 2 gives a list of trees and shrubs that can be sourced for combinations for effective and protective green walls along roads. It was estimated that the amount of dust that can potentially be captured by such belts per kilometer of a road (Fig. 1), was impressively high (Pandit, 1995). Maharashtra State Road Development Corporation (MSRDC) authorities preferred to plant three rows of trees on either side of the Mumbai-Pune Express Highway. Tallest foliage trees, medium-size flowering trees and smaller fruit trees, was the order of plantation starting from the edge of the road. The median (divider) was planted with ornamental and/or medicinal bushes (Fig. 2). Chaphekar *et al.* (2000) have recommended that while planting trees on city footpaths, elongated pits of land to accommodate two to four trees in a row should replace the present practice of only 0.5 x 0.5 meter pits for trees. Such elongated pits would ensure availability of water to tree roots in otherwise compact ground.

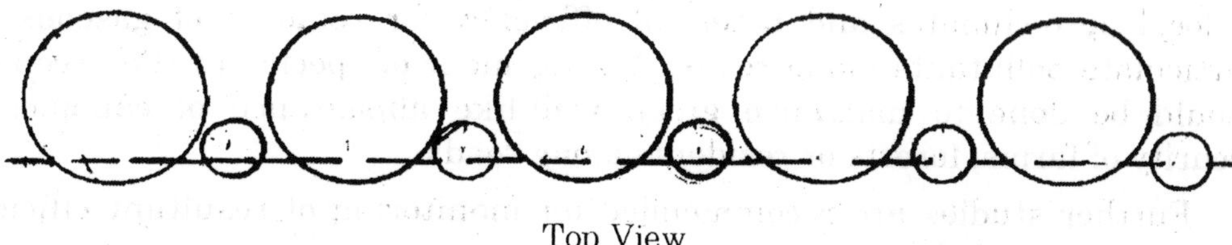

Top View

Road

Profile

Potential Dust Removal - 23 kg per kilometer

Fig. 1. Suggested plantation along roadside - Bigger *Syzygium cumini* and smaller *Bauhinia variegata*

2.3 CONCLUSIONS

Green belts recommended for pollution absorption along major roads require research-based attention to ensure optimization of their efficiency.

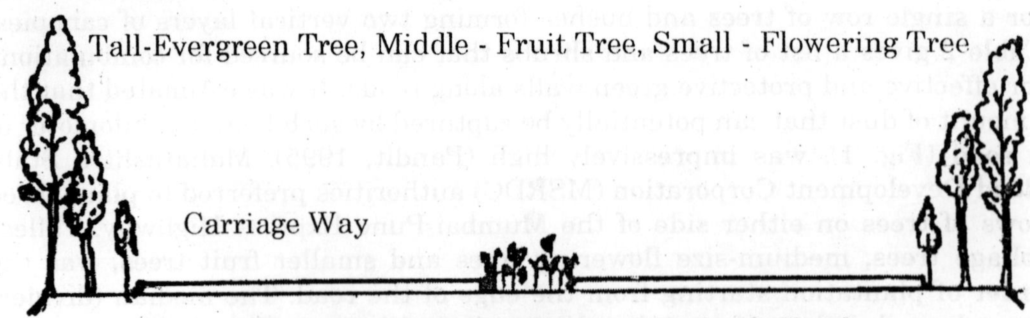

Fig. 2. MSRDC's Plantation Plan Along Mumbai-Pune Expressway, within ROW

Selection of plant species for green belts needs to be based on information about their relationship to the climate and potential for adjustment according to local agro-climates and potential efficiency for removal of gaseous and particulate pollutants along roads. Arrangement of species in the green belt should be done to maximize green wall-like appearance of canopies for security of farms, forests or residents along roads.

Further studies are recommended for monitoring of resultant efficiency of 'in situ' green belts.

Acknowledgements

The authors thank the Central Pollution Control Board, New Delhi, for financial support to the project, to Dr. R.K. Kapoor of Nuclear Power Corporation and Shri V.K. Gupta of Atomic Energy Regulatory Board, Mumbai, for extensive discussion sessions, to Dr. Chandrakant Lattoo and Dr. Seema Ghate for help and to Pune University, for facilities provided.

REFERENCES

Agrawal, M. 2005. Effects of air pollution on agriculture: an issue of national concern. Source and Technology Development and Policy Issues. 28(3&4): 93-104.

Alagh, Y.K. 1990. (Summary by Wadia, F.K.) Agro-climatic Regional Planning: Overview of a new approach. *Jour. Ind. School of Political Economy*. 2(1): 81-111.

Carter, D.B. 1954. Climates of Africa and India according to Thorntwaite's (1948) Classification. John Hopkins University Publication. Climatology, 7(4).

Chaphekar, S.B. 1972. Effects of atmospheric pollutants on plants in Bombay. *J. Biol. Science*. 15: 1-6.

Chaphekar, S.B., 1982. Air Pollution and Plants. *Indian Review of Life Sciences*, 2:41-66.

Chaphekar, S.B. 1995. Green belts. In: *Landmarks of Botany in India*. (Ed. P. Soni & A.K. Srivastava), Surya Internatl. Publ.. Dehradun. pp. 17-23.

Chaphekar, S.B., Kapoor, R.K. and Gupta, V.K. 1995. Final Report on "Guidelines for Development of Green Belts", for Central Pollution Control Board, New Delhi.

Chaphekar, S.B., Lattoo, C.S., Khalap, K.P. and Madav, R.P. 2000. Health of Greenery in Mumbai and suggestions for Feasible Remedies. Report of MMRDA-sponsored project.

Gupta, V.K. and Kapoor, R.K. 1985. Reducing the consequences of reactor accidents, with a green belt. *Nuclear Technology*, 70 (2): 204-214.

Indian Institute of Environmental Medicine (IIEM), 2006. Project proposal to MPCB, Mumbai.

Krishnan, A. 1988. Delineation of soil climatic zones of India and its practical application in agriculture. *Fertilizer News*, 33 (4): 11-19.

National Environmental Engineering Research Institute (NEERI), Nagpur. 2005. Periodic Reports and Lecture Notes.

Pandit, A. 1995. Sink Capacity of Common Roadside Trees in Pune. M.Phil. Dissertation to Pune University.

Smith, W.H. 1981. Air Pollution and Forests: Interactions Between Air Contaminants and Forest Ecosystems. Springer-Verlaag, New York.

Treshow, M. 1970. *Environment and Plant Response.* Mc Graw Hill, New York.

Singh, N. and Rao, D.N. 1980. Growth responses of wheat plant to cement dust polluted environment. *Proc. Indian Natl. Sci. Acad.* 3: 325-330.

CHAPTER - 3

Air pollution control through use of appropriate technology

H.V. Bhavnani, M.P. Keshwani, M.M. Gor, Jigar Padaria and Diptesh Jalul

Parul Institute of Engineering and Technology, Vadodara

ABSTRACT

Nanotechnology is the technology where we can handle individual atoms and molecules and arrange them in desired complex structures. In the same way we can have stronger materials than ever found and this will make possible manufacturing machines / equipments or structures of incredibly huge size, but at the same time it will also permit us to build machines of the nanoscopic level. With nanotechnology we can make materials with atomic precision. We will be able to put together the fundamental building blocks of nature easily, inexpensively and in most of the ways permitted by the laws of physics. Thus it is apparent that it has its applications in all the fields of science and we can have variety of the applications such as biotech, ultra precision machining, nano computing, molecular modeling, etc. In India the transport system is based on petroleum fuel, but its availability coupled with environmental pollution is a matter of growing concern. With the enhancement of the rapid industrialization air pollution is the major concern in our country. Through nonotech it is feasible to meet challenges for the control of air pollution on global level. The paper discusses at length, Eco-friendly environment *THROUGH NANOTECHNOLOGY.*

Key words: Nanotechnology, Air Pollution control, Nano-materials, Eco-friendly vehicles

3.1 NANOTECHNOLOGY : INTRODUCTION

A nanometer is one billionth of a meter (10^{-9} m)—about one ten-thousandth the diameter of a human hair, a thousand times smaller than a red blood cell, or about half the size of the diameter of DNA. Nanotechnology is defined as research and technology development at the atomic, molecular, or macromolecular levels using a length scale of approximately one to one hundred nanometers in any dimension thus creating and use of structures, devices and systems that have novel properties and functions because of their small size, and the ability to control or manipulate matter on an atomic scale. Nanotechnologies with information, biological, and cognitive technologies will lead to products we can only now vaguely imagine.

As applications of nanotechnology develop over time, they have the potential to help the shrink age of the human footprint on the environment. This is important, because over the next 50 years the world's population is expected to grow 50%, global economic activity is expected to grow 500%, and global energy and materials use is expected to grow 300%. This has been true for municipal waste generation, as well as for environmental impacts associated with vehicle travel, groundwater pollution, and agricultural runoff. This paper describes how nanotechnology can create materials and products that will not only directly advance our ability to detect, monitor, and clean-up environmental contaminants, but also help us avoid creating air pollution in the first place. By more effectively using atoms and energy throughout a product lifecycle, nanotechnology may contribute to reducing air pollution, by reducing their "volume effect".

For example, nano-sized cerium oxide has been developed to decrease diesel emissions, and iron nanoparticles can remove contaminants from soil and ground water. Nanotechnology also has the potential to improve the environment, both through direct applications of nanomaterials to detect, prevent, and remove air pollutants, as well as indirectly by using nanotechnology to design cleaner industrial processes and create environment friendly products.

3.1.1 Precautions on implementation of this technology

It may be true that in most cases nanomaterials will not be of human health or ecological concern. However, at this point not enough information exists to assess environmental exposure for most engineered nanomaterials. This information is important, because we need a sound scientific basis for assessing and managing any unforeseen future impacts resulting from the introduction of nanoparticles and nanomaterials into the environment. Inhaled nanoparticles may become lodged in the lung, and the high durability and reactivity of some nanomaterials raises issues of their fate in the environment. A challenge for environmental protection is to allow full realization of the societal benefits of nanotechnology, while identifying and minimizing any adverse impacts to humans or ecosystems from exposure to nanomaterials. In addition, we need to understand how to best apply nanotechnology for air pollution prevention in current manufacturing processes and in the manufacture of new nanomaterials and nanoproducts, as well as in environmental detection, monitoring, and clean-up.

3.2 CONTROL OF AIR POLLUTION BY APPLICATION OF NANOTECHNOLOGY

3.2.1 Alternate Fuels

There has been lot of research and development work going on in the field of alternate fuels (non-conventional fuels). Some of the examples are propane, butane, hydrogen etc. Among these hydrogen is the one the alternate fuel which can be used in controlling air pollution with the application of nanotechnology.

HYDROGEN AS NON POLLUTING FUEL

Hydrogen stores energy more effectively than batteries, burns twice as efficiently in a fuel cell as gasoline does in an engine, and produces a single waste product, water. But this system requires more space and is to 0 costly. The amount of hydrogen that would be needed for a hydrogen economy is about five times what we are producing today. So we would have to find much better techniques to produce more hydrogen economically. Ironically, most of the nine million tons of hydrogen we produce today is extracted from fossil fuels. In order to generate hydrogen energy, we first need to develop inexpensive, renewable sources of electricity like wind or solar power. Some nanotechnologists are already working on more efficient solar cells, which would help provide inexpensive electric power to produce more hydrogen at low cost.

Besides producing more hydrogen, we must increase the storage volume of hydrogen fuel tanks in cars, so that consumers can drive at least 400 miles before refueling. For example, one way to harness the power of hydrogen is through its gaseous state. But because the gas takes up a lot of room, it would have to be compressed, and the tank that holds it would need to be able to withstand the resulting high pressure. Nanotechnology could provide materials which are strong, light, and cheap enough to build such a tank. Another approach would be to have a tank filled with a solid material that absorbs hydrogen at fill-up time, and releases it while driving. Nanotechnology could produce that very porous solid material.

3.2.2 Eco-friendly Vehicles

Catalytic converters using nanotech will almost eliminate the pollutants from the exhaust or even make useful material out of it. Nano engineered catalysts can replace platinum and palladium, two precious metals most commonly used in catalytic converters. Platinum and palladium are expensive and in short supply worldwide. Because catalysis only occurs on the surface of the metals, most of the weight - which drives up the price - is essentially just filler. Nanotechnology can create catalysts that will provide entire surface area. The carbon nanotube based capacitors also support this concept because they use a very clean and effective way to replace car batteries which are rather causing many problems to the environment. As mentioned above hydrogen fuel cells will also support this notion.

3.2.3 Sensors for detecting Pollutions

Sensor development and application based on nanoscale science and technology is growing rapidly due to the advancements in the microelectronics industry and the increasing availability of nanoscale processing and manufacturing technologies. One of the near-term research products of nanotechnology for environmental applications is the development of new and enhanced sensors to detect biological and chemical contaminants in air. Nanotechnology offers the potential to improve exposure assessment by facilitating collection of large number of measurements at a lower cost and improved specificity. It will be soon possible to develop micro- and nanoscale sensor arrays that can detect specific sets of harmful agents in the environment at very low concentrations. On the other hand, the environmental measurement field is a cost sensitive arena with less available resources for leading-edge development. Therefore, environmental nano-sensor technology likely will evolve by leveraging the investment in nanosensor research by related fields.

Manufacturing process control, compliance, ecosystem monitoring, and environmental decision making would be significantly improved if more sensitive and less costly techniques for contaminant detection were available. Recently, chemical sensors based on individual single walled nanotubes have been demonstrated. Altered electrical resistance of semi conducting single walled nanotubes through exposure to gaseous molecules, such as nitrogen dioxide or ammonia, forms the basis for these sensors.

The nanotube sensors also exhibited fast responses at room temperature to the gases and a substantially higher sensitivity than existing solid state sensors. Boron-doped silicon nanowires (SiNWs) have been used to create highly sensitive, real-time electrically based sensors for biological and chemical species in air. These mechanisms could apply for detecting pathogens and chemical and biological agents in air. For instance, nanobarcodes-highly encodable and chemically diverse nanoparticles-may allow for high-level multiplexing of small sample volumes. Evaluating the exposure of complex systems in the environment to toxic pollutants often has such requirements for sample analysis.

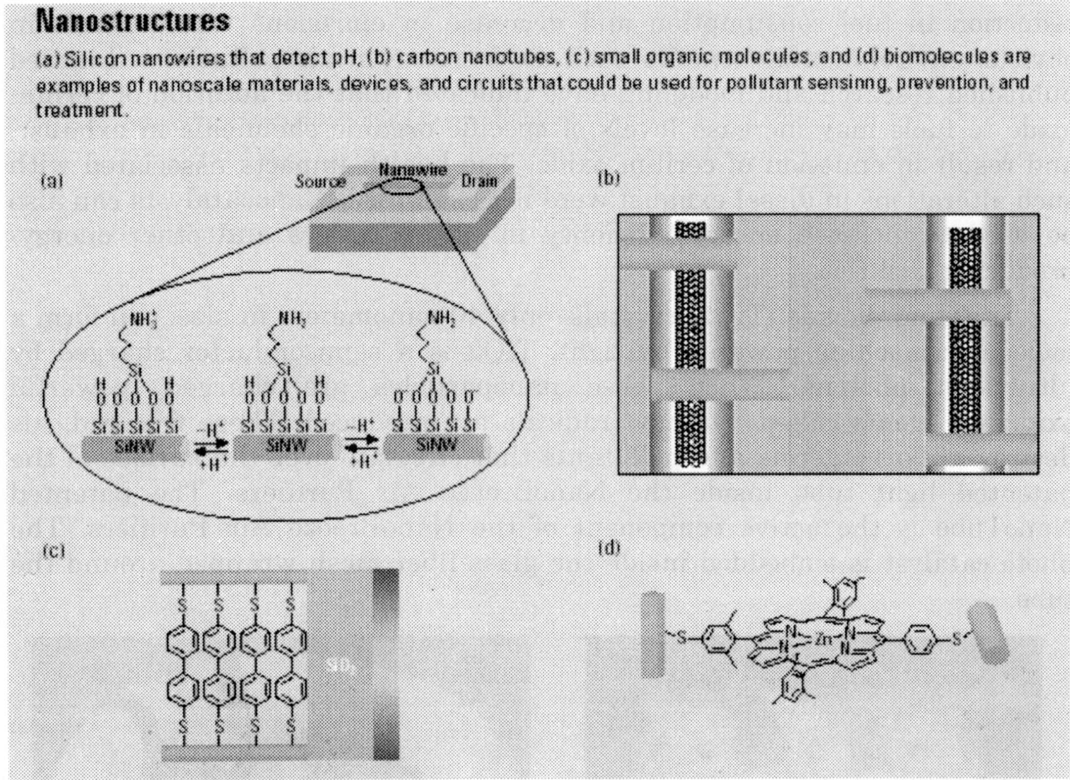

Figure 1.

Nanobarcodes can greatly enhance our ability to identify the source and strength of contaminants, determine the route and mechanism of environmental fate and bioavailability, and assess the effectiveness of treatment and remediation techniques in controlling air pollutants.

3.2.4 Nano-Materials

Nanotechnology may also lead to more efficient and effective use of materials. For example, nanotechnology may improve the functionality of catalytic converters and reduce by up to 95% the mass of platinum group metals required. Because platinum group metals occur in low concentration in ore, this reduction in use may reduce ecological impacts from mining thus reducing air pollution. Nanostructured catalysts can also increase yield (and therefore reduce materials use) at the process level.

Nanomaterials also show potential as fuel additives and automotive catalysts. For example, nano-cerium oxide particles are being employed in the United Kingdom as on- and off-road diesel fuel additives to decrease emissions. These manufacturers also claim a more than 5% decrease in fuel consumption with an associated decrease in vehicle emissions. Such a reduction in fuel consumption and decrease in emissions would result in obvious environmental benefits such as decrease in air pollutants. Limited published research and modeling have indicated that the addition of cerium oxide to fuels may increase levels of specific organic chemicals in exhaust, and result in emission of cerium oxide. The health impacts associated with such alterations in diesel exhaust were not examined. Nanocatalysts can also be used to increase energy efficiency in utility boilers and other energy-producing facilities.

Titanium dioxide (TiO2) crystals, only 40 nanometers in size, can form a molecular machine powered by light. TiO2 is a semiconductor charged by ultraviolet photons. When these nanoparticles are charged, powerful oxidizing agents called hydroxyl radicals are produced. These free radicals destroy airborne germs and pollutants that circulate over the surface of the patented light tube inside the NanoBreeze Air Purifiers. The patented NanoTube is the active component of the NanoBreeze Air Purifiers. The photo catalyst is embedded inside the glass fiber mesh wrapped around the tube.

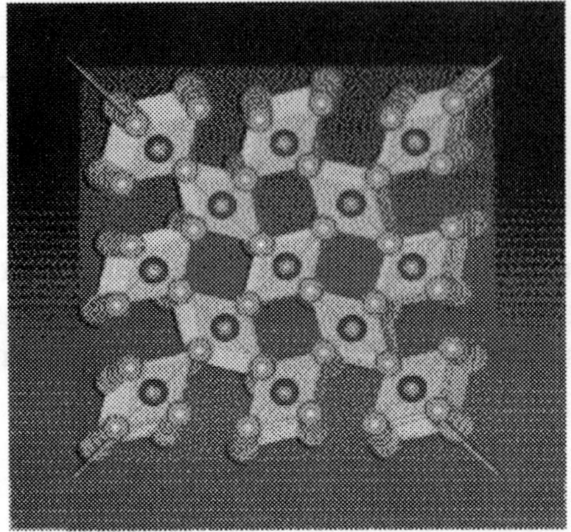

Figure 2. Crystal nanoparticle of TiO2

Figure 3. The patented NanoTube is an air cleaning filter activated by light.

3.3 RISK ASSESSMENT IN IMPLEMENTING NANOTECHNOLOGY FOR AIR POLLUTION

A challenge in evaluating risk associated with the manufacture and use of nanomaterials is the diversity and complexity of the types of materials available and under development as well as the seemingly limitless potential uses of these materials. A risk assessment becomes prime requirement to understand the hazardous properties of environmental agents the extent of exposure of humans or environmental receptors to those agents. It is important to note that nanomaterials have large surface areas per unit of volume, and novel electrical and magnetic properties relative to conventional chemicals. Some of the special properties that make nanomaterials useful may pose hazards to humans and the environment, under specific conditions in different form. Thus risk assessment becomes the key feature in implementing nanotech in air pollution.

3.3.1 Chemical identification and characterization of Nanomaterials

The identification and characterization of chemical substances and materials is an important first step in assessing their risk. Understanding the physical and chemical properties in particular is necessary in the evaluation of hazard (both toxicological and ecological) and exposure (to all routes). Chemical properties that are important in the characterization of discrete chemical substances include, but are not limited to, molecular weight, melting point, boiling point, vapor pressure, water solubility, reactivity, and stability etc.

The diversity and complexity of nanomaterials make chemical identification and characterization not only more important but also more difficult. A broader spectrum of properties will be needed to sufficiently characterize a given nanomaterial for the purposes of evaluating hazards and assessing risk for air pollution. Chemical properties such as those listed above may be important for some nanomaterials, but other properties such as particle size and distribution, surface/volume ratio, magnetic properties, coatings, and conductivity are expected to be more important for the majority of nanoparticles in understanding their reaction to the air pollutants. For example, single-walled carbon nanotubes can be produced by four different processes that can generate products with different physical-chemical properties (e.g., size, shape, composition) and potentially different ecological and toxicological properties. It is not clear whether existing physical-chemical property test methods are adequate for sufficiently characterizing various nanomaterials in order to evaluate their hazard and exposure and assess their risk. It is clear that chemical properties such as boiling point and vapor pressure are insufficient. Because of the current state of development of chemical identification and characterization, current chemical representation and nomenclature conventions may not be adequate for some nanomaterials. Nomenclature conventions are important to eliminate ambiguity when communicating differences between nanomaterials and bulk materials and in reporting for regulatory purposes in studying the effects for air pollutions.

3.3.2 Fate of Nanomaterials in Air

Several processes influence the fate of airborne nanomaterials in addition to their initial dimensional and chemical characteristics: the length of time the particles remain airborne, the nature of their interaction with other airborne particles or molecules, and the distance that they may travel prior to deposition.

With respect to the length of time particles remain airborne, particles with aerodynamic diameters in the nanoscale range (<100 nm) follow the laws of gaseous diffusion when released to air. The rate of diffusion is inversely proportional to particle diameter, while the rate of gravitational settling is proportional to particle diameter. Airborne particles can be classified by size and behavior into three general groups:

1. Small particles (diameters <80 nm) are described as being in the agglomeration mode; they are short-lived because they rapidly agglomerate to form larger particles.
1. Large particles (>2000 nm) are described as being in the coarse mode and are subject to gravitational settling.
1. Intermediate-sized particles (>80 nm and < 2000 nm) are described as being in the accumulation mode and can remain suspended in air for the longest time, days to weeks, and can be removed from air via dry or wet deposition.

Note that these generalizations apply to environmental conditions and do not preclude the possibility that humans and other organisms may be exposed to large as well as smaller particles by inhalation. Additionally, intentionally produced nanomaterials may or may not behave in similar fashion to the ultra fine particles described in the above citations.

Deposited nanoparticles are typically not easily resuspended in the air or reaerosolized. Because physical particle size is a critical property of nanomaterials, maintaining particle size during the handling and use of nanomaterials is a priority. Current research is underway to produce carbon nanotubes that do not form clumps either by functionalizing the tubes themselves, or by treatment with a coating or dispersing agent, so future materials may be more easily dispersed. Many nano-sized particles are reported to be photoactive, but their susceptibility to photo degradation in the atmosphere has not been studied. Nanomaterials are also known to readily adsorb a variety of materials, and many act as catalysts. However, no studies are currently available that examine the interaction of nano-sized adsorbants and chemicals absorbed to them, and how this interaction might influence their respective atmospheric chemistries. Although most nanomaterials do not demonstrate high vapor pressures indicative of volatile chemicals, when released to the atmosphere, nanomaterials behave like a gas or vapor demonstrating Brownian motion. Many nanosized particles are reported to be photoactive, but their susceptibility to photo degradation in the atmosphere has not been studied.

3.4 CONCLUSIONS

Environmental air pollution is the major, increasing concern in the modern technological world. Nanotechnology is one of the emerging science which is one of the best alternative solution for controlling air pollution on the global level. However, it is in the stage of the development, but surely on the plot of the information presented in above text we can say that with adequate researches we can implement this technology for controlling air pollution.

Acknowledgements

We are thankful to all those who helped us for the successful completion and providing the valuable guidance throughout for presenting this matter. Our sincere thanks are due to the management and Prof B.B. Mistry, the Principal of Parul Institute of Engg. & Tech. Vadodara for their continuous encouragement and support.

REFERENCES

Engines of Creation by Eric Drexler.

Global Environment & Technology Foundation. 2000. Final Report: The US EPA Environmental Management System Pilot Program for Local Government Entities.http://www.epa.gov/owm/iso14001/emsrepor.pdf

Health Effects Institute, Communication 9, August 2001. www.healtheffects.org/pubscomm.htm.

National Institute for Occupational Health and Safety. 2004. Nanotechnology Workplace Safety and Health. http://www.cdc.gov/niosh/topics/nanotech/ default.html

Nanotoxicology Workshop: Developing Experimental Approaches for the Evaluation of Toxicological Interactions of Nanoscale Materials, University of Florida, Nov. 3-4, 2004. www.nanotoxicology.ufl.edu.

CHAPTER - 4

Our children's Earth - Exposed to air pollutions, kids' lungs show later function loss

V. S. Patel

President IAAPC, Consulting Engineer and Visiting Professor,
The Maharaja Sayajirao University of Baroda, Vadodara 390 002
e mail: vpatel_usa@hotmail.com

ABSTRACT

Children and infants are at greater risk from air pollution than adults for many reasons. Children tend to breathe more rapidly and inhale more pollutants per pound of body weight. Babies' and children's bodies, lungs and immune systems are still developing and more prone to air pollution's ill effects. Children also spend more time exercising outside, especially during summer months when smog levels are highest. Heavy exercising also means that children breathe through their mouths more. The nose provides a natural way to filter particulates from reaching the lungs, the mouth does not. This is a problem for anyone exercising outside.

Air pollution can affect our health in many ways with both *short-term* and *long-term* effects. Different groups of individuals are affected by air pollution in different ways. Some individuals are much more sensitive to pollutants than are others. Young children and elderly people often suffer more from the effects of air pollution. People with health problems such as asthma, heart and lung disease may also suffer more when the air is polluted. The extent to which an individual is harmed by air pollution usually depends on the total exposure to the damaging chemicals, i.e., the *duration of exposure* and the *concentration of the chemicals.*

Key words: Planet earth, Air pollutants, Children, Lungs, Asthma, Respiratory care.

Many children exposed to polluted air have poorly developed lungs at the age of 18 - a deficit that may put them at risk for respiratory illness and other health problems later in life, a study says. Scientists have known that smog can sicken or, in extreme cases, even kill people. But the new study, published in *The New England Journal of Medicine*, suggests that daily exposure to polluted air could leave teens with lungs that will never function they way they should. "We found some significant deficits in growth for kids breathing the most polluted air," says W. James Gauderman at the University of Southern California, who with his colleagues studied 759 fourth-graders living in 12 Los Angeles suburbs in 1993. For eight years, the team kept track of lung development and the amount of smog in these areas.

The team found that teens living in the most polluted areas were nearly five times more likely to have very low lung function: These kids had less than 80% of the lung function expected for their age. In most cases, the teens have enough lung capacity to play soccer and do most other daily activities without any problems. But the deficit when added to the normal age-related decline in lung function could cause trouble years later, says C. Arden Pope, an air pollution expert at Brigham Young University. "When I look at these data, yeah, I am concerned," says Pope, who says people with immature lungs might develop asthma or emphysema.

People with poor lung function often can't get over a cold or other respiratory infections as easily as people with healthy lungs, a problem that could put them at risk of getting severely ill or even dying from an illness that others shrug off, Gauderman says.

In this study, kids living in communities polluted with nitrogen dioxide, acid vapor, particulate matter and elemental carbon had the greatest risk of having underdeveloped lungs. Most pollutants come from car exhaust in the Los Angeles area, but the burning of fossil fuel, such as coal, can contribute to air pollution elsewhere.

The pollution could get worse if vehicles continue to clog the nation's highways in record numbers, Gauderman says. The findings likely would apply to any urban area in the USA with lots of car exhaust and other forms of pollution, says Jonathan Samet, a lung specialist at Johns Hopkins University. Some of the California towns in the study are the most polluted in the nation. No one really knows how breathing polluted air might hamper lung development. Gauderman suspects that inhaling tiny particles might trigger inflammation in the lungs, a process that might slow down the normal growth of the air tubes.

The researchers plan to keep track of the teens, as they grow older to see if the deficits really do persist and cause health problems down the line.

No Deep Breathing: Air pollution impedes lung development

– Ben Harder

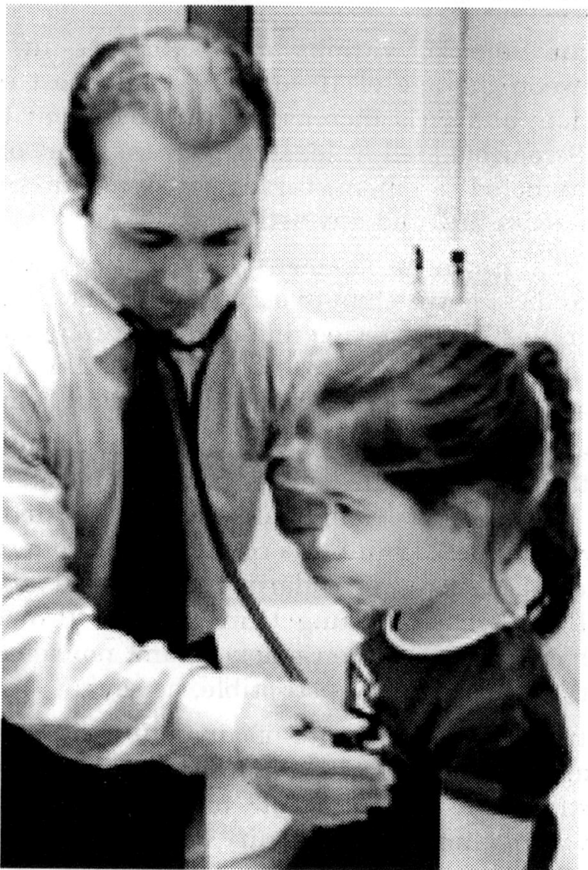

Fig. 1. A doctor examining a child

Spending one's childhood in a community with polluted air stalls lung development roughly as much as does having a mother who smokes, according to a study of children growing up in southern California.

That finding lengthens the list of negative effects on health that stem from bad air (SN: 8/2/03, p. 72: http://www.sciencenews.org/articles/20030802/bob8.asp), but it also suggests that antipollution measures could prevent deficits in pulmonary development. Adolescence is prime growing time for lungs. To evaluate the effects of air pollution during this important period, W. James Gauderman of the University of Southern California in Los Angeles and his colleagues in 1993 tested the lung capacity of 1,759 fourth graders in a dozen southern Californian communities, including some notably clean areas and some of the most polluted towns in the United States.

Then, through 2001, the researchers annually repeated the test on as many of the original volunteers as they could locate. Moreover, they recorded information on each child's health and exposure to tobacco smoke. Gauderman and his colleagues also took air samples, which they used to estimate concentrations of various air pollutants in each participating community. Between the age of 10 and 18 years, most of the boys doubled their lung function, and girls increased theirs by about two-thirds. About 4 percent of the 18-year-olds fell at least 20 percent short of the value that the scientists had predicted for them, taking into account such factors as their sex, size, tobacco exposure, and any asthma.

Teens who had grown up in towns with the dirtiest air were up to five times as likely to have lower-than-expected lung function as were kids in the cleanest communities. The average differences in lung capacity between the most-polluted and cleanest-air towns were similar to those previously observed among children whose mothers did and didn't smoke, respectively.

Airborne carbon, nitrogen dioxide, acid vapors, and particulate matter contributed to the diminished lung capacity, Gauderman and his colleagues report in the Sept. 9 New England Journal of Medicine. In southern California, they note, these pollutants come mainly from cars and trucks. The entire population faces long-term harm from modest air pollution, says C. Arden Pope III of Brigham Young University in Provo, Utah. The study indicates that boys and girls, teen smokers and nonsmokers, and kids with and without asthma are equally susceptible, he says.

"Even relatively small amounts of air pollution have fairly large, long-term effects on pulmonary health," Pope says. The harm to young lungs appears proportional to the pollutant" concentrations, no matter how low they fall? Any worsening of air quality, even within currently permitted ranges, is thus likely to harm public health, he says.

On the other hand, air-quality improvements could produce health benefits. "Extremely high concentrations of air pollution remain in many areas of the world, and decreasing these concentrations offer substantial opportunities for disease prevention," Pope says.

4.2 SHORT TERM AND LONG TERM HEALTH EFFECTS

Air pollution can affect our health in many ways with both short-term and long-term effects. Different groups of individuals are affected by air pollution in different ways. Some individuals are much more sensitive to pollutants than are others. Young children and elderly people often suffer more from the effects of air pollution. People with health problems such as asthma, heart and lung disease may also suffer more when the air is polluted. The extent to which an individual is harmed by air pollution usually depends on the total exposure to the damaging chemicals, i.e., the duration of exposure and the concentration of the chemicals must be taken into account.

Examples of short-term effects include irritation to the eyes, nose and throat, and upper respiratory infections such as bronchitis and pneumonia.

Other symptoms can include headaches, nausea, and allergic reactions. Short-term air pollution can aggravate the medical conditions of individuals with asthma and emphysema. In the great "Smog Disaster" in London in 1952, four thousand people died in a few days due to the high concentrations of pollution.

Long-term health effects can include chronic respiratory disease, lung cancer, heart disease, and even damage to the brain, nerves, liver, or kidneys. Continual exposure to air pollution affects the lungs of growing children and may aggravate or complicate medical conditions in the elderly. It is estimated that half a million people die prematurely every year in the United States as a result of smoking cigarettes.

Research into the health effects of air pollution is ongoing. Medical conditions arising from air pollution can be very expensive. Healthcare costs, lost productivity in the workplace, and human welfare impacts cost billions of dollars each year. In many countries in the world, steps are being taken to stop the damage to our environment from air pollution. Scientific groups study the damaging effects on plant, animal and human life. Legislative bodies write laws to control emissions. **Educators** in schools and universities teach students, beginning at very young ages, about the effects of air pollution. The first step to solve air pollution is assessment. Researchers have investigated outdoor air pollution and have developed standards for measuring the type and amount of some serious air pollutants.

4.3 SCIENTISTS MUST THEN DETERMINE *HOW MUCH* EXPOSURE TO POLLUTANTS IS HARMFUL ?

Once exposure levels have been set, steps can be undertaken to reduce exposure to air pollution. These can be accomplished by regulation of man-made pollution through legislation. Many countries have set controls on pollution emissions for transportation vehicles and industry. This is usually done through a variety of coordinating agencies, which monitor the air and the environment. At the United Nations, the Atmosphere Management Program carries out world wide environmental projects. In the United States, the primary federal agency is the Environmental Protection Agency. Many state and local organizations also participate in monitoring and controlling the environment. These include the San Francisco Bay Area's Air Quality Management District (B.A.A.Q.M.D.), the Air Quality Management District in southern California, the Environmental Protection Agency of California, Smog Busters of southern California, and the Galveston-Houston Association for Smog Prevention (GHASP).

Prevention is another key to controlling air pollution. The regulatory agencies mentioned above play an essential role in reducing and preventing air pollution in the environment.

In addition, it is possible to prevent many types of air pollution that are not regulated through personal, careful attention to our interactions with the environment. In the United States, most household products come with instructions about safe use. Additional information about product chemical safety are available in an article and a fact sheet from Enviroene, and in Material Safety Data Sheets (MSDS).

Building materials should be reviewed for potential harmful effects. Information on building materials can be found in LBNL's Center for Building Science newsletter. Additional information is available from EcoTech.

Adequate ventilation is also a key to controlling exposure to indoor air pollution. Home and work environments should be monitored for adequate airflow and proper exhaust systems installed. Additional information is available in a book titled *Understanding Ventilation*.

One of the most dangerous air pollutants is cigarette smoke. Restricting smoking is an important key to a healthier environment. Legislation to control smoking is in effect in some locations, but personal exposure should be monitored and limited wherever possible. Additional information about the effects of "secondhand" cigarette smoke is available from the American Association for Respiratory Care (AARC) and Medicine On-line.

Only through the efforts of scientists, business leaders, legislators, and individuals can we reduce the amount of air pollution on the planet. This challenge must be met by all of us in order to assure that a healthy environment will exist for our children and us. Find out "What you can do to reduce air pollution"?

4. 4 HEALTH EFFECTS IN CHILDREN OF DHAKA

A case study in Dhaka, Bangladesh is presented by **Manzoor Hussain. Air** pollution can affect the health of all human, regardless of age, but the greatest impact is likely to be among urban children. Increased urbanization and rapid industrialization in developing countries is causing a virtual epidemic of diseases related to air pollution among younger children. Most children who live in large cities in the developing world breathe air that is equivalent of smoking two packs of cigarette a day.

Respiratory disease is now the leading cause of deaths in children worldwide and air pollution in the developing world is responsible for at least 50 million cases of chronic cough. Children living in cities in developing countries face double jeopardy from both living in poverty and from exposures to degraded environment. Many children in cities in developing countries already suffer from hunger, malnutrition and infections, and do not have access to basic medical care. Air pollution only worsens their burden by aggravating diseases like bronchitis, asthma and other lung diseases. Medical practitioners have not been trained to relate specific illness or symptoms with air quality problems. So, connection between environmental causes and health effect is not easy to arrive at.

Dhaka city is the capital of Bangladesh, where about nine million people live in an area of 1353 sq.km. and over thousand immigrants join to swell the population of Dhaka for livelihood everyday from different parts of the country. Fifty-five per cent of the inhabitants live under the absolute poverty line and standard of living of about 2.5 million slum dwellers is precariously low in Dhaka city.

There is considerable evidence that the present level of atmospheric contaminants that exist in Dhaka is sufficient to cause discomfort or significantly impair child health. Sky of Dhaka gets overcast with smoke that forms due to uncontrolled emission of exhaust from 70-80 per cent of the total 175 thousand automobiles operating in the city.

From automobile emission - about 3,700 tones of suspended particles are added to the air of Dhaka yearly. Hundreds of brick kilns around the city are also belching smoke in the atmosphere. It is claimed that the density of lead in the air of Dhaka is the highest in the world and about 50 tones of lead is being mixed with air annually. Studies showed that children in Dhaka have blood lead level in excess of WHO guidelines.

Children of Dhaka are more at risk than the adults because their bodies are still developing. The lung of a child grows most rapidly in the first two years of life. They also tend to absorb pollutants more readily than adults do and retain them in the body for longer period of time. Because, they're breathing, rates are substantially higher than the adults and the children are exposed to greater levels of pollutants. Children with diet deficient of micronutrients have also weaker immune system. For this reason children below six are highly vulnerable to the toxic effects of air pollution.

On an average of 450 to 500 patients attend at the Out-patient Department including emergency room of Dhaka Shishu (Children) Hospital for treatment. Out of them, 350 to 400 patients attend for medical consultation. Disease pattern of out patient department of Dhaka Shishu Hospital shows that respiratory problems topped the list. Even ARI prevention programme launched in 1990 could not reduce the hospital attendance due to ARI in Dhaka. From November to February 1998, 52 per cent of total 13,025 patients were suffering from ARI and 33 per cent had wheezing who attended the out patient department of Dhaka Shishu Hospital.

Long term low level exposure to pollutants in the environment depresses immune system. The same may be responsible for vague undesirable illnesses. Common complaints of headaches, sneezing, coughing, nausea, fatigue, skin rashes, irritation of noses, throat and eye, and difficulty in concentration may be nature's way of informing us that there is a problem in the air quality. The ill effect of air pollution is most obvious during acute episodes of unusually high pollution. So, connection between environmental cause and health effect is not easy to arrive at when the children are exposed to long term low-level exposure to the pollutants. As little data is available — pediatricians are beginning to realise that cumulative effect of long-term exposure to low-level pollutants may, infect, be responsible for those vague undesirable illnesses or symptoms. Important pollutants and their effect on child-health is briefly discussed below:

Emission of smoke from motorized vehicle is the main culprit for impairing health status of a child. Smoke inhalation altars lung function resulting in chronic symptoms and significantly decrease performance endurance. Immediate effect of smoke inhalation causes headache, vertigo, burning sensation of the eyes, sneezing, nausea, tiredness, cough etc. Long-term effect of chronic smoke inhalation may cause asthma, recurrent respiratory tract infection, wheezy bronchitis, bronchitis, triggers asthma attacks. The effects are manifestations of body response to irritant materials present in the smoke that interfere with lung functions.

Sub-clinical toxicity has been reported in dense traffic situation. Symptoms consist of headache, lethargy, nausea, and vomiting and depressed sensorium. Carbon monoxide gas combines with hemoglobin to form carboxy-haemoglobin, which fails to carry oxygen and results in tissue anoxia (lack of oxygenation).

Children below six of Dhaka city are exposed to lead poisoning due to use of leaded petrol in the vehicle. Studies in USA indicate that 5 per cent to 15 per cent of their children have increased lead absorption. We need such studies to understand our situation. Lead poisoning is actually a chronic disorder that may be punctuated by acute episodes. We do not know the exact incidence. The clinical diagnosis tends not to be suspected until the onset of nervous system symptoms or evidence of sideroblastic anemia in the blood film. Chronic lead poisoning may present as non-specific development delay, behavioral and attention disorder or a peripheral neuropathy where weakness begins in the foot muscles. Mild sensory loss may accompany. Eventually hand muscles are affected. General symptoms include apathy, decreased play, clumsiness and intermittent vomiting.

The dust particles in the air collect bio-aerosols (microscopic living organisms or fragments of living thing). Bio-aerosols floating about our homes consists of viruses, bacteria, mold spores, fungi, yeast, animal and insect allergens, pieces of dead skin, microscopic dust mites and their excrement. Breathing these various elements all the time obviously puts pressure on our immune systems. Allergies, asthma, colds and flue are some of the health effects of these pollutants. Bio-aerosols can cause severe health problems during December to March period when total suspended particulate level of air of Dhaka remains high. During this period incidence of respiratory problems sharply rises among the children of Dhaka.

The recent increasing trend of the incidences of respiratory disease in Dhaka signals the acute necessity of combating air pollution in a befitting manner. Lead is a serious environmental pollutant. One can hope for cleaner air if persons in policy making positions recognize the problem like leaded gasoline and move quickly to enforce strict emission standards for automobiles; only then children will be able to breathe improved air. Author has surveyed few slum quarters at Vadodara city. Most of families are living in small unventilated room having size less then 900 sq. cm. The cooking is carried out half burn kerosene stove. The new born babies sleep on floor where pollution is very high. Most of the babies in slum area are suffering from asthma, colds and flue. On construction site, same problem, the new born infants are brought by working women on site where these children are exposed to the dust particles, Carbon monoxide and other pollutants.

We are pitifully behind the time-not just scientifically and technologically but more potentially in our ability to grasp the true nature of the problem in time. This time, if we fail to face the challenges of air pollution then there will be a man made disaster — the cost of morbidity will be unbearable for both the families and the nation.

This is our children's Earth. Think clearly about our future before polluting of more.

Reference

http://www.sciencenews.org/articles/20030802/bob8.asp

CHAPTER - 5

Significance of metals as air pollutants

A. D. Galgale and Bhavin Patel

*Civil Engineering Department, Faculty of Technology and Engineering,
The Maharaja Sayajirao University of Baroda, Vadodara 390 001*

ABSTRACT

Pollution due to metals is well known. Some metal contaminants are Cadmium (Cd), Chromium (Cr), Lead (Pb) and Mercury (Hg). Heavy metals are non biodegradable and tend to accumulate in living beings, causing various diseases and disorders. Metals in higher concentration are harmful to plants as well. The harmful effects of particulates are associated with their chemical nature. Heavy metals occur, in atmosphere, basically in particulate form. Hence the air borne particles to land or water surfaces by dry, wet and occult deposition constitutes the first stage of accumulation of heavy metals. The paper describes sources of metal pollutants and their harmful effects on human health. The harmful effects of heavy metal pollutants like Lead, Mercury, Cadmium, and Chromium are described for human beings. Maximum permissible limits for Lead is 1.5 $\mu g / m^3$. It may be produced by motor vehicle exhaust, lead smelters and battery plants.

Key words: Metals, Air pollution, Industrial emissions.

Particulates form a major part of emissions of air pollutants and come from diverse sources such as vehicles, materials handling and various other industrial processes. In general the term particulate refers to all atmospheric substances that are not gases, they can be suspended droplets or solid particles that are of mixture of two. Particulates may be composed of inert or extremely reactive materials ranging in size from 100 μm down to 0.1 μm and less.

The chemical composition of particulate pollutants varies over a wide range. The actual composition is very much dependent upon the origin of particulates. Particulate matters are measured as their mass concentration in the atmospheric air. The various standards related to particulate matter are available for the permissible concentration of the particulates in terms of mass concentration. The available standards give only physical quantities of the particulates. The harmful effects of particulates are associated with their chemical nature.

With the increasing use of various metals in industry and in our daily life problems arises due to metal pollution of environment have assumed serious dimensions. Another face of the problem of metal is in occupational health area when an industrial worker handling metal is exposed at high concentration leading to toxic manifestations and deterioration to its health status. The emission factor of various metals gives an average estimate at which a metal is released to atmosphere as a result of some activities.

5.1. LEAD

In the past, automotive sources were the major contributor of lead emissions to the atmosphere. As a result of regulatory efforts to reduce the content of lead in gasoline, however, the contribution of air emissions of lead from the transportation sector, and particularly the automotive sector, has greatly declined. Today, industrial processes, primarily metals processing, are the major source of lead emissions to the atmosphere. Lead is used as an industrial raw material for storage batteries manufacture, printing, and pigments and in fuel. It is one of the most widely used, non-ferrous metals. The highest air concentrations of lead are usually found in the vicinity of smelters and battery manufacturers.

Table 1. Lead Emissions Factors (Khan, 2001)

1.	Metallurgical industry	0.3175 Kg/Tone of product (controlled)
2.	Storage batteries	0.5896 Kg/Tone of Lead process (controlled)
3.	Municipal incineration	0.0907 Kg/Tone of charged (controlled)
4.	Sewage and sludge incineration	0.2721 Kg/Tone of charged (controlled)
5.	Pigments	4.5360 Kg/Tone of Lead process (Uncontrolled)

5.2 MERCURY

Because of its high volatability emissions of mercury emanate from any source where mercury will expose to atmosphere, or where any material bearing mercury is processed. Mercury and its compounds enter the atmosphere through emissions in the form of vapors and particulates from industries and also are evaporation from soil and water. Once in the atmosphere mercury widely returns to the earth surface as settleable particulates, but most of it returns with the rainfall.

The majority of mercury emissions are produced from fossil fuel combustion and solid waste incineration. These emissions are transported through the atmosphere as vapor and particulates and may be deposited on land and in water. EPA is in the process of developing new regulations that will significantly reduce mercury emissions from power plants.

Evaporation takes place as the ores are mined, transported, processed and used. Many metal ores contains mercury from percentage quantities to trace amounts. Coal in particular contents significant quantities of mercury. Power plant using 1000 of tones of coal / day can release large quantities of mercury even if present in trace amount. One study indicates that the concentration of mercury in air over non – mineralized areas ranges from 0.6 to 50 μg per cu. m.

Table 2. Emission factor for consumptive uses of mercury and its compounds (Khan, 2001)

Paint	650 Kg / 1000 Kg of contained Hg
Pharmaceutical	200 Kg / 1000 Kg of Hg applied
General Laboratory handling	40 Kg / 1000 Kg of Hg used

Table 3. The approximate estimate of Chromium containing emissions (Khan, 2001)

	Uncontrolled emission factor	Emission of Chromium after control in Tones per year
Electroplating	0.024 Kg / 1000 Kg	Negligible
Chemical processing (Dye Chromate)	15 Kg / 1000 Kg	92
Steel & Alloys, Super Alloys & Alloys	12 Kg / 1000 Kg	33

Table 4. Emission factors for cadmium from various Industrial sources (Khan, 2001)

Source	Emission factor
Pigments	8 Kg/1000 Kg of Cd charged
Plastic stabilizers	3 Kg/1000 Kg of Cd charged
Batteries (Ni-Cd)	1 Kg/1000 Kg of Cd charged
Rubber tire wear	0.003 Kg/106 Km
Super phosphate fertilizers	0.0001 Kg/103 Kg
Municipal incinerator (Uncontrolled solid waste incinerator)	0.0002 Kg/103 Kg waste incinerated

Fly ash particles are composed of stable elements or compounds that are usually not considered directly toxic in concentration found in ambient air. From the stand point of specific toxicity the chief importance of fly ash may be the tendency of trace metallic element vapors to condense on it as an effluent gas stream cools. These trace elements include many of known biological activities e.g. Cadmium, Vanadium, Nickel, Mercury and Manganese. Solid particles may initiate toxicity when inhaled or by the alimentary route when toxic compounds accumulate on forage, food or in nasal or bronchial mucus i.e. ingested.

Table 5. Concentration of Certain Metallic Elements in Airborne Fly Ash According to Particle Size (Ref. 3)

Particle Diameter (μm)	Element concentration (μg / g)									Sulfur Concentration (wt %)
	Pb	Tl	Sb	Cd	Se	As	Ni	Cr	Zn	
11.3	1100	29	17	13	13	680	460	740	8100	8.3
7.3-11.3	1200	40	27	15	11	800	400	290	9000	-
4.7-7.3	1500	62	34	18	16	1000	440	460	6600	7.9
3.3-4.7	1550	67	34	22	16	900	540	470	3800	8.3
2.1-3.3	1500	65	37	26	19	1200	900	1500	15000	8.3
1.1-2.1	1600	76	53	35	59	1700	1600	3300	13000	8.3
0.65-1.1	-	-	-	-	-	-	-	-	-	8.3

5.3. EFFECTS ON HUMAN HEALTH

5.3.1. Lead

Exposure to lead can occur from breathing contaminated workplace air or house dust or eating lead-based paint chips or contaminated dirt. Lead is a very toxic element, causing variety of effects at low dose levels. The major source of exposure to lead is from food and water, with estimated intake of 0.12-0.35 mg/day. But only 5-10% of ingested lead may be metabolically absorbed in adults, whereas about 20 to 50% of the lead inhaled from community air pollution may be metabolically absorbed into the body.

The percentage absorbed tends to be higher with smaller particle sizes. Acute forms of toxicity are manifested when blood levels in the adult males are greater than 75 μg/100gm. Three types of toxicity are documented gastrointestinal cramps, central and peripheral nervous system effects and anemia.

Lead particulates may settle on the objects and handling of such objects, lead paints etc. specially by the children, cause sufficient ingestion of lead resulting into anemia and effects on the brain.

Children are particularly sensitive to the chronic effects of lead, with slowed cognitive development, reduced growth and other effects reported.

Table 6. Lead concentration in Air and in Blood of the primary school's children in two comparable communities

Community	Lead concentration in Air μg/m³	Lead in Blood of children (μg/100gm)	
		Male	Female
A	3.27 ± 1.59	23.3 ± 4.70	20.4 ± 2.91
B	1.87 ± 1.37	16.8 ± 4.01	17.1 ± 4.37

5.3.2. Mercury

Mercury occurs in the environment in several physical and chemical forms and the most common forms are in the inorganic forms and organo-mercurials. "Organic mercury compounds" are those in which mercury is linked. The organic mercuric compounds are further subdivided into the alkyl, aryl and alkoxyalkyl groups. There is a great variation in the toxicity among the organic forms.

Metallic mercury enters the body primarily via inhalation. When elemental mercury is spilled in the work place or the home, vapor inhalation can occur. In workmen occupationally exposed to mercury, some mercury may be found in the exhaled breath samples, and skin and clothing may absorb mercury and disseminate it after the workman come home to his family. Thus, an exposed workman may become a source of pollution in his own home.

Elemental mercury is not absorbed to any significant degree by the gastrointestinal tract. Elemental mercury can easily diffuse from blood into tissues. Experiments in animals have shown high uptake of mercury by the brain after exposures to mercury vapor. Following absorption in the blood, metallic mercury is oxidized mercury joins complexes with sulfhydryl groups and other ligaments in the body or tissues.

Mercury is initially transported from lungs to tissue (as this process will take some time) mostly as physically dissolved mercury vapor. After oxidation the blood cells and plasma contained almost equal amounts of mercury. The placental transfer of elemental mercury has not been studied in detail as elemental mercury penetrates membranes easily. A considerable amount penetrates the blood–brain barrier. The brain is the critical organ after exposure to metallic mercury.

Aerosols of inorganic mercury compounds are absorbed via respiratory tract to a considerably lesser degree than the mercury vapor. The kidney, which generally contains the higher concentration of mercury, is the critical organ for exposure to inorganic mercury compounds.

5.3.3. Cadmium

The normal concentration of cadmium in air is about 0.001 μg/m^3 and at this concentration there would not be any significant daily intake of cadmium. The cadmium concentration in air around cadmium factories, however, may be several hundred times greater. Absorption via inhalation will vary with different chemical forms of cadmium and different particle sizes.

Once absorbed, cadmium will be found in blood, organs and excreta. With chromic exposure, the cadmium in the blood will be found primarily in the red cells. Inhalation of high concentration of cadmium dust or fumes in the industry may lead to kidney damage and lung damage. Cadmium may also give rise to dysfunction of the liver and to anemia. Industrial exposure to cadmium has been associated with an increase of lung and prostate cancer.

5.3.4. Chromium

Occupational exposure to hexavalent chromium compounds, which are both irritant and corrosive, has caused symptoms in both the respiratory tract and the skin. Nasal perforation has been a common finding in workers exposed to chromates. It is also well established that lung cancer may result from occupational exposure to the chromates.

In one study of effects on general population of emissions from a chromate producing plant in Japan, "light" cases of paryngitis were found at a concentration of chromic acid of < 1.5 μg/m^3. Exposure to trivalent chromium compounds has not been found to cause any significant effects. Whereas, in most human body organs chromium levels do not change or even decrease during a lifetime, there is an accumulation with age in the lungs, indicating that chromium has a long biological half time in that organ. In the Soviet Union, the threshold concentration for chromium to cause nasal irritation and reflex activity has been found to be 2.5 μg/m^3.

5.4. STANDARDS FOR METALS IN AMBIENT AIR

National Ambient Air Quality standards give permissible concentration for lead only. For other metals standards are not available for ambient concentrations.

The following tables can serve as a guideline to decide the permissible concentrations in ambient air for metals such as Cadmium, Chromium, Mercury and other metals.

Table 7. Permissible Levels of different Metals in work environment (The second Schedule appended to Factories Act, 1948) (Athar & Vohra)

Sr. No.	Metals and compounds		Permissible limits in mg/m^3 On average concentration (8 Hr)
1	Arsenic and soluble compounds	(as As)	0.20
2	Cadmium dusts and salts	(as Cd)	0.05
3	Copper Fumes		0.20
4	Chromic acid and Chromates	(as Cr)	0.05
5	Lead, inorganic, dusts & fumes	(as Pb)	0.15
6	Manganese fumes	(as Mn)	1.00
7	Manganese fumes	(as Mn)	1.00
	Mercury	(as Hg)	
8	Alkyl compound		0.01
	All forms except alkyl vapour		0.05
9	Nickel carbonyl	(as Ni)	0.35
10	Tetraethyl Lead	(as Pb)	0.10
	Zinc Oxide		
11	Fumes		5.0
	Dust		10.0

Table 8. Exposure Levels of Heavy Metals (µg / day) (Athar & Vohra)

In Air	Cadmium	Lead	Chromium	Nickel
Average	0.005-.010	0.16-10.2	0.16-2.1	0.08-1.04
Medium	0.005-.003	0.16-3.7	0.08-1.1	0.008-0.46
P (90)	0.01-.020	0.37-20.1	0.3-3.2	0.12-1.92
WHO value	0.25	4-16	0.2	0.4

REFERENCES

Athar, M. and Vohra, S.B. Heavy metals and environment.

Elsom, D. Atmospheric pollution, causes, effects and control policies.

Environment Protection Agency (EPA) website.

Khan, R.R. 2001 Environment and metal pollution, I[st] Edition, 2001.

Sittig M. Environmental sources and Emissions Handbook.

Stern, A.C. Air pollutants, their transformation and transport, Third Edition, volume-I, The effects of Air pollution, volume-II.

CHAPTER - 6

Mitigating urban air pollution through town planning practices

Preety Shah

D.C. Patel School of Architecture, Vallabh Vidyanagar, Anand 388 120
e mail: preetyarch@rediffmail.com

ABSTRACT

Urban atmosphere is affected by pollution generated by processes of urbanization and industrialization. Air pollution has a number of biological, economic and meteorological implications. But the major source of concern is the harmful effects created due to introduction of man-made pollutants in the atmosphere. The pollutants can cause health effects like chronic bronchitis, asthma, cardiovascular diseases, eye and skin irritations and damage to animals and plants. The cities have high suspended particulate matters which demands immediate attention besides $PM_{2.5}$ particles. A comprehensive urban air quality management strategy is required using proper urban planning. Urban planners can contribute to air pollution control through proper land use planning and landscape planning. This paper examines in depth, of how judicious urban planning strategies including land-use planning, transportation planning, landscape planning, building bye-laws etc. can help to generate and maintain a healthful urban environment.

Key words: Urban air pollution, Mitigation, Town planning, Landscape planning.

Man, until recently, has largely ignored the fact that locally and globally, he has overloaded the natural self-cleansing capacity of the atmosphere so as to produce very high and perhaps dangerously high concentrates of air pollutants. Because man's socio-economic organizations have concentrated the waste forming activities in towns, it is here that the most serious problems which the human race is witnessing, are generated.

The freshness of the air in one's environment has a most fundamental and direct impact on the quality and length of one's life. Air is more a necessity of life than either food or water. Although air pollution is only one of the many environmental hazards in urban centers of the world, along with water contamination, hazardous wastes, overcrowding, congestion, and so on, it is a unique problem as it affects each and every resident, without any doubt.

The city is a physical as well as a social setting. When these two factors interact and when the choices made by urban dwellers overtax the environment's ability to cope effectively with those choices, environmental problems arise. Like all problems, environmental difficulties are intensified in the city because it concentrates a very large number of people in such a relatively small space.

6.1 AIR - VITAL ELEMENT OF ENVIRONMENT

Air is the swirling mixture of gases and vapour that surrounds the planet Earth. The flowing air mass around us is extremely susceptible to contamination which is rapidly dissipated into the surrounding atmosphere for an area extending many square kilometers and to a height of many thousands of meter.

Air pollution is the introduction into the air, any substances different from its natural constituents. Major source of concern is the harmful effects created due to introduction of man-made pollutants in the atmosphere. Air pollution in its many forms is a killer and devastator. To untold thousands of humans it brings sufferings and premature death. It is debilitating to livestock. In addition, it causes extensive economic loss through damage to materials-eroding stone, corroding metals, making rubber crack and nylons run. It shortens the useful life of everything that it touches.

Each year we add 270 million tons of highly toxic carbon monoxide to the atmosphere. Of the total Air – pollution, about 70% is generated through vehicles and automobiles. It is believed by many that air pollution is the price we must pay for progress. This is far from the truth. First of all, what is progress unless it includes the assurance of a wholesome place in which to live? Moreover, there are case histories to prove that many communities and industrial cities have learned to achieve increasing prosperity while providing an attractive environment in which people can breathe clean air.

6.2. CAN AN URBAN PLANNER PLAY A ROLE IN CONTROLLING AIR POLLUTION?

The answer to this question is a capital YES! Urban Planners can contribute to air pollution control through judicious Land-Use Planning involving zoning, proposing suitable transportation networks, appropriate Landscaping and well distributed green spaces and suggesting proper urban design and building bye-laws.

6.2.1. Landuse Planning

Large cities generate their own traffic and unless growth of cities is checked, increasing urbanization throughout the world will result in even more movement within cities in proportion to the population. **While planning new areas or re-planning the existing areas, it is possible to consider the relative location of different types of land use in relation to accessibility.**

This means that changes should be strategically utilized to strengthen environmentally-friendly city-structures and transport systems, improve the physical environment and promote health-enhancing activities. This can be accomplished through long-term preventive planning and more-short term measures. **A core strategy would be to develop more concentrated cities and city structures in order to reduce the pressure on important areas for recreation and outdoor activities. Such a pattern of development will reduce the need to travel and increase the efficiency of public transport.** Denser cities will also give the possibilities for people to travel by foot or bicycle, and thus contribute to a better environment and better individual health.

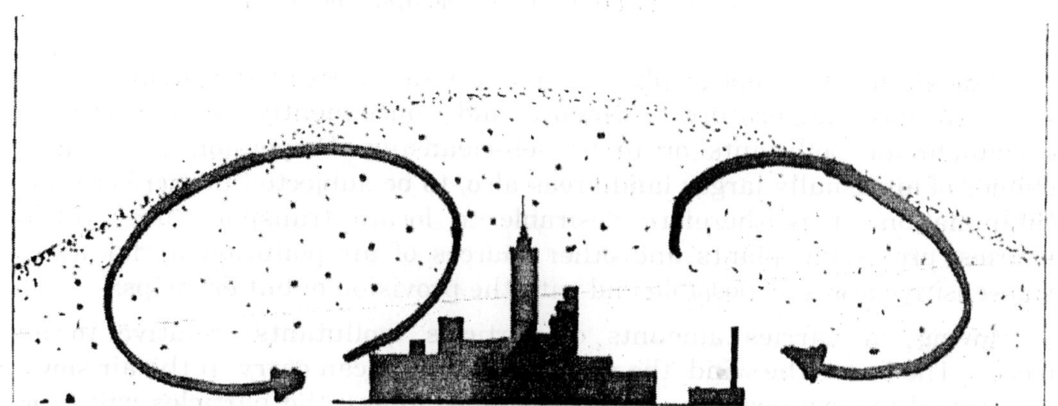

Figure 1

DUST DOME: Dust dome takes shape periodically over large cities because of the particles of dust and smoke that enter the air as a result of activities in the city. Air tends to rise over the warmer central part of the city, and to settle over the cooler environs, so that a circulatory system develops. 80% of the solid contaminants remain suspended for several days in still air. The dust dome is likely to persist, significantly affecting the city's climate, until a strong wind or a heavy rain disperses it.

Physical planning criteria, the distribution of land uses, provision for movement routes and designed form must derive, from and be closely related to environmental conditions. A close fit between these and the other environmental factors enables the city to be regarded as part of an organic ecosystem which includes landform, air, water, the nature of the soil and the climatic conditions. The slope and orientation of the ground affects wind exposure, solar gain, differences in temperature, movement and humidity.

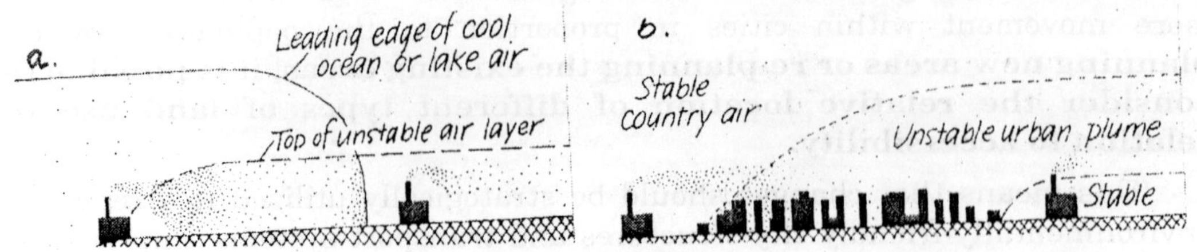

Figures 2:a,b.

Suggests the pattern of distribution of polluted air(plume) in near sea/lake and in urban areas.

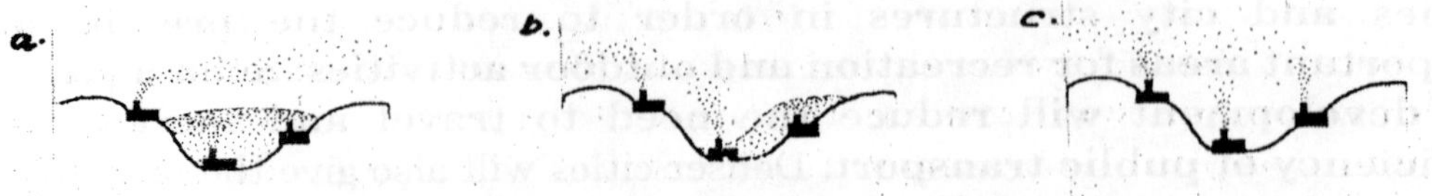

Figures 3:a,b,c.

Plume behaviour depends on topographic position.

The shape of the topography also has a vital effect in the occurrence of fog, frost and temperature inversion, and consequently its capacity to accumulate air pollutants or to be self-cleansing. Dispersion into larger volumes of air usually larger land areas also, to be subjected to increased air contamination. It is therefore desirable to locate transportation routes, factories, processing plants and other sources of air pollution as far away from sensitive areas as possible and with the provision of buffer strips.

Moving air carries amounts of particles (pollutants) relative to its velocity. The faster the wind, the more pollutants it can carry. If the air slows (or is forced to slow down by a fence or wind break), the particles will drop out and be deposited.

6.2.2 Landscape Planning

On its way into the city the quality of the incoming air will be increasingly lowered, accumulating a higher proportion of particles and gases and becoming progressively deoxygenated. This effect will be most marked near the city center. **Parks, especially those with trees, which interrupt**

the inflowing air, will alter the wind flow and improve the quality of air. An effective distribution of adequate numbers of trees in the city can be thus deliberately arranged to put the normal convection cycle to useful purpose not merely in providing additional oxygen and cleansing the air, but also to introduce cooler, fresh air into the city.

In polluted environments grit and dust particles are deposited and accumulate where local turbulence occurs close to ground level. The edges of highway curbs, if unclean, frequently harbour a wide band of fine dust and large gritty particles. The disturbance of this from vehicle movement can cause great local discomfort in addition to the noise and exhaust pollution from urban highways. Extensive areas of our cities, therefore, become virtually uninhabitable.

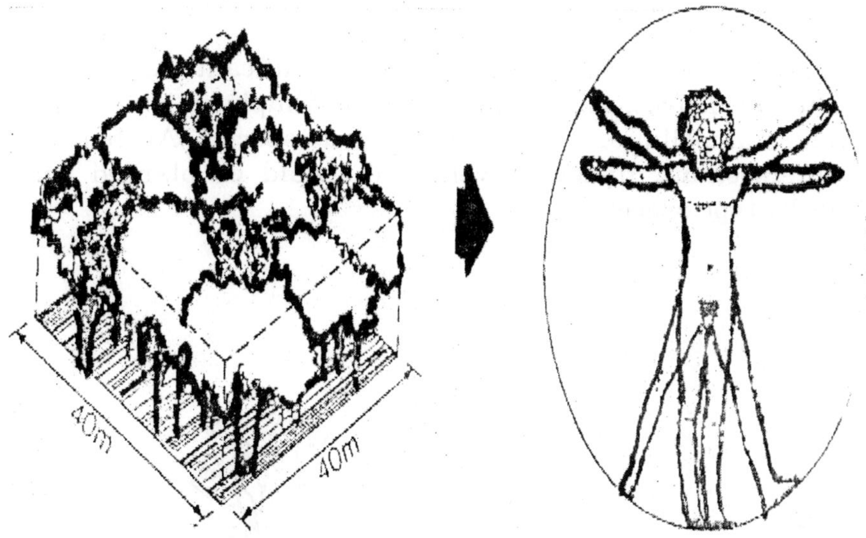

Figure 2

Trees Oxygen Producing Machines.

The removal by plants, during photosynthesis, of carbon dioxide, and replenishment of tree oxygen in the air is of great significance. The presence of plants in polluted atmospheres dilutes the level of pollution by increasing the oxygen content.

An area of trees 30 × 40 metres supplies the oxygen requirements for one man. The city is also a prodigious generator of heat. Consumption of fuel energy both for heating buildings and particularly from vehicles, are a major source of heat production. But plants can ameliorate the temperature, airflow and humidity.

The role of grass areas, planting trees, most particularly in controlling and ameliorating many of these environmental problems, is of the greatest significance in both the planning and the detailed design of the external spaces in our cities. Plants have the ability to mitigate air pollution. It is said that a tree can collect on an average, 5 to 10 kg of dust per day and act as an effective screen against the movement of dust particles/sand by wind. In replenishment of atmospheric oxygen, in transpiration and in modifying solar radiation, the plants control the air movement caused by both wind and convection; they therefore, influence humidity and precipitation. In thus balancing climatic extremes, all vegetation, but trees most particularly, have a role in ameliorating the city climate which is not merely important but also quantifiable.

Pollution level is generally very high at road intersections due to high volume of traffic and congestion causing adverse effects on the environmental quality of urban areas, particularly in and around the heavy traffic intersections. Planting suitable species in the polluted urban areas, bus stops/ terminals, along avenues and intersections etc. would help mitigate pollution to a reasonable amount.

The value of trees in cleansing the air is significant. Airborne particles are intercepted on leaves, stems and branches. Emphasis should be laid on selecting vehicular pollution tolerant indigenous species. All bare spaces at junctions, intersections, under flyovers, etc should be planted with trees, shrubs or bushy type species.

Figure 3a.
Streets without trees:10,000 to 12,000 dust particles per litre

Figure 3b.
Streets with trees:1,000 to 3,000 dust particles per litre.

Dust pollution – the value of trees.

Plants cleanse the air by collecting air borne particles; a mature tree can intercept several hundred weights of dust on its leaves. This is periodically removed by heavy rainfall. The difference in dust pollution between streets with and streets without trees can be quite dramatic.

Trees like Neem, Amaltas, Ashok, Arjuna, Pipal, Sisham, Tamarind etc.. and shrubs like *Bougainvillea* and oleander are suitable plants for pollution control.

6.3 TRANSPORTATION PLANNING

Motor vehicle traffic is a major source of air pollution in large- mega cities. In some of the cities, it is the single most important source. It is a major source of four of the six major air pollutants-carbon monoxide, nitrogen oxides, hydrocarbons, lead and contributes to the SPM concentration as well. Within the next twenty years, it is said that the number of vehicles in the world is likely to double from its present total of 650 million vehicles. As cities expand into mega-cities, more people will drive more vehicle greater distance and for longer times. In the absence of controls, the automotive emissions will likewise increase. In cities where a substantial proportion of the motor vehicle fleet is diesel-powered there are additional problems of black smoke and greater particulate emissions. **No doubt, the need for automotive emissions controls is paramount, but more important is the judicious transportation planning which will relieve the town and city of the unwanted air pollutants. In the long term, preventive measures must necessarily be incorporated in new industrial and urban developments.** Proper urban and transportation planning can achieve significant improvements in air quality, by the use of improved modern technology. Urban sprawl leads to greater travel distance functions and poses an increased demand for transport services.

Since the exhaust fumes from automobiles are a principal source of air pollution, the arterial roads have to be planned as controlled-access roadways. The fact that they will be substantially free of stop-and-go traffic will materially reduce the volume of emissions over the conditions that prevail on conventional grid street patterns. In addition, the provision of an extensive open green space framework will provide natural ventilation for the building groupings. Even the prevailing winds will help to protect the air quality since the predominant direction of air flow will be away from populated areas.

A well-defined hierarchy of streets and motorways would contribute a lot to abating urban air pollution. Transportation routes and highways particularly are presently the major source of air pollution. Improvement can be done in several ways: their roadways may be reduced in scale as by the reduction in the number of lanes.

Gradient of the road also affects the work imposed on an engine. Thus, hazards are likely to occur at busy junctions and are relevant only to people working and living in buildings close to the point at which vehicles stop and start. Although the only way to control atmospheric pollution from vehicles is at the vehicle itself, some amelioration of the effects can be derived by changes in the highway pattern and by traffic management *e.g.* a by-pass or relief road will divert traffic away from streets where such traffic has no business and so reduce the area affected.

Proposal, promotion and provision of an effective mass transport system to reduce the use of personal vehicles will be a great asset in keeping the air pollution in cities under a check.

6.4 CONCLUSIONS

To sum up, following recommendations emerge as ways and means by which air pollution can be controlled.

1. Land-use Planning should take into consideration also the topography of the land as well as all the natural processes like heat islands, temperature inversion, wind velocity, radiation etc.
1. Major vehicular street sections would have to be designed in such a manner that the projections of upper floors do not interrupt the dilution and dispersion of polluted air over larger areas at the top.
1. Buildings abutting main roads with heavy traffic should have limited fenestration and other openings directly into the said streets.
1. Major roads must have enough vegetation to collect dust particles and pollutants over its leaves which can be cleaned everyday by a wash of water from the tip of the trees.
1. Properly spaced system of green areas cutting through built up area will help in dispersion of polluted air by funneling effect.
1. Selection of tree species and other plant material will drastically influence conditions for air pollution control.
1. Main road network should serve the public transport nodes within and outside the cities. Vehicular traffic should be avoided on narrow street sections; instead pedestrianization should be promoted. Also all cities must rely and lay stress on a good public transport system.

We cannot do away with vehicles; they are now a necessity and a very important part and parcel of urban life. But the above recommendations would definitely work towards abatement of pollution of air in urban areas.

REFERENCES

1. Shah P. 2003 Paper presented at National Seminar: *Urban Air Pollution: Issues and Solutions*, The M.S. University of Baroda Vadodara, March 2003.
2. Gurumukhi, J.S. 2003 *Plants for abatement of vehicular pollution: A case study in Delhi*, Spatio-Economic Development Record, May-June 2003
3. Lenihan J. and Fletcher W.W. 1978. *The Built Environment*, Blackie, Glasgow and London.

CHAPTER - 7

Auto fuel policy - Government of India

V.S. Patel[1] and Ravi Dhiman[2]

[1]President IAAPC,
Consulting Engineer and Visiting Professor,
The Maharaja Sayajirao University of Baroda, Vadodara 390 002
email : vpatel_usa@hotmail.com
[2]Environment Science Programme
The Maharaja Sayajirao University of Baroda, Vadodara - 390 002

ABSTRACT

There were several agencies who tried to find fault with energy sources being used and provided solutions without any strategies. When many individuals/organizations attempted to virtually force their views based on a series of untruths, and half-truths, and affecting government work. The Prime Minister's Office had had to intervene and set up an expert committee, headed by Dr R.A. Mashelkar, Director-General, CSIR, to device an "auto fuel policy' for the country. This committee, consisting of eminent technologists, environmental scientists, automobile experts and economists, submitted its interim report which was approved by Cabinet. The panel has recommended an auto fuel policy which targets vehicular emission standards rather than fuel type, and permits the use of merit-based fuel and engine technologies. The report categorically suggests that mandating single fuel is neither a technically-sound nor scientifically- or strategically-wise option. It suggests that no single fuel be made mandatory and there should be a choice between gaseous and other liquid fuels for public transportation. Thus, the expert committee report, by recommending multi-fuel choice, has sought to align India with practices followed in the developed countries. The chapter describes the emission standards for petrol, diesel and CNG which are commonly used as fuels in the country.

Key words: Diesel, Petrol, Auto fuel, Policy, Government of India.

Use of different fuels to drive the vehicles has been tried from time to time. Still we have not found a pollution free fuel. The current reliance on petro-based fuels and chemicals is not sustainable as the energy demand is projected to grow by more than 50% by 2025 (Ragauskas, 2006). Multifaceted approach has resulted at large to think a number of alternatives like nuclear, solar, hydrogen, wind, electricity, fossil fuels (petrol, diesel, LPG, CNG) and biofuel etc. Rudolph Diesel in 1895 demonstrated the use of oil obtained from different varieties of plants. Brazil is the only nation where majority of vehicles are driven on cane – derived ethanol (Koonin, 2006). The majority of petroleum has its origin in algae, which were grown using CO_2 as a sole carbon source. To find zero pollution fuel, scientists are working to grow micro algae in bio-reactors. Algae contain up to 60% fat. The researches have found that after pressing the algae to get the oil the remaining waste can be processed to produce ethanol. Per hectare yield of oil rich algae is more than that of sunflower or rapeseed (Daniel 2005). Efforts have been done to grow Cyanobacteria for hydrogen production which will surely replace most of present day fuels. The need to set standards for different fuels is very essentials. Government of India has proposed standards for common fuels used in the country like Petrol (Table 2) Diesel (Table 1) CNG (Table 3).

One eco-friendly fuel is ethyl tertiary butyl ether (ETBE), which is derived by reacting C4 olefinic petroleum stream with ethyl alcohol. Existing MTBE plants can be retrofitted to produce ETBE. Use of biofuels has a special significance in India. They not only promise a better and clean environment but also open growth opportunities for millions engaged in agro cash crop plantations and its processing industry. If its use can be coupled as blends alcohol / biodiesel, it can provide an answer to much required energy security needs of the country. It has been well documented that oxygenates such as alcohol, ETBE and bio-diesel, significantly improve emission quality even in old vehicles, including life of after exhaust treatment devices. These eco-friendly products are now already in use in the developed countries, including Europe and the US.

Talking about long-term needs for clean fuels, hydrogen gas would be an undisputed best choice, say, by 2020 either as such in the gaseous form or through fuel cell route. It is 100 per cent eco-friendly, as its combustion product is simply water. Hydrogen can be generated electrolytically, chemically or biotechnically and India has large potentials in this area. Efforts have been done to grow Cynobacteria for hydrogen production. In this regard, mention must be made of Benaras Hindu University (BHU) and some other academic bodies which have done pioneering work in this area and brought technology to the take-off stage. They have demonstrated the feasibility of their novel technology for generation as well as hydrogen storage technology based on reversible hydrides using indigenously available materials. The BHU's interdisciplinary group on hydrogen energy has demonstrated successfully the possibility of hydrogen use to fuel two-wheelers and other devices such as home cookers and the gen-sets. India should take an initiative and if due time and priority is given, could show the way to others. Hydrogen generation has a synergy with refining and fertilizer industry, which may find their reformers for large-scale hydrogen production, storage in hydride form and its marketing is an attractive commercial venture.

Table 1. Indian Diesel specifications required to meet Euro equivalent norms

Characteristics	Unit	Requirements for Euro II or Bharat II Emission Norms	Requirement for Euro III or Bharat III Emission Norms	Requirement for Euro IV or Bharat IV Emission Norms
Acidity inorganic	-	Nil	Nil	Nil
Acidity, total, mg or KOH/g max	mg of KOH/g	0.2	Nil	Nil
Ash, max	% mass	0.01	0.01	0.01
Carbon residue (Ramsbottom) on 10% Residue, max	% mass	0.3 without additives	0.3 without additives	0.3 without additives
Cetane number (CN) min		48*	51	51
Cetane index (CI) min		46*	46	48
Pour point, max as per OCC Directive				
Winter (Nov - Feb) max	°C	3	6	Nil
Summer, max	°C	15	Nil	Nil
Copper strip corrosion for 3 hrs @100°C max & for euro 3 & 4 @ 50° max		Not worse than No 1	Rating class 1	Class 1
Distillation :				
(A) at 350°C, min recovery	% volume	85	Nil	Nil
(B) at 370°C min recovery	% volume	95	Nil	Nil
Flash point				
Abel, min	°C	35	35	35
Kinematics viscosity @ 40°C	cst	2.0-5.0	2.0-4.5	2.0-4.5
Sediment, max	% mass	0.05	Nil	Nil
Density @ 15°C	kg/m^3	820-860 (820-870*)	Nil	Nil
Total Sulphur, max	% max	0.05	350 mg/kg	50

Water content, max	% volume	0.05	200mg/kg	200
Cold filter plugging point (CFPP)				
Winter (Nov - Feb) max	°C	6	18	18
Summer, max	°C	18	6	6
Total sediments, max	mg/100ml	1.6	Nil	Nil
Lubricity, corrected wear scar diameter (wsd1, 4) @ 60°C, max	micron	460	460	460
Distillation 95% vol recovery at °C max	°C	Nil	360	360
Total contaminations, max	mg/kg	Nil	24	24
Oxidation stability, max	gm^3	Nil	25	25
Polycyclic Aromatic Hydrocarbon (PAH) max	% mass	Nil	11	11

* For diesel processed from Assam crude, either CN of 45 min and density of 820-870 shall be applicable

Center for Science and Environment (CSE) is not satisfied with the set standards. Despite noting at the outset that "public health is a prime concern", the report offers nothing to address this concern. Even in 2005 diesel will have sulphur content as high as 350 ppm with no cap on polycyclic aromatic hydrocarbons (PAH) that makes diesel fume such a strong carcinogen. Petrol will be rich in aromatic content that is extremely toxic. Indiscriminate use of unsafe oxygenates like MTBE will continue (www.cseindia.org/campaign/apc/press _20020108.htm).

"It is a farce to even talk about letting emissions standards govern when we do not even have a legally enforceable system to ensure compliance with the regional air quality targets," says Narain. She cites the recent case in Maryland, USA, where federal grants were cut as the state had failed to meet the air quality target. In the US, the federal clean air act specifies that if regional air quality targets are not met, the states can lose federal highway funds. Monetary penalties and fear of court action keep the states in line. "It is meaningless to talk about emissions standards and air quality objectives without a legally enforceable compliance clause" says Narain (www.cseindia.org/campaign/apc/press_20020108.htm). This multi-fuel policy option is open to spirit of innovation and would promote competitiveness. This is also in line with the practice followed in other parts of the world, and is ideally suited for India. Opportunity exists for new alternate environment-friendly technologies, which also need to be considered.

Table 2. Indian Petrol specifications required to meet Euro equivalent norms

S.N.	Characteristics	Unit	Requirement for Euro II		Requirements for Euro III or Bharat III		Requirements for Euro IV or Bharat IV	
			Un-leaded	Un-leaded	Un-leaded	Un-leaded	Un-leaded	Un-leaded
			regular	Premium	Regular	Premium	Regular	Premium
01	Color, visual		Orange	Red	Orange	Red	Orange	Red
02	Density @ 15°C	Kg/m³	710-770	710-770	720-775	720-775	720-775	720-775
03	Distillation:							
	(a) Recovery up to 70°C (E70)	%Volume	10-45	10-45	10-45	10-45	10-45	10-45
	(b) Recovery up to 100°C (E100)	%Volume	40-70	40-70	40-70	40-70	40-70	40-70
	(c) Recovery up to 180°C (E180) for euro III & IV upto 150° (E 150) min	%Volume	90	90	75 min	75 min	75 min	75 min
	(d) Final Boiling Point (FBP)Max°	C	215	215	210	210	210	210
	(e) Residue, max.% Volume		2	2	2	2	2	2
04	Research Octane number (RON), min		88	93	91	95	91	95
05	Anti-Knock Index (AKI), min		84	88	Nil	Nil	Nil	Nil
06	Existent Gum, max	g/m³	40	40	Nil	Nil	Nil	Nil
07	Potential Gum, max	g/m³	50	50	Nil	Nil	Nil	Nil
08	Sulphur, total, max.	%mass	0.05	0.05	150mg/kg	150mg/kg	50mg/kg	50mg/kg
09	Lead content (aspb), max	g/l	0.013	0.013	0.005	0.005	0.005	0.005
10	Reid Vapor pressure (RVP), max	Kpa	35-60	35-60	60	60	60	60
11	Vapor Lock Index (VLI =10RVP+7E70)							
	a) Summer, max		750	750	750	750	750	750
	b) Other months, max		950	950	950	950	950	950

12	Benzene Content, max	%Volume			1	1	1	1
	a) For the Metros		3.0	3.0	Nil	Nil	Nil	Nil
	b) For the rest		5.0	5.0	Nil	Nil	Nil	Nil
13	Copper strip corrosion for 3 hrs@ 50°C, max	Rating	Not more than 1	Not more than 1	Class1	Class1	Class 1	Class 1
14	Water tolerance of Gasoline-alcohol separation							
	a) Summer, max	°C	10	10	Nil	Nil	Nil	Nil
	b) Winter, max	°C	0	0	Nil	Nil	Nil	Nil
15	Engine intake system cleanliness	MFA used	To report	To report	Nil	Nil	Nil	Nil
16	Olefin content, max	%Volume	Nil	Nil	21	18	21	18
17	Aromatics content, max	%Volume	Nil	Nil	42	42	35	35
18	Oxygenates Content	% mass	Nil	Nil				
	(a) Methanol, max	%Volume	Nil	Nil	3	3	3	3
	(b) Ethanol, max	%Volume	Nil	Nil	5	5	5	5
	(c) Iso propyl alcohol, max	%Volume	Nil	Nil	10	10	10	10
	(d) Tertiary butyl alcohol, max	%Volume	Nil	Nil	7	7	7	7
	(e) Ethers containing 5 or more carbon atoms per molecule, max	%Volume	Nil	Nil	15	15	15	15
	(f) Other oxygenates, max	%Volume	Nil	Nil	8	8	8	8
	(g) Iso butyl alcohol, max	%Volume	Nil	Nil	10	10	10	10
19	Oxygen content, max	%Volume	Nil	Nil	2.7	2.7	8	8
20	Gum content (solvent washed) max	Mg/100	Nil	Nil	5	5	5	5
21	Oxidation stability, min	min	Nil	Nil	360	360	360	360

Table 3. Proposed specifications for CNG as auto – Fuel in India

S.N.	Constituents	Value	Tolerance
1	Wobber Number	1350	+ / - 20%
2	Water, lbs / million ft^2	0.5	Maximum
3	Hydrogen Sulphide, grains/100 ft^3	0.1	Maximum
4	Other Soluble Sulphide, grains/100 ft^3	0.1	Maximum

5	Carbon Dioxide, vol. %	3.0	Maximum
6	Oxygen, vol. %	1.0	Maximum
7	Hydrocarbons (% of Total Organic Carbon Present)		
	(i) Methane	80	Maximum
	(ii) Ethane	10	Maximum
	(iii) C_3 and Higher HC	5.0	Maximum
	(iv) C_6 and Higher HC	1.0	Maximum
	(v) Total Unsaturated HC	1.0	Maximum
	Other Species (mole %)		
8	Hydrogen	0.1	Maximum
9	Carbon Monoxide	0.1	Maximum
10	Other Requirements Free from liquids over the entire range of temperatures and pressures encountered in the engine and fuel system. Free from solid particulate matter.		

Table 4. B I S Specifications for automotive LPG (IS 14861 – 2000)

S.N.	Characteristics	Requirement
1	Vapour pressure (gauge) @ 40°C, kpa	
	Min	520*
	Max	1050
2	C_5 Hydrocarbons and Heavier mol %, max	2.0
3	Dienes (as 1.3 Butadiene) mol %, max	0.5
4	Total volatile sulphur (after stanching) ppm, max	150
5	Copper stirp corrosion @40°C for 1 hour, max	Class 1
6	Hydrogen sulphide	Pass the test
7	Evaporation residue, mg/kg, max	100
8	Free water content	Nil**
9	Motor octane number (MON), min	88
10	Odour	Unpleasant and Distinctive down to 20% Lower Explosive limit***

* In winter, the gauge vapour pressure requirement shall be minimum 700 kpa at 40°C Winter period shall be from November to 15th February

** The water contained shall be determined at the Refinery First Dispatching location.

*** Product shall contain 20 ppm, min. ethyl mercaptan at first dispatching location to insure detection of leakage by odour.

REFERENCES

1. Daniel O. (2005) *Biofuture,* 6:57
2. Koonin S.E. (2006) *Science*, 311:432
3. Ragauskas A.J. *et al.* (2006) *Science,* 311: 484
4. Report of the Expert Committee on Auto Fuel Policy, Government of India August 2002
5. www.cseindia.org/campaign/apc/press_20020108.htm
6. www.thehindubusinessline.com/2002/06/07/stories.

CHAPTER - 8

Alternate transport fuels : A solution to control air pollution

Arun Arya[1] and J.H. Bhavnani[2]

Botany Department, Faculty of Science,
The Maharaja Sayajirao University of Baroda, Vadodara 390 002
e mail : aryaarunarya@rediffmail.com; bhavananimsu@yahoo.com
[2]Faculty of Technology and Engineering,
Kala Bhawan, The Maharaja Sayajirao University of Baroda, Vadodara - 390 001

ABSTRACT

Fossilized algae and plants are the source of petrol (gasoline) and diesel used as fuel in vehicles. The impending depletion of these non-renewable resources of energy and emission of toxic fumes containing CO, NO_x SO_2 and aromatics has forced to find out renewable alternatives such as CNG, LPG, biofuel, petro-crops, liquid resins, vegetable oils, alcohol, etc. When Rudolph Diesel run his first diesel engine on peanut oil in 1895 no one thought that one day it will be the better alternative than diesel. The non edible oil obtained from *Jatropha curcas*, *Pongamia pinnata*, *Calophyllum lanigerum* and *Ricinus communis* are used as a motor fuel as they are convertd to plant oil methyl ester (PME). Biodiesel is biodegradable and non toxic fuel free from sulphur and aromatics. ICRISAT in India has developed a variety of sweet sorghum which will yield biofuel - ethanol from sorghum stalks. The grains are used as food and the crops can be grown any time in the year. If hydrogen is used as fuel the problem of pollution can be solved another alternative is use of biological fuel cells. A fuel cell is an electro-chemical device that continuously converts chemical energy to electrical energy as long as fuel and oxidant are supplied to it. In biological fuel cells the catalyst is a microorganism.

Key words : Alternate fuels, LPG, CNG, Biofuel, Petro crops, Solar power, Biological fuel cells, Air Pollution.

Plants perform photosynthesis and are important source of renewable energy. All over world 2.5 billion tonnes of wood is burnt directly to get heat. To drive the vehicles energy may be obtained by liquid fuels. The exploitation of plants to obtain the liquid fuels requires physical, chemical and biological techniques. The technology involved in this conversion may be simple or difficult depending on the nature of the biomass source and the kind of liquid fuel finally produced. Three types of liquid fuels can be derived from plants, viz., (1) the vegetable or fatty oils, (2) alcohol and (3) liquid hydrocarbons.

8.1. BIOMASS AS ENERGY SOURCE

Biomass is a local resource which can contribute to the diversification of energy supply. It creates opportunities of employment for cultivation, harvesting transport and preparations. Well managed, biomass allows for carbon neutral fuels when substituted to fossil fuel. Amongst renewable energy sources, biomass appears to be the most important in terms of technical and economic feasibility. Biomass that can be converted in biofuels are of four different origins:

(i) Conventional agricultural products like oily seeds and sugar or starch rich crops

(ii) Lignocellulosic products and residues

(iii) Waste cooking oil

(iv) Biofuel from car exhaust

8.1.1 Conventional agricultural products

Ethanol is traditionally produced from the fermentation of glucose by yeast, therefore, sugarcane and sugar beets which contain a substantial amount of sugar constitute the main biomass source of ethanol production worldwide, even if the US production from maize is nowadays important in volume as from sugar cane in the world. Other suitable sources of sugar are molasses (a by product of sugar industry and sweet sorghum). Brazil is largest producer of ethanol. Many developing countries are producing sugar cane. Availability of land is main limiting factor. Sugar beet is another crop used in France. To produce one tonne of bio–ethanol from wheat, 3.3 tonnes of grain are needed. The demand for grain bio–ethanol production will reach 17.8 Mt in 2010 – 2011: 4.6 Mt maize; 8.7 Mt wheat; 1.1 Mt barley; 2 Mt rye; 1.3 Mt unspecified grains (Ziggers 2007).

Sweet Biofuels

The sugar found in fruits such as apples and oranges can be converted into a new type of carbon fuel for cars, US scientist have said the fuel made from fructose, contains far more energy than ethanol, according to the report in Nature. University of Wisconsin – Madison researches say that sugar can be converted into a fuel that has many advantages over ethanol. Called dimethylfuran – it can store 40% more than ethanol, does not evaporate as easily and is less volatile. However, more work needs to be done to assess the environment impact of this new fuel.

The development of fermentation process using economical carbon sources is important for the biofuel ethanol production on a commercial scale. Fermentation utilizing strains of *Zymomonas mobilis,* in place of traditional yeasts, have been proposed due to their ethanol yields being close to theoretical. The TRS (total reducing sugars) concentrations in molasses, temperature, agitation and culture time effects were studied simultaneously through factorial design. The best conditions for ethanol production were 200 gL^{-1} of total reducing sugars in the molasses, temp. 30°C and static culture and time of fermentation of 48h, achieving 55.8 gL^{-1} (Cazetta *et al.* 2007).

Starch present in maize, wheat, potato, cassava, sweet potato is converted into ethanol by yeast and enzymes. Cassava (*Manihot esculenta*) could be an interesting feed stock for developing countries as its cultivation has increased now. The international Crops research Institute for the semi arid tropics (ICRISAT) and Rusni Distilleries are implementing a pilot project to produce ethanol from sweet sorghum. The first batch of ethanol has been produced at the distillery in Andhra Pradesh. The project is being implemented with the help of Aakririt Agriculture Associates of India (AAI). It generates ethanol as biofuel from the sugar – rich juice extracted from sweet sorghum stalks. The provides the resource farmers of the dry lands with a source of additional income even while they do not lose out food security since the farmers can continue to use the grain for food. It is relatively easy and do not require rich soil as sugarcane. Contaminating organism *Lactobacillus* bacteria were found to consume arabinose, a sugar not utilized by the yeast, producing acetic acid and lactic acid that has detrimental effect on fermentation. The infection can be controlled by antibiotic virginiamycin, but routine use of antibiotics is cost prohibitive (Schell *et al.*, 2007).

Soyabean as biodiesel

Biodiesel production is set for global boost as a Cargill a major international company has just announced the intention to invest in a new 2 lakh tonnes biodiesel facility in Mainz in Germany. The facility will compliment Cargills existing oil seeds crushing and refining operations in the region and is expected to commence production in 2006. The company says it is facing increasing demands from customers from biofuels and Mainz is ideally situated for the production and supply of biodiesel – both in terms of its geographical locations and in the availability of raw materials such as rapeseed oil and other vegetable oil, which can be supplied via Rhine.

The news follows an earlier announcement that Cargill will construct 37.5 million gallons per year in biodiesel in Lowa falls, USA, adjacent to its existing soyabean crush facility. Initially the plant will produce biodiesel exclusively from soyabean oil, but Cargill officials expect to have the capability of using waste grease for biodiesel production in the future.

Sweet Sorghum as Biofuel

In order to enable sweet sorghum to be available all year round, ICRISAT sorghum breeders have developed hybrids that can be planted any time of the year. The sorghum biofuel project accommodates the farmers food needs with an additional income source and enables a biofuel strategy that does not jeopardize food security in at least in sorghum eating areas. Countries sensitized to food security threat of biofuels are taking policy measures that will prevent conversion of land available for food crops for growing biofuel crops. The Xinhua News Agency reported that the Chinese Government has asked biofuel crops growers to switch to crops such as sweet sorghum for their projects. The Philippines Government has also decided not to use crop for bioethanol production and has invested in research and development of other crops such as sweet sorghum and cassava. Sweet sorghum has other benefits over sugarcane and maize as feed stock for ethanol production. It only requires one half of the water as compared to sugar cane; its cost of cultivation is around one fifth of the cost for growing sugarcane.

Sweet sorghum is also said to be a carbon neutral crop. This means that the amount of carbon dioxide that sweet sorghum fixes during its growing period is equal to the amount it emits during crop growth, conversion to ethanol and combustion of ethanol.

Table 1. Different crop resources for bio-ethanol and their ethanol production

Crops	Ethanol production potential (litres/ tonne)
Sugar crops	
Sugar cane	70
Sugar beet	110
Sweet sorghum	60
Starchy sources	
Sweet potato	125
Potato	110

Casava	180
Maize	360
Rice	430
Barley	250
Wheat	340
Cellulosic biomass	
Forest thinnings	257
Bagasse and other agro residues	351

Nigam and Agarwal 2004 and DoE 2005

8.1.2 Lignocellulosic products and residues

The largest part of a plant is not sugar or starch but rather cellulose, hemicellulose and lignin (particularly for woody biomass). Cellulose and hemicellulose can be converted into ethanol through a first step which consists in converting them into sugar. Unfortunately, it is not possible with lignin. The cellulose component of woody and herbaceous biomass ranges from 30% to 70%. The remaining organic part is lignin which can be used to generate the energy for the process or can be considered as raw material for biosyn-fuel routes.

Large quantities of lignocellulosic material in the form of agricultural and forest residues are available in India. For example, India's production of wheat straw and rice straw is approximately 200 million tonnes per year. Although straw has low protein content, it is rich in fermentable carbohydrates. Similarly, hemicellulose rich waste materials like corn-cob or coconut pith and shell are valuable resource. Production of coconuts in India is as high as 6000 million nuts per year (Lachke 1989).

The major constituents of lignocellulosic substrates are cellulose, hemicellulose and lignin which are present in an average proportion of 4:3:3. This proportion varies significantly in their native states. Some of the commonly occurring hemicelluloses are arabinoxylan, glucoxylans and arabinogalactans. Depending on the biomass source, the hemicellulose fraction can be as much as 85% xylan. Xylan on hydrolysis yields D-xylose as the major product.

Hemicelloses have a relatively open structure as compared to cellulose. This facilitates diffusion of acid into the polymer resulting in accelerated hydrolysis of xylans. Thus, hemicellulosic sugars can be obtained more readily, with better yield (80-90%) from xylan by acid or enzymic hydrolysis as compared to the D-glucose yield (50%) from cellulose.

Glucose is the well-known fermentable six carbon sugar and can be easily used for the production of fuel. However, about 40% of the total carbohydrates obtained from plant materials consists of hemi-cellulose, five carbon sugars. Thus, D-xylose is the second most abundant sugar in the biosphere offering us a potential feed stock for generating fuel.

The enzymic hydrolysis of xylan to produce monomeric sugars requires synergistic action of at least two enzymes, namely, Beta-D-xylanase and Beta-D-xylosidase. Using microbial xylanases the hydrolysis of xylans to monomeric sugars has been achieved by several investigators. A few spp. of *Streptomyces* and fungi (e.g. *Sclerotium rolfsii* and *Penicillium funiculosum*) have been reported to produce enhanced levels of xylanolytic enzymes.

During last few years, yeasts namely *Pachysolen tannophilus, Candida shehatae, Pichia stipitis* and *C. tropicalis* have been identified for ethanol production from D-xylose in appreciable yields under well defined aerobic or anaerobic conditions (Gong 1983). Some fungi such as *Fusarium, Mucor* and *Rhizopus* are also known to ferment D-xylose to ethanol. Many bacteria are able to utilize xylose; especially members of the genera *Clostridium* and *Bacillus* carry out mixed fermentation in which ethanol is a major product. The main problems in the use of bacteria are low ethanol tolerance and the formation of products other than ethanol.

The rates and yields of ethanol production from D-xylose depend upon several process variables such as substrate concentration, nitrogen source and aeration and agitation. For example, in the case of *P. tannophilus* it is observed that oxygen is required for cell growth but not for ethanol production. The cell growth and alcohol production is better at 32°C and at pH 2.5. The improvement in the yields has been achieved by strain improvement through mutation studies.

The Indian Government has accepted the plan to start with a blending proportion of 5 per cent (5% biofuel with 95% petroleum) by 2012, 10 per cent by 2017. The research focus has shifted to improve oil content of crop yields and reduce the environmental impact of biofuel's use. According to Sahai (2007a) *J. curcas* is low yielding; giving one tone seed price of Rs. 5 per kg, the farmer would get Rs. 5000 per hectare per year. This makes it a loss making venture. To produce a litre of ethanol from corn 6,597 kilo calories are required. It gives 5,130 kilo calories of energy, which is a 22 per cent deficient. On top of this, biofuels are more expensive than petrol. In the US, a litre of petrol costs roughly 33 cents to produce, a litre of ethanol can cost upto $ 1.88. At present, this difference is covered by subsidies.

The UN is urging governments to beaware of human and environmental consequences of the agro-fuel trend, some of which could be irreversible. They warn that current agro-fuel route will lead to deforestation, push small farmers off the land, and lead to serious food shortages and increased poverty. The statistical figures suggest that even if high yield bio-energy crops were grown on all the arable land on earth, the biofuel produced would cater to only 20 per cent of current demands. The biotechnologists are developing genetically engineered maize, cassava and other crops to produce agro fuel. A company in New Zealand has successfully developed a system for using sewage waste as a substrate for algae and then producing biodiesel.

8.1.3 Vehicle fuel from waste cooking oil

A new system has been developed by Green Fuels Ltd. Cheltenham U.K. to create biodiesel from cooking oil in closed units. The small scale units produce 150 litres biodiesel per day. The transesterification process is used to separate glycerine from vegetable oil. The process results in biodiesel, mono-alkyl ester and glycerine, which is valuable by product used in soaps and other products (www.greenfuels.co.uk).

8.1.4 Biofuel from car exhaust

An organic chemist and two engineers from Britain have join hands to make biofuel from car exhaust, turning car emissions environmentally safe in the bargain. The three have developed a box, which replaces the exhaust in a car. This box can trap the exhaust fumes containing carbon dioxide and nitrous oxide and emit mostly water vapour. The captured gases can be further processed using genetically modified algae, to create a biofuel.

The technology developed by organic chemist Derek Palmer and engineers Ian Houstan and John Jones is dubbed the "Greenbox". The trio claims it can be used for small and big vehicles and perhaps even industry and power plants. The idea emerged while experimenting the carbon dioxide to help boost algae growth for fish farming. At present, the prototype is about the size of a bar stool, but the inventors say they can build one small enough to replace a car exhaust that will last for a full tank of petrol.

The carbon dioxide is held in a safe, inert state and can be handled, transported and released into a controlled environment easily, using a minimal amount of energy. Through a chemical which can be crushed to produce bio-oil. This extract can be converted to produce a biodiesel almost identical to normal diesel. More than 130 tests carried out on GM algae over two years have yielded a capture rate between 85 and 95 per cent. If the systems take off, drivers with a Greenbox can replace it when they fill up their cars and it would go to bioreactor to be emptied. The biodiesel produced can be fed back into a diesel engine, the emptied Greenbox can be affixed to the car and the cycle can begin again. It is estimated that the algae required to capture all the auto emissions of vehicles in Britain would take up around

1,000 acres. The conversion process also yields methane gas and fertilizer. Both of which can be captured separately (Sahai 2007b).

Table 2. Annual production of Non – edible oil Seeds in India

Type	Production (MT)	Oil %
Neem	500	30
Karanja	200	27 – 39
Kusum	80	34
Pilu	50	33
Ratanjot	-	30 – 40
Jojoba	-	50
Bhikal	-	37
Wild Walnut	-	60 -70
Thumba	100	21

Table 3. Seed and oil yields and equivalent Bio-diesel.

Crops	Seed yield t\ha	Oil content %	Litres-biodiesel per ha
Soybean	2.67	18	524
Cotton seed	1.05	19	216
Canola	1.54	40	665
Sun flower	1.52	40	657
Peanuts	3.40	25	920
Rape seed	1.47	40	638
Mustard (spice)	1.04	40	452
Jatropha	2	37	701
Palm oil	20 (fresh fruit bunches)	49	3000

Table 4. Physical properties of Biodiesel

Properties	Values
Specific gravity	0.88
Viscosity @ 20C(centistokes)	7.5
Cetane Index	49
Cold filter Plugging Point (C)	-12
Net Heating Value (Kilojoules/Liter)	33,300

Table 5. Biodiesel emissions compared to conventional Diesel

Emissions	B100	B20
Regulated Emissions		
Total Unburnt Hydrocarbons	-93%	-30%
Carbon Monoxide	-50%	-20%
Particulate Matter	-30%	-22%
NO_x +13%	+2%	
Non-Regulated Emissions		
Sulphates	-100%	-20%*
Polycyclic Aromatic Hydrocarbons (PAH)**	-80%	-13%
NPAH(Nitrated PAHs)**	-90%	-50%***
Ozone Potential of Speciated HC	-50%	-10%
Life-Cycle Emissions		
Carbon Dioxide (LCA)	-80%	
Sulphur Dioxide (LCA)	-100%	

*Estimated from B100 results. **Average reduction across all compounds measured.
***2-nitroflourine results were within test method variability.

Table 6: Oil present in seeds of other plants

Sr. No.	Name of plants	Percentage of Oil
1	*Jatropha gossypifolia* Linn.	24.68
2	*Croton sparsiflorus* Dalz.	10.68
3	*Chrozophora brachiata* Juss.	24.4
4	*Datura innoxia* Mill.	9.14
5	*Gynandropsis pentaphylla* DC.	20.88
6	*Cassia renigera* Benth.	1.18
7	*Grewia villosa* Willd.	1.61
8	*Abutilon indicum* Sweet.	11.5
9	*Cordia sebastena* Linn.	40.59
10	*Ricinus communis* Linn.	14.5
11	*Citrus limon* Burm f.	33.78
12	*Cassia occidentalis* Linn.	1.59
13	*Cassia tora* Linn.	5.66
14	*Cassia fistula* Linn.	2.02
15	*Balanites roxburghii* Planch.	43.34

16	*Corchorus acutangula* Lam.	5.97
17	*Hibiscus* sp.	12.0
18	*Ipomea* sp.	10.16
19	*Hibiscus* sp. 2	12.6
20	*Anisomeles ovata* Br.	24.75
21	*Lagerstroemia flos-reginae* Roxb.	1.30
22	*Caesalpinia crista* Linn.	20.50
23	*Acacia catechu* Willd.	3.27
24	*Eucalyptus hybrida*	1.54
25	*Jatropha curcas* Linn.	50.25
26	*Melothria maderaspatana* Cogn.	15.93
27	*Salavadora persica* Linn.	17.06
28	*Acacia chundra* DC.	3.64
29	*Butea monosperma* Roxb.	16.09
30	*Cassia leucophloa* Willd.	1.22
31	*Bauhinia racemosa* Lam.	0.81
32	*Nyctanthes arbor-tristis* Linn.	5.6
33	*Mucuna prunita* Hook.	2.9
34	*Bauhinia purpurea* Linn.	4.0
35	*Casuarina equisetifolia* Forst.	15.24
36	*Dalbergia sissoo* Roxb.	1.39
37	*Lagerstroemia parviflora* Roxb.	1.3
38	*Diospyros melanoxylon* Roxb.	1.22
39	*Adina cordifolia* Benth. and Hook.	4.42
40	*Dalbergia latifolia* Roxb.	7.44
41	*Haplophragma adenophyllum* Wall.	1.16
42	*Sida acuta* Burm.	21
43	*Abutilon indicum* Linn.	2.67
44	*Lawsonia inermis* Linn.	7.9
45	*Nicotiana tabaccum* Linn.	13.28
46	*Clitoria ternatea* Linn.	7.71
47	*Canna indica* Linn.	0.67
48	*Teramnus labialis* (L.) Spreng	2.22
49	*Mimusops elengi* Linn.	3.02
50	*Tephrosa villosa* Pers.	7.20

51	*Anona squamosa* Linn.	11
52	*Merremia dissecta* Hall.	2.73
53	*Leucaena leucocephala* Benth.	4.31
54	*Bridelia retusa* Spr.	0.897
55	*Albizzia lebbeck* Benth.	2.37
56	*Acacia nilotica* Willd.	2.75
57	*Tamarindus indica* Linn.	3.85
58	*Peltophorum ferrugineum* Benth.	3.4
59	*Samanea saman* (Jacq.) Merrill	6.44
60	*Acacia senegal* Willd.	2.64
61	*Pithecellobium dulce* Benth.	11.75
62	*Cassia auriculata* Linn.	4.57

Source : (Nair, 2007)

8.2 LIQUEFIED PETROLEUM GAS (LPG)

According to an estimate there were 172,806 LPG fueled vehicles in use in United States which increased to 190,438 in 2003 and 194,389 in 2004. Liquefied Petroleum Gas (LPG) is the generic name for commercial Propane and Butane, which form liquid when pressurized for storage. It is usually kept in liquid state for easier storage and transport. In its liquid state it is 250 times denser than in the gaseous state. It is an excellent alternative fuel for normal petrol vehicles. It is said that even bio-diesel is not safe as its use can have large carbon emissions which can effect the global climatic change and can result into heavy rains and floods. LPG is cheaper than petrol and LPG driven vehicles have 75% less CO, 85% less hydrocarbons, 40% less of nitrous oxide. LPG causes 87% less ozone depletion as compared to petrol. LPG is 84% more fuel efficient as compared to petrol. Likewise, compared to diesel LPG contains 90% less of particulate matter and HC and 50% less of nitrous oxide. Not only LPG is economical its use means a cleaner engine with extended periods between oil changes, and reduced carbon deposits. The development of LPG systems is as follows:

Generation 1: For Carburetor Systems

This early type mixer system introduces the LPG to the engine via a mixer ring mounted either ahead or under the carburetor. It had crude screw adjustments to idle, part and full load fuelling. This early system could produce poor derivability and caused backfire if mixtures were set lean. When fuelling was rich to the point of reasonable derivability to prevent back fire, the resulting LPG consumption was usually high, reducing the benefits of installing it in the first place.

Generation 2: For injected and Carburetor vehicles

Vast improvements were made to Gen. 1 systems by introducing Lambda control to it. In later mixer type systems ECU electronics have further improved the derivability and resulting emissions. Some systems have lambda ECU controlled stepper motors mounted on the vaporizer unit. Some of these systems have base settings, after which they are controlled by lambda, others are totally lambda controlled. Although much improved the Gen. 2 systems can still suffer from induction backfire, which is eliminated by the application of generation 3 multi-point system.

Generation 3: Single and multi-point Systems

The single point and multi point systems almost completely mimic the vehicles existing single and multi point petrol system, and utilize most of the Petrol ECU's input signals. The advanced mixer system for single point vehicles now has all of the generation 2 advantages, plus an advanced self learning map ability that is always adjusting the system to optimum emissions, but again induction backfire could occur if fitted to very late vehicles. The initial set up is made by the dealer using advanced equipment to establish the basic settings. The multipoint system has a large volume LPG injector that supplies simultaneously LPG to pressure values mounted in each manifold leg close to the cylinder head. This system does not suffer from induction backfire reduces power losses to almost zero, and affords the most accurate control of LPG.

Generation 4:

The generation 4 system from AG Auto Gas now offer sequential control of LPG injection. The system takes most of its control signals directly from the OEM manufacturers ECU, this enables the SGI system to accurately mimic the manufacturers own calibrations. The result is less wastage of LPG, lower emissions and greatly reduced development period.

According to Patel (2005) if we have a modern fuel injected vehicle we can choose generation 3 or 4, while system 2 is good for older vehicles. An auto LPG dispensing station has to confirm to strict safety guidelines set up by oil industry directorate, as well as department of explosives of the government of India. It is not advisable to use cooking LPG cylinders without safety measures.

8.3 COMPRESSED NATURAL GAS: A NATURAL CHOICE

CNG is a mixture of hydrocarbons, especially methane, which is compressed to 80 atmospheric pressure. It is a cost effective fuel. It is cleaner than petrol and diesel, which play a major role in air pollution abatement. It also costs 70 and 30% less than petrol and diesel respectively. CNG is free from toxic pollutants like sulphur, benzene, lead and polycyclic aromatic hydrocarbons. Vehicles running on CNG emit 90% less carbon monoxide, 60% less hydrocarbons and 50% less nitrous oxide as compared to traditional fuel. Petrol and diesel exhausts contain high level of particulate matter. Emissions of diesel and petrol exhaust are a major cause of heart and lung diseases and they also contain carcinogens.

You can cook with it heat your house with it or you can power a car or truck with it. Like oil (petroleum), this common fuel comes from underground. Natural gas is 95% methane and 5% other gases like butane, propane and ethane. Several vehicles are available today such as the Honda civic CGX and Ford crown Victoria that operate on compressed natural gas. Some run on natural gas only and others can run on natural gas or gasoline called bi-fuel vehicles. CNG has an excellent anti-knocking property, which results in better fuel efficiency. As CNG vapours are lighter than air, they can easily dissipate into atmosphere in the case of an accidental leakage and thus the risk of catching fire is minimal. CNG cannot be adulterated.

8.4 METHANOL AS FUEL

Rising energy prices and environmental problems have led to increased interest in alcohol as a fuel. The first four aliphatic alcohols (methanol, ethanol, propanol and butanol) are of interest as fuels because they can be synthesized biologically and they have characteristic which allow them to be used in current engines. One advantage shared by all alcohols is octane rating. Until today production of all four alcohols from petroleum is cheaper than fermentation and extraction from biomass, but this is expected to change as fermentation and extraction processes become more efficient while petroleum becomes more expensive.

Methanol is more corrosive than ethanol. Its energy per litre is 55% lower than that of petrol. High compression ratios and corrosion resistant materials can overcome these issues, but require extensive engine modification. The main hazard of methanol is its toxicity. Methanol fuel is widely used in motor sport. According to an estimate there were 18,319 vehicles using methanol as fuel in 1995 which reduced to 4,592 in 2004.

8.5 BIODIESEL: FUEL FOR THE FUTURE

Biodiesel is the name for a variety of ester based oxygenated fuels derived from natural, renewable biological sources such as vegetable oils. Biodiesel operates in compression ignition engines like petroleum diesel therapy requiring no essential engine modifications. Moreover it can maintain the payload capacity and range of conventional diesel.

Biodiesel (mono alkyl esters) can be made from new or used vegetable oils and animal fats. Unlike fossil diesel, pure biodiesel is biodegradable, nontoxic and essentially free of sulphur and aromatics. The concept of using vegetable oil as a fuel dates back to 1895 when Dr. Rudolf Diesel developed the first diesel engine to run on vegetable oil obtained from Groundnut (*Arachis hypogaea*). Vegetable oils can be chemically reacted with an alcohol (methanol) to produce esters. Biodiesel is the name given to these esters when they are intended for use as fuel. Vegetable oil is first filtered, then processed with alkali to remove free fatty acids. It is then mixed with methanol and a catalyst usually sodium or potassium hydroxide. The oil's triglycerides react to form esters and glycerol, which are then separated from each other and purified. Soybean, peanuts, cotton seed, sunflower seeds and canola are some of the potential feedstock for biodiesel. Some of the physical characteristics of biodiesel are given in table 4. Emission results for pure biodiesel (B100) and mixed biodiesel (B20-20% biodiesel and 80% petrodiesel) compared to conventional diesel are given in table 3.

8.5.1. Why to use biodiesel?

The production and use of biodiesel produces approximately 80% less carbon dioxide and almost 100% less sulphur dioxide as compared to conventional diesel. Use of 1 kg of biodiesel leads to esterification of some 3 kg of CO_2. The emissions of NO_x from pure biodiesel increased in this test by 13%. However, biodiesel's lack of sulphur allows use of NO_x control technologies that cannot be used with conventional diesel. So, biodiesel NO_x emissions can be effectively managed and efficiently eliminated to make it a safe alternative fuel.

8.5.2 Algae as bio-diesel source

Algae can be used as source of biodiesel. The Sundarban delta, an archipelago of some 100 islands spread over 4262 sq km. on the Indian side of the Bay of Bengal, is becoming the incubator of eco-friendly energy source which are now lighting up the lives of many people who never get access to conventional forms of electricity. Efforts are on to cultivate algae to extract biodiesel. Chennai based Bio Max is keen to set up an algae cultivation project on one of the uninhabited Sundarban islands.

8.5.3 *Jatropha* as bio-diesel source

Jatropha curcas Linn. (Ratanjot, Plate I Fig. A) a non-edible oil-bearing and drought hardy shrub with ecological advantages, belonging to the Euphorbiaceae family, was found to be the most appropriate renewable alternative source of biodiesel. The seeds of this perennial shrub are used to extract oil with 30 to 40% recovery (Kandpal and Madan 1995). The extracted oil could not be used directly in diesel engines because of its high viscosity. High viscosity of pure vegetable oils would reduce the fuel atomization and increase fuel spray penetration, which would be responsible for high engine deposits and thickening of lubricating oil (Silvio *et al.* 2002). *Jatropha* can be exploited effectively in combating desertification. It will help to improve rural economy. In Egypt *Jatropha* is being cultivated on a 500,000 hectares of desert land which is irrigated by treated sewage water (Ackom and Ertel 2005). Geography department of Pune University has digitized a map through use of remote sensing to help plantation of *Jatropha*. Kaushik *et al.* (2007) studied the genetic variability of 24 accessions. In general phenotypic coefficient of variation was higher than genotypic coefficient of variation indicating the predominant role of environment. Seed weight has positive correlation with seed length, breadth, thickness, and oil content.

The stoichiometric equation requires one mole of triglyceride and three moles of alcohol to form three moles of methyl ester and one mole of glycerol in the presence of a strong base or acid (Muniyappa *et al.* 1996). According to Chitra *et al.* (2005) optimum concentration of methanol required for effective trans-esterification of *Jatropha* oil was 20%. The concentration of NaOH required was 1.0%. The fatty acid composition of *Jatropha* classifies it as a linoleic or oleic acid type, which are unsaturated fatty acids. Myristic, palmitic, stearic, arachidic, oleic and linoleic acids. The seeds and oil are toxic due to the presence of cursive and curcasive (Pramanik, 2003). Trials conducted at Regional engineering college Warangal (Pramanik 2003) established that upto 50% *Jatropha curcas* oil can be substituted for diesel for use in a C.I. engine without any major operational difficulties. Pre-treatment of oil can further improve the quality of oil. Biodiesel has a higher Cetane rating than diesel, which improves engine performance. It has energy content 39.6-41.8 MJ/Kg as compared to 42.6-45 MJ/Kg in diesel. The sulphur content is 0.13% as compared to 1-1.2% in diesel. Agarwal and Agarawal (2007) have reduced the viscosity of oil by increasing the fuel temperature (using waste heat of the exhaust gases) and thereby eliminating its effect on combustion and emission characteristics of the engine. While operating the engine on *Jatropha* oil (preheated and blends), performance and emission parameters were found to be very close to mineral diesel for lower blend concentrations.

Table 6. Physical and chemical properties of diesel and Jatropha oil blend

Properties	Diesel	*Jatropha curcas* oil
Density (g/cc), 30°C	0.836 – 0.850	0.9329
Kinematic viscosity (cSt), 30°C	4 – 8	52.76
Cetane No.	40 – 55	38
Flash point, °C	45 – 60	210
Calorific value, MJ/kg	42 – 46	38.20
Saponification value	-	198.00
Iodine No.	-	94.00

8.5.4 Castor oil as biodiesel

This work presents the transesterification of castor oil with ethanol in the presence of sodium ethoxide as catalyst. This is an exceptional option for the biodiesel production, since the castor nut is quite available in the country. Castor (Plate I Fig. B) oil is the oil derived from castor beans. Castor oil is one of the hard oils, where the oil content in the seed is relatively high. In the case of castor seed, the oil content is close to 50% of the total by weight – the castor bean contains 50-55% oil. The oil itself contains a number of fatty acids similar to those in cooking oils, such as oleic acid, stearic acid and palmitic acid. No other vegetable oil contains so high a proportion of fatty hydroxyl acids. Castor oils unsaturated bond, high molecular weight (298), low melting point (5°C) and very low solidification point (-12°C to -8°C) make it industrially useful, most of all for the highest and most stable viscosity of any vegetable oil.

Castor oil is unique among all fats and oils. It has an unusual chemical composition of a triglyceride of fatty acids. It is the only source of 18–carbon hydroxylated fatty acid with one double bond. The product uniformity and consistency of castor oil are significantly high for a naturally occurring material.

The chemical composition of castor oil is:

- Ricinoleic Acid – 89.5%
- Linoleic Acid – 4.2%
- Oleic Acid – 3%
- Stearic Acid – 1%

8.5.5 Biofuel from *Pongamia* (Honge oil)

Most of the common oil seeds contain 20-30% of oil by weight, and in some cases up to 42%.

SuTRA [Sustainable Transformation of Rural Areas] a programme unit of SID [The Society for Innovation and Development], Indian Institute of Science, Bangalore has been popularizing the direct use of oil in diesel engines for a variety of applications and of the starch in biogas plants. Oils of honge (*Pongamia pinnata*), Neem and *Madhuca indica* have been commonly available in most part of India for a long time as fuels for illumination purposes. However, to use them (a) as diesel substitutes in transport vehicles these oils need to be a degummed to remove the drying components in them, (b) thoroughly filtered to remove the particulates and c] warmed before being feed into the fuel injector to reduce the viscosity to acceptable levels. The efforts have been made by SuTRA to used *Pongamia* oil in engine pump sets. Engine generators consume about 10% honge oil by volume as compare to diesel (Nagendrappa 2000).

Table 7. Comparison of *Pongamia* oil with diesel

Sl. No	Particulars	Honge (unfiltered)	Honge (degummed)	Diesel	Neem
1.	Saponification value	178	186	Nil	196
2.	Iodine value	116	95	38	110
3.	Acid value	2.3	1.6	Nil	1.4
4.	Colour in 1 inch cell (Y + 5 R)	50	50	102	50
5.	Specific gravity	0.92	0.93	0.84	0.92
6.	Refractive index	1.48	1.48	1.47	1.48
7.	Moisture and volatile matter	0.05	0.2	25	0.04
8.	Viscosity at 25°C (centi strokes)	85	89	8.5	86
9.	Cloud point (°C)	15	15	13	13
10.	Pour point (°C)	-1	-1	1	1
11.	Gross calorific value (k Cal/kg)	9340	-	10,500	9,000

8.5.6 Sweet Biofuels

`The sugar found in fruits such as apples and oranges can be converted into a low new type of carbon fuel for cars', US scientists have said. The fuel made from fructose, contains far more energy than ethanol, according to the report in Nature. University of Wisconsin- Madison researches say that sugar can be converted into a fuel that has many advantages over ethanol. Called dimethylfuran – it can store 40% more than ethanol, does not evaporate as easily and is less volatile. However, more work needs to be done to assess the environment impact of this new fuel (www.bti-eneflash.com/ bionews22).

8.5.7 Hydrocarbons as Biofuel

The plants that are promising to yield "bio-crude" are the members of the family Euphorbiaceae. Many of these species are being cultivated on experimental basis in USA, Japan and India and few other countries. The succulent species like *Euphorbia tirucalli* (Plate I Fig. D), *E. antiquorum, E. nevalia, E. milli, Pedilanthus tithymaloides* (Plate I Fig. C) are grown in Karnataka and other parts of South India as hedge or ornamental plants. They are recommended to be grown as petro crops. Non-succulent *Manihot esculenta* is a tuber and latex yielding plant, it can, therefore, be a source of both alcohol and hydrocarbons. Its tuber can be used in alcohol production and other parts for hydrocarbon processing. The extractable hydrocarbon content that is useful as petrol substitute in these plants varies from about 3 – 9%.

8.6 BIOLOGICAL FUEL CELLS

In about 200 yrs. from now, vehicles will not have petrol tanks because our petroleum reserves will be depleted. An alternative, which is less wasteful and cleaner, is to power vehicles directly with carbohydrates using biological fuel cells (Bennetto 1987). A fuel cell is a device that converts electrochemical energy to electrical energy for as long as fuel and oxidant are supplied to it.

A fuel cell is an electrochemical device that continuously converts chemical energy to electrical energy for as long as fuel and oxidant are supplied to it. Beginning with the space programme of the seventies, we have seen a truly substantial improvement in generic fuel cell technologies. Albeit their high operational-efficiency, the advantage of these fuel-cells is partially offset by the high cost of catalysts, high operational-temperature, corrosive electrolytes employed with them. In this regard, biological fuel cells are both attractive and permissive.

Unlike chemical fuel cells, biological fuel operates and the mild reaction condition, namely ambient operational temperature and pressure. They also employed neutral electrolyte and used in expensive catalyst such as platinum. In biological fuel cells, the catalyst is either a microorganism as simple as baker's yeast or enzyme. For sanctuaries, microorganism; which transforms food into a electron flow, were on biological curiosity; but scientists have made it possible to use them in watches and cameras as power sources. Luigl Galvani, who noticed twitching of isolated frog leg, when a brief electrical discharge was passed through it, was to first observe the bioelectric phenomena as early as in 1790 and the term bioelectricity was coined after that observation. In 1910, Michael Cresse Potter, Professor of

Botany at the University of Durhan, U.K. demonstrated that organisms could generate voltage and deliver current. Cohen at Cambridge revised Potter's idea in 1931. He described how batch of biological fuel cells produce more than 35 V. But biological fuel cells became popular in 1960, when the national aeronautics and space administration evinced interest in turning organic wastes into electricity on its long-haul space flights. Algae and bacteria were the first organisms used in biological fuel cells. During the mid 19th century; first time Rohrbach *et al.* designed a biological fuel cell in which *Clostridium butyricum* was used as biological material to generate hydrogen by glucose fermentation. In 1963; biological fuel cells were already commercially available for use as a power source in radios, signal lights and other appliances at sea. However, these fuel cells were not commercial successes and soon these disappeared from markets. With the successful development of technical alternatives; example solar photovoltaic for the energy supply on space flights; biological fuel cells suffered a short setback. Later during the oil crisis of 70s and 80s the interest in the development of biological fuel cells was revived. In1966, Williams showed that biological fuel cells powered by rice husk produced 40 mA at 6v. Rice husk is a potential source of ligno-celluloses, which on fermentation yield many useful enzymes and biofuels like ethanol that could be used in biological fuel cells. In 1969, Yao *et al.* showed that glucose could be used as a fuel in the presence of platinum black. Later Karube *et al.* reported the generation of about 300 mA electric current from an *Anabaena* spp. based biological fuel cell in which phosphoric acid was used as a electrolyte. *Aspergillus niger, Trachyderma tsunodae, Polyporus versicolor* are used in biofuel cells (Tayhas *et al.* 1999 and Mano *et al.* 2002).

Bennetto and co-workers have made noteworthy contribution to biological fuel cell. They have developed and demonstrated improved biological fuel cell using various micro organisms and mediator system. They showed that the mediator could enhance both the efficiency of electron transfer and reaction rate. Redox mediator, 2,2-azinobis (3-ethylbenzo-thiazoline-6-sulfonate), was used by Tayhas *et al.* (1999) to reduce dioxygen to water using the fungal enzyme, laccase, as the catalyst for homogenous electron transfer. Recently Chaudhary and Lovely have reported that micro organism *R. ferrireducens* can recover an electron from glucose oxidation in the presence of Fe 3+ up to 83% without a mediator. Presently, efforts are being expanded to improve the performance of mediator less biological fuel cells as well as on finding an effective route to wire the microorganisms to the electrode so as to promote the efficiency of electron transfer.

Plate I. Fig. A. Plant of *Jatropha curcas*; a good source of biodiesel; **Fig. B.** Castor plant inflorescence of *Ricinus communis;* **Fig. C.** Plant of *Pedilanthus tithymaloides*; a petro-crop; **Fig. D.** A hydrocarbon yielding plant of *Euphorbia terucalli*

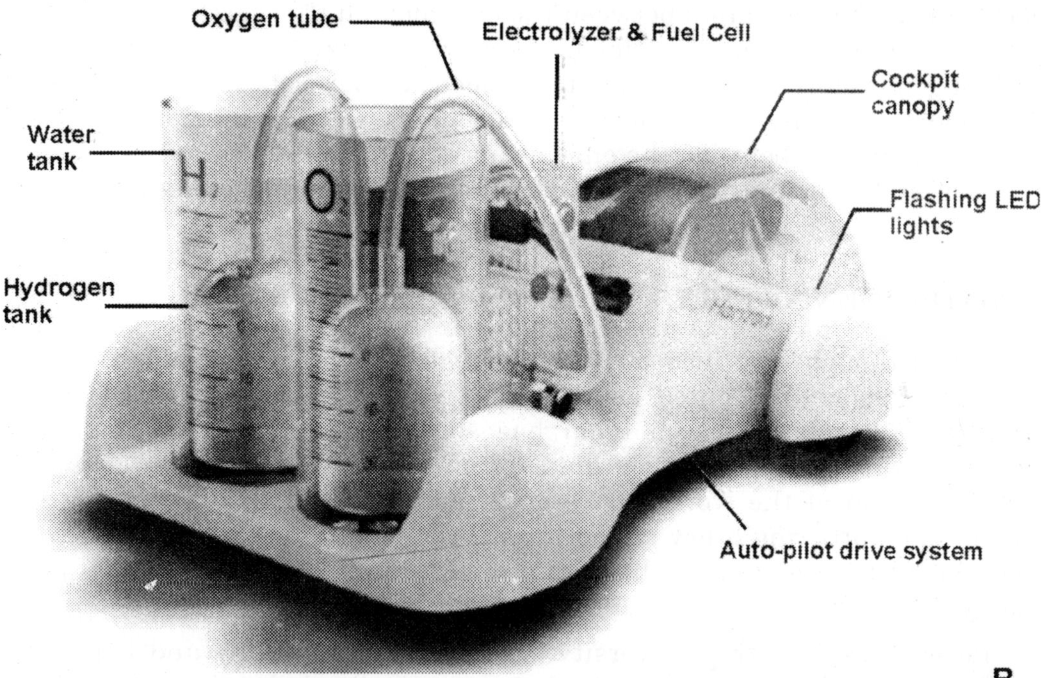

Plate II. Fig. A. A solar car; better car for future without pollution; **Fig. B** hydrogen powered car; a vehicle with great promises

8.6.1 Types of biological fuel cells

There are 3 types of biological fuel cells, namely microbiological fuel cells where the entire organism is used as the catalyst, photomicrobial fuel cell where a photosynthetic organism is used as the catalyst and an enzymatic fuel cell where enzymes, both in solution or immobilized, are used as the catalyst.

1. Microbial fuel cells

The use of microorganisms in biological fuel cells eliminates the isolation of individual enzymes; there by providing cheaper substrate for biological fuel cells, microorganisms are used in 4 ways to producing electrical energy.

2. The enzymatic fuel cells

Enzymes have a complex structure comprising proteins. The electron transferring unit of the enzyme, namely the apoenzyme and cofactor, is deeply buried inside its complex structure. Hence, efficient electrical communication between the electrode substrate and the enzyme biocatalyst is difficult. Mediated bioelectro catalyses use small redox molecules having electron transfer mediators between the enzyme and the electrodes. They are useful for study electron transfer kinetics of the enzyme and there is a key reaction in aping the enzymes to biological fuel cells. In the case of quinohaemoprotiens, direct electron transfer as well as mediated electron transfer is possible between the enzyme and an electrode, where the haeme is in the protein functions as a built in mediator in an enzyme electrochemical reaction.

8.7 HYDROGEN AS FUEL

If you thought the idea of breeding hydrogen producing algae was far fetched think again. Touted as the green energy breakthrough of future, Germany–Australia research team has succeeded in breeding algae that produce hydrogen in completely unheard of quantity. Amid growing fears of the decimation of the worse oil reserves and peak oil theory making new ground and gathering a few supporters, A genetically altered algae provides nourishing vision and possibilities of a ready supply of environment friendly energy.

Researches from the University of Bielefeld in Germany and University of Queensland, Brisbane, Australia have genetically changed single cell alga *Chlamydomonas reinhardii*, in such a way that it produces an especially large amount of hydrogen. This gas can then be used to produce energy. 'In contrast to use of fossil based fuel such as oil, petroleum and natural gas, no carbon – dioxide is generated. Water is the only residue.

According to Bielfeld biologist Olaf Kruse, the long standing belief that certain algae can produce oxygen during photosynthetic process has now been proved beyond doubt. The catch, as always, is efficiency. One litre of algae can produce approximately 100 ml of hydrogen and then they collapsed. The genetically spruced up variant can boost this up to half of litre of nitrogen. These algae can produce 5 times the volume made by the wild form.

Economic feasibility of algae becomes consideration only when the energy efficiency – the conversion of sunlight into hydrogen reaches 7-10%, but algae in its natural form achieves efficiency of a meager 0.1%. The new 'turbo alga' has now managed to achieve efficiency of 1.6-2%. The final goal according to researcher has not been reached yet. At the end of the development process there could be a biological fuel cell in which the alga produces the necessary hydrogen at the site of consumption. A motorist could than instead of stop at petrol pump, need only to have an alga power plant on board.

This is naturally a futuristic vision. Olaf Kruse and Ben Ankamer have been working on such a turbo alga' at the institute of molecular bioscience at the University of Queensland since 2001. From a pool of 20,000 random mutation, the scientist have selected 20 algae, ultimately coming up with a genetically altered mutation called as Stm6. Meanwhile at the Technical University in Karlsruhe, a prototype of a bioreactor containing 500-1000 litres of algae cultures is being developed. The reactor is to be used to prove the economic feasibility of the system in the next 5 years. On the basis of calculation, a reactor shaped like a cube measuring 3metres per side and filled with algae could fulfil the energy needs of a 2 person household.

Burned or used in fuel cells, Hydrogen is an appealing option for powering future automotive vehicles for several reasons. Domestic industries can make it from a range of chemicals feedstock and energy sources (for instance, from renewable, nuclear and fossil fuel sources), and the non-toxic gas could serve as a virtually pollution – free energy carrier for machines of many kind when it burns it releases no carbon – dioxide, a potent green house gas. And if hydrogen is fed into a fuel – cell – stack – a battery like device that generates electricity from hydrogen and oxygen – it can propel a electric car or truck with only water and heat as bio products (Satyapal *et al.* 2007). Hydrogen could therefore, help in pressing environmental and social problems, including air pollution and its health hazards, global climate change and dependence on foreign oil imports. Yet barriers to gasing up cars with hydrogen are important. Kilograms for kilogram hydrogen contains 3 times energy or gasoline but today it is impossible to store hydrogen gas as compactly and simply as conventional liquid fuel. Picture in Plate II shows a hydrogen driven car with two cylinders (Plate II, Fig. B).

One of the most challenging technical issues is how to efficiently and safely store enough hydrogen on board to provide the driving range and performance that motorists demand. Researches must find the "Goldilocks", storage solutions that are "just right". Storage devices should hold sufficient hydrogen to support today's minimum acceptable travel range – 500 kms on a tank of fuel in volume of a space that does not compromise passenger or luggage room. They should release it at the required flow rates for accelerat-ion on highways and operate at a practical temperature. They should be refil-led or recharged in a few minutes. Today a 20-gallon tank containing hydro-gen gas at atmospheric pressure would propel a standard car only about 500 feet so engineers are trying to increase the energy of hydrogen in fuel tank.

8.8 HITHANE FOR VEHICLES

Indian Automobile Manufacturers Society has planned for a project in which CNG and hydrogen will be used for running the vehicles. The project costs Rs. 120 million and would yield results within 2 years. The centre has agreed to bear about 30 million of the cost with a similar amount being invested from IOC. The rest would come from the 5 automobile majors, Tata motors, Bajaj auto, Mahindra and Mahindra, Ashok Leyland and Eicher motors have also come together for this project. The idea is to use various combinations of CNG and Hydrogen- called Hithane-in existing vehicles to determine an optimum mix, which not only ensures fuel efficiency but also allows minimum engine modification.

8.9 ELECTRIC FUEL

Electricity is the unique among the alternative fuels in that mechanical power are derived directly from it, whereas the other alternative fuels release stored chemical energy through combustion to provide mechanical power. Batteries commonly provide electricity used to power vehicles, but fuel cells are also being explored. Batteries are energy storage devices but unlike batteries, Fuel cells convert chemical energy to electricity. A large number of various types of batteries are being tested for use in EVs nickel iron, nickel zinc, nickel air, aluminum air. The first benefit of using electric fuel some emissions that can be attributed to EV s the emissions that are generated in the electricity production process at the power plants. The maintenance costs for EVs is less. EVs have fewer moving parts to service and replace *i.e.* Hydro-gen, natural Gas, P-series, Solar fuel, CNG Cylinder, Vapor Bag Assembly.

8.10 SOLAR FUEL

Recently there has been efforts to evaluate how solar energy may be used to power vehicles, however, the possibility of operating a vehicle on solar energy is very slim. Solar energy is derived from sun. In order to collect this energy and use it to fuel a vehicle, photovoltaic cells are used. The research on solar power is slowly shifting to the idea of using the solar energy to run certain auxillary systems of vehicle. Plate II, Fig.A shows a solar powered car.

8.11 FUEL EFFICIENCY

There is a need to increase fuel efficiency. The Petroleum Conservation and Research Association in collaboration with the Bureau of Energy Efficiency in developing fuel economy standards under the Energy conservation Act, 2001.

According to International Council on Clean Transportation (ICCT), worldwide, regulations are set either to control fuel consumption per unit of distance travelled or to regulate direct emissions of total greenhouse gas or CO_2 emissions. Both are directly linked with fuel consumption levels of the vehicles (Anonymous 2007). The EU auto industry has set CO_2 reduction target 120 g CO_2 /km in 2012 and Japan 125 g CO_2/km in 2015. Among the developing Asian countries China and Korea has taken steps to reform passenger vehicle tax to encourage smaller engine vehicles.

REFERENCES

Ackom, E.K. and Ertel 2005 An alternative energy approach to combating desertification and promotion of sustainable development in drought regions. Forum der Forschung 18: 74-78.

Agarwal D. and Agarwal A.K. 2007 Performance and emissions characteristics of *Jatropha* oil (preheated and blends) in a direct injection compression ignition engine. *App. Thermal Engineering* 27 (13): 2314-2323.

Anonymous 2007 News: Fuel efficiency, India in step with global rules. *Down to Earth* 16:20

Cazetta M.L. Celligoi, M.A.P.C. Buzato, J.B. and Scarmino I.S. 2007 Fermentation of molasses by *Zymomonas mobilis* : effects of temperature and sugar concentration on ethanol production. *Bioresource Technology* 98(15) 2824-2828.

Chitra P. Venkatchalam P. and Sampatrajan A. 2005. Optimisation of experimental conditions for biodiesel production from alkali catalyze transesterification of *Jatropha curcas* oil Energy for Sustainable development Sept IX 3 :13 – 18

DoE (Department of Energy) USA. 2005 Theoretical ethanol yield calculator. Details available at http://www.eere.energy.gov.biomass/ethanol_yiled_calculator.html, last assessed on 2nd January 2005

Dwary A. and Pramanick M. 2006 *Jatropha* – A biodiesel for future, *Everyman's Science* XL (6) 430 – 432

Gong C.S. 1983 Recent advances in D-xylose conversion by yeasts. In: Ann. Rep. on Fermentation Processes vol 6. Acad. Press N.Y. p. 253.

Kandpal J.B. and Madan M. 1995. "*Jatropha curcas* : a renewable source of energy for meeting future energy needs." *Renewable Energy*, 6(2): 159-160.

Kaushik, N. Kumar, K. Kumar S. Kaushik N. and Roy S. 2007 Genetic variability and divergence studies in seed traits and oil content of Jatropha (*Jatropha curcas* L.) accessions. *Biomass and Biotechnology* 31(1) 497-502.

Lachke A.H. 1989. Liquid fuel from solid waste. *Science Reporter* 26 (6):318-321.

Mano N., Mao, F. and Heller A. 2002 A miniature biofuel cell operating in physiological buffer. *J. Am. Chem. Soc.* 124: 12962-12963.

Muniyappa P.R. Brammer S.C. and Noureddini H. 1996 Improved conversion of plant oils and animal fats into biodiesel and co-product. *Bioresource Technology* 56: 19-24.

Nagendrappa G. 2000 Liquid fuel from plants. *Resonance,* 5(11): 47-55.

Nair D.R., Patel P. A. and Daniel M. 2007 Some new vegetable oils as sources of Biofuel. In: *Herbal technology : Recent trends and progress* (eds) M. Daniel, A. Arya and V.M. Raole Scientific Pub. Jodhpur 297-300.

Nigam P.B and Agarwal P.K 2004 Ethanol manufacture from cane juice and molasses : A new diversification for sugar industry in India. Proceedings of International Conference of Biofuels : Perspectives and Prospects, New Delhi, Sept 2004, Winrock International India, New-Delhi.

Patel V.S. 2005 LPG: the greener fuel alternative. *IAAPC News letter* 9:1-2

Pramanik K. 2003 Properties and use of *Jatropha curcas* oil and diesel fuel blends in compressor ignition engine. *Renewable energy* 28: 239 - 248

Sahai S. 2007a Biofuel vs. Food Security *Gene News* 2: 2 – 3

Sahai S. 2007b Biofuel from car exhaust *Gene News* 2: 5

Satyapal S. Petrovic J. and Thomas G. 2007 Hydrogen. *Scientific American* April 2007: 62 – 69

Schell, D.J. Dowe, N. Ibsen K.N. Riley C.J. Ruth, M.F. and Lumpkin R.E. 2007 Contamination occurrence, identification, and control in a pilot scale corn fiber to ethanol conversion process. *Bioresource Technology* 98(15) 2942-2948.

Silvio C.A., de A. Belchior, C.R. Nascimento M.V.G. Vieira L. dos S.R. and Flueury G. 2002. Performance of a diesel generator fuelled with palm oil. *Fuel* 81: 2092-2102

Tayhas G., Palmore, R. and Kim H. 1999 Electro-enzymatic reduction of dioxygen to water in the cathode compartment of a biofuel cell. *Journal of Electro-analytical Chemistry,* 464 : 110-117.

www.greenfuels.co.uk

www.bti-eneflash.com/bionews22.

CHAPTER - 9

Problems of biodeterioration affecting valuable cultural heritage

Arun Arya[1] and A.R. Shah[2]

[1]*Department of Botany, Faculty of Science,*
The Maharaja Sayajirao University of Baroda, Vadodara – 390 002.
e-mail: aryaarunarya@rediffmail.com
[2]*Baroda Museum and Picture Gallery, Sayajibaug, Vadodara – 390 008.*

ABSTRACT

In simple terms biodeterioration means any alteration or change in a given object or building on account of biological agencies. Biodeterioration by microbes include damage caused by algae, fungi, bacteria actinomycetes etc. Fungi are major contributors to the deterioration of cultural property in the tropical countries like India, Sri Lanka, Malaysia and Thailand etc. Fungi are one of the major bio-pollutants present in air. Their number per cubic meter of air may further increase in urban air due to presence of favourable substrates and environmental conditions. Present paper deals with deterioration caused by fungi present in indoor air. These fungi get entry into museums through doors, windows, visitors or objects kept in museum without proper treatments.

Qualitative and quantitative estimation has been done earlier to detect the load of colony forming units per unit area. Fungi were isolated from different deteriorating objects, displayed in Baroda museum. When isolations were made from 7 miniature paintings. A total of 20 different colonies were observed belonging to 9 different genera. The fungus *Aspergillus* was present in 5 paintings. Total 4 spp. of **Aspergillii** were observed. All the members of *Aspergillus* are potential allergens. Other fungi isolated are phytopathogens and may have got entry from surrounding garden.

Key words : Baroda Museum, Biodeterioration, Miniature paintings, *Aspergillus* spp.

"A good museum attracts, entertains, arouses curiosity, leads to questioning and this promotes learning. It is an educational institution that is set up and kept in motion that it may help the members of the community to become happier, wiser and more effective human beings."

– (Dana, 1909) Foster, 1986

The museum today is not merely a repository of objects, it effectively helps in conservation, documentation and presentation of the national and cultural heritage (Baxi and Dwivedi, 1973). The word **Museum** is derived from the Greek 'Mouseion' which was a place of contemplation, a philosophical institution or a temple of Muses. ICOM (1990) defines museum as a non-profit making, permanent institution in the service of society and of its developments, communicates and exhibits for purpose of study, education and enjoyment, material evidences of people and their environment.

Museum's primary function is to be a source of inspiration because the 'Muses' whose house it is, are goddesses of liberal art and are eternal inspiration of men. A museum should be able to give its visitor a lasting impression, some experience which could introduce a new element into his life. Hooper-Greenhil (1991) commented, "a museum presents probably the only effectual means of educating the adult, who cannot be expected to go to school like the youth...."

9.1 MUSEUMS - CENTERS OF EDUCATION AND VALUABLE CULTURAL WEALTH

Museum collection comprising of materials of cultural, artistic, historical and scientific significance from different sources serves as an index a civilization between the present and the past. India is one of those few nations which have a very rich ancient culture. Its several thousand years old cultural history shows a strong thread of continuity. Gujarat has 33 museums. Of the different museums of Vadodara two museums are noteworthy i.e. Baroda Museum and Picture Gallery and Fatehsingh Museum.

The Baroda Museum and Picture Gallery is famous for its collection of European oil Paintings all over Asia. The buildings of the museum are situated in the pastoral environment of Sayajipark. The area inhabited by Kamati tribe was built into large public garden under the leadership of Colonel Carls in 1875. Construction of museum started in 1887 by Maharaja Sayajirao Gaikwad III and thrown open to the public in 1894. Picture gallery building begun in 1908 and was completed in 1914. This museum is built in the Indo-Saracenic style of architecture. A new building is added to this museum recently. Vadodara city gallery was inaugurated by the Chief Minister Shri Narendra Modi on 1st May 2003. Museum office is also shifted to this new building.

The cultural heritage is perishable. It does not regenerate itself like natural heritage. Owing to passage and ravages of time, it disintegrates. The damage caused to it by human agency is much more than natural forces. According to Baxi and Dwivedi (1973)' museum objects like textiles, paintings etc. will be eaten up by insects if left to themselves". In the words of Coremans (1960), "There is inexorable law which spare no work, past or present; all living matter is subject to decay and one day must perish; like doctors we can only do our best to delay the deadly process", Nair (1972) defined biodeterioration as, "any undesirable change in the properties of the materials caused by vital activities of organisms." Biodeterioration has direct link with our economy as it is related to the survival of object material and to the maintenance of their quality. Museum exhibits of Baroda Museum and Picture gallery include European oil paintings, portraits of Royal family members, carved gold, silver ornaments, stone sculptures, bronze sculptures, metalware, wooden temple, chariot, gold Gaekwadi turban, embroided cloths. Science gallery has specimens of animals and birds and a skeleton of large blue whale. Egypto-Babylonian gallery has mummy of a lady placed in a glass chamber is center of attraction. It is one of the 7 mummies present in Indian museums.

9.2 BIODETERIORATION OF MUSEUM OBJECTS

In museums and Libraries, paper manuscripts, miniature paintings and palm leaf manuscripts are extremely vulnerable to infestation by insects. Agrawal (1991) found palm leaf manuscripts and paper documents damaged by insects. He observed devastating and quick action of fungus on art objects. Numerous paintings received by Lalit Kala Academy developed 'black or grey layer' on the surface, when paintings were examined they were affected by fungus. Within a short span of time 4-5 days such damage occurred because these paintings were stored in a basement, where humidity was very high.

Indian miniature paintings probably constitute one of the richest treasures of Indian cultural heritage. A miniature painting being composed of multilayered material, numerous agencies decay it rapidly. Biodeterioration is one of the most commonly encountered problems faced by miniature paintings. This is primarily due to favourable climatic conditions and the constituent materials of the paintings (Uniyal 1991).

Not only is the type of decay varied, but the mode of attack by different biological agents also differs and is sometimes very selective. The exact nature and type of damage caused by living organisms upon different kinds of materials in general, and on miniatures in particular, the species of fungus or insect responsible for damage need more study to evolve appropriate treatment of the objects.

Plate I. Fig A. A Rajasthani Miniature painting showing Biodeterioration by fungi and silver fish

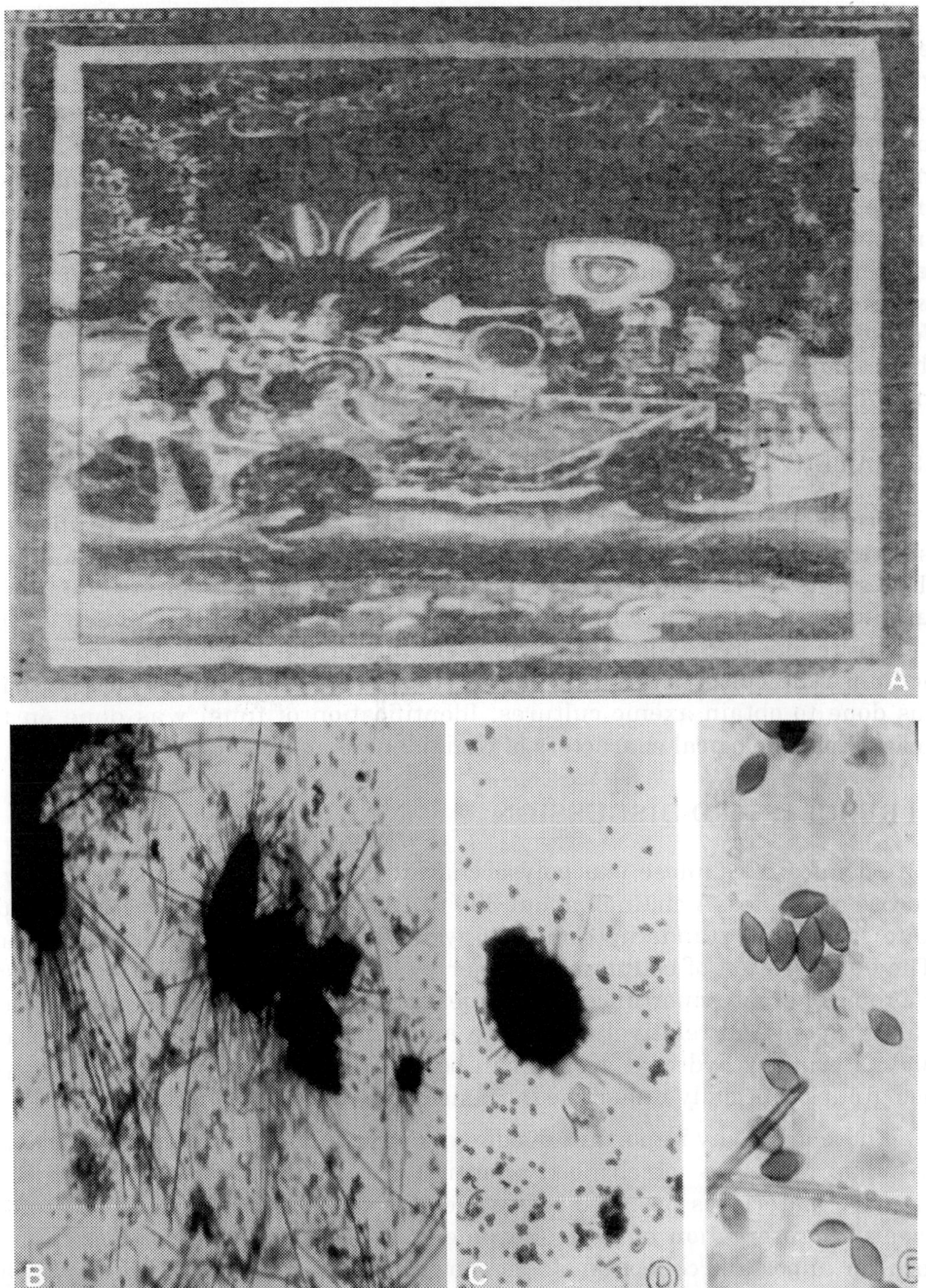

Plate II. Fig. A. Rajasthani miniature painting showing damage by *Chaetomium gangligerum;* **Fig. B.** Perithecia of *Chaetomium gangligerum*; note the straight hairs on the wall; **Fig. C.** Perithecia and ascospores of *Chaetomium cymbeforme.*

Cellulolytic fungi are the most common organisms responsible for the deterioration of various cultural properties viz., manuscripts, books, various types of paintings, wall hangings, wood craft etc. Cellulolytic fungi have the ability to attack cellulose which is then broken down into glucose molecules and utilized. Thus, they grow well on these substrates. However, the growth and distribution of these organisms is greatly influenced by the association of simple sugars, nitrogen content, moisture content and growth promoting substances with the substrates. Common cellulolytic fungi reported by Lakshmikant and Mathur (1991) from Gorakhpur were *Alternaria*, *Aschochyta*, *Aspergillus*, *Chaetomium*, *Cladosporium*, *Curvularia*, *Humicola*, *Memnoniella*, *Penicillium*, *Scopulariopsis*, *Trichoderma* and *Trichothecium*, Thakre and Bhajbhuje (1991) found *Aspergillus*, *Chaetomium*, *Nigrospora* and *Fusarium* with deteriorating books in Nagpur.

9.3 MATERIALS AND METHODS

Problems of biodeterioration due to fungi are acute and it is difficult to ascertain an individual fungus attacking an object. Isolations were made on PDA slants from miniature paintings. Later on the area of miniature paintings was cleaned with 90% ethanol and after 10 days when fungal growth reappeared, the isolations were made on Potato Dextrose Agar medium. The cultures were incubated at 25 ± 2°C for 8 days. Subculturing was done to obtain axenic cultures. Identification of fungi was done on the basis of morphological characters of conidiophore and conidia.

9.4 RESULTS AND DISCUSSION

Chemically cellulose is a polysaccharide chain consisting of a number of glucose molecules. A chain of about 300 to 1000 glucose molecules is normally quite stable and stands upto much physical strains and chemical reactions. "The deterioration of cellulose, whether by oxidation or hydrolytic processes, is a topic of great importance to the question of the preservation of books and manuscripts. It is relatively stable to autoxidation at ordinary temperature but the primary hydroxyl groups can be progressively oxidized to carboxylic acid groups especially under the influence of light" (Mills and White 1987).

There are many important documents written in ancient times preserved in Indian Museums which need scientific preservation. Paper being a most sensitive material is prone to many agencies of deterioration and, therefore, scientific preservation of paper documents is a prime need to stop ruins of such an important collection. It is known that in many libraries books are damaged severely by insects, fungi or other ways. In museums these agencies may cause severe and irreparable damage to archival collections. Fungi which induce biodeterioration on paper can be divided into two groups. Those which belong to the first group initiate the degradation process and are mostly saprophytes including representative of genera *Penicillium*, *Aspergillus*, *Alternaria* and *Fusarium*. The second group consists of fungi with adaptivity to definite substrates such as cellulose products e.g. *Trichoderma viride* and *T. reesei*. Most of the members of genus *Chaetomium* are cellulolytic and contribute significantly to paper damage (Szczepanowska 1986).

The cellulolytic enzyme system consists of three types of activities: Cellobiohydrolase (CBH) activity, endo ß–glucanase (EG) activity and ß–glycosidase (BG) activity. These activities act synergistically to hydrolyse cellulose (Persson *et al*. 1991). A hypothesis for the synergistic action between the different enzyme activities has been proposed by Enari and coworkers (1987). Each activity consists of several enzymes. Cellobiohydrolase I and II hydrolyse insoluble cellulose into soluble cellodextrins and cellobiose. Endo - ß-glucanase I and II act on the cellodextrins and hydrolyse them into cellobiose units. The cellbiose is then split into two glucose units by ß–glucosidase. Unless purified it is difficult to determine the activity of the individual cellulolytic enzymes due to the synergistic action between the enzymes. Filter paper assay is used to compare cellulolytic enzyme activity. However, this assay does not necessarily show the true hydrolyzing capacity of a cellulolytic enzyme system since the activity is also dependent on the substrate. Shah and Arya (2000) studied biodegradation of different kinds of paper. Percentage loss of paper was observed after 25 days. It was observed that two organisms *Cochliobolus verruculosus* and *Aspergillus ustus* had a choice for Japanese tissue paper while third one (*Chaetomium cymbiforme*, Plate II, Fig. C). had a choice of brown paper and the other (*Penicillium citrinum*) had a choice for card board. Hence, different fungi had different choice of paper.

Studies were undertaken on five miniature paintings displayed in Baroda Museum and Picture gallery due to fungal growth. These paintings belong to 18th century A.D. The paintings are depicting consort of Malkaush and other Ragas, a type of classical music prevailing in India. Isolations from these Kanhara Ragini paintings resulted in pure cultures of *Aspergillus fumigatus*, *A. niger*, *Alternaria alternata*, *Cladosporium herbarum*, *Curvularia lunata*, *Helminthosporium* sp. and *Paecilomyces puntoni*. The discolouration of paintings was clearly visible. Infestation of silver fish was also noticed on borders of paintings (Plate I). Earlier Arya *et al*. (1988) found *Aspergillus terreus*, *A. versicolor* and *Chaetomium gangligerum* (Plate II, Fig. B) from Rajasthani miniature painting placed in Baroda Museum. (Plate II, Fig. A) Agrawal *et al*. (1989) described mechanism of microbial deterioration of paintings:

a. Attack by microorganisms on paintings mostly result in deterioration of the whole of the object, generally starting from the support and then penetrating through all the layers.

a. Appearance of microbial growth on the surface of paintings results in the formation of filaments spreading over the surface changing their design and colour.

a. Wazny and Rudniewski (1972) reported that hyphae growing inside the painting cause the structure of the layers to become dislodged. A similar effect may be brought about by fungal fruiting bodies, which are frequently formed under the surface of the painting layers.

a. Extra cellular enzymes enabling to penetrate inside the paintings cause deterioration of all components of the object. Enzymatic activity causes the painting layers to loosen or crack. Advanced microbial activity brings about powdering of the surface of paintings (Ross, 1963).

a. Ross *et al.* (1968) reported that hyphae growing inside the paintings excrete metabolic products like organic acids into the environment.

a. Fungal metabolism may also produce coloured compounds, which stain the paintings. The stains that appear as a result of fungal activity (especially the Dematiaceae) are almost impossible to remove from the surface of the paintings.

a. Ross (1963) observed that development of fungi, particularly in their vegetative stages, promotes the settling of dust on paintings. Hyphae readily grow through the dust particles, consolidating them and utilizing them as nutrients.

Rajasthani miniature painting affected by *Chaetomium gangligerum* produced green velvety growth. Fungi like *Alternaria alternata*, *Curvularia lunata*, *Cladosporium herbarum* and *Helminthosporium* sp. can produce black pigmentation and destroy the beauty of miniature paintings.

REFERENCES

Agrawal O.P. 1991 An overview of studies on biodeterioration of cultural property. In : *Biodeterioration of cultural property* (eds. Agrawal O.P. and Dhwana S.) Pub. By Macmillan India Limited.

Agrawal O.P., Dhawan S. and Garg K.L. 1989 *Microbial deterioration of paintings : A review* Pub. By Intach Conservation Centre, Lucknow.

Arya A., Bhowmik S.K. and Shah A.R. 1988 Biodeterioration of paintings. *Geobios new Reports* 7 : 179-180.

Baxi, S.J. and Dwivedi, V.P. 1973. *Modern Museums-organisation and practice in India.* Abhinan Publications. New Delhi 208 pp.

Coremans P. 1960. The Museum laboratory, In: *The organization of Museums, Practical Advice* UNESCO. 93-118.

Enari, T.M. and Niku-Paavola, M.J. 1987 CRC Critical Reviews in Biotechnology 5(1) : 67.

Foster, R.A. 1986. The larger provincial museums and art galleries. In, *Manual of curatorship* Butterworths, London : 66-75.

Hooper- Greenhil E. 1991. *Museum and gallery education*. Leicester Univ. Press, Leicester and London pp. 18.

ICOM 1990. ICOM Statutes code of Professional ethics, Paris, 23-24.

Lakshmikant and Mathur S.N. 1991. Biodeterioration of cultural property by cellulolytic fungi in", *Biodet.of Cul. Prop.*" (eds. Agrawal, O.P. and Dhawan, S.) Macmillan India Ltd., New Delhi 168-172.

Mills, J.S. and White R. 1987. *The organic chemistry of museum objects*. Butterworths, London. pp. 165.

Nair, S.M. 1972. Biodeterioration of museum materials in tropical countries, In, "Conservation in tropics". Proc. of Asia Pacific seminars on Con. of Cul. Prop. (ed. Agrawal O.P.) The International Centre for the Preservation and the restoration of Cul. Prop., Rome : 150-158.

Persson I., Tjerneld F. and Hahn-Hagerdal B. 1991. Fungal Cellulolytic enzyme production: A Review. *Proc. Biochemistry* 26: 65-74.

Ross, R.T. 1963. *Advances in Applied Microbiology* **5**: 217-234.

Ross, R.T., Sladen J.B. and Wienert, L.A. 1968 In, "Biodetrioration of Materials" eds. Walter, A.H. and Elphick, J.J. Elsevier, Amsterdam. pp. 317-325.

Shah, N.R. and Arya, A. 2000 *Studies on some fungal biodeteriogens*. Bhartiya Kala Prakashan, Delhi pp. 202.

Szczepanowska, H. 1986. Biodeterioration of art objects on paper. The paper conservator 10: 31-39.

Thakre, R.P. and Bhajbhuje 1991. Biodeterioration of books and journals. In, "*Biodet. of Cul. Prop.*" (eds. Agrawal, O.P. and Dhawan, S.) Macmillan India Ltd., New Delhi :139-147.

Uniyal, C.P. 1991. Biodeterioration of Indian miniature paintings-problems and prospects. In, "*Biodet of Cul. Prop.*" (eds. Agrawal O.P. and Dhawan, S.) Macmillan India Ltd., New Delhi : 91-97.

Wazny, J.A and Rudniewsku P. (1972). *Material and Organismen* **7**(2) : 81-92.

CHAPTER - 10

Awareness programme regarding shifting cultivation among tribal society

Alok Chantia

Department of Anthropology, Sri Jai Narain Post Graduate College, Lucknow 226 001
E-mail alokchantia@rediffmail.com

ABSTRACT

India (2001 census) has 698 tribal groups living across the length and the breadth of the country, each with their individual and characteristic culture. A peculiar cultural trait regarding agriculture is shifting cultivation, which is directly related with deforestation and air pollution. Eighty tribal groups are engaged in this cultivation across the country. Mainly Northeastern tribes (Adi, Aka, Apatani, Bokar, Bori, Deori, Galung, Karbi, Khampti, Libo, Memba, Miji, Mikuir, Miri, Mishmi, Monpa, Nishi, Sulung, Tagin, tangsa tribes of Arunanchal Pradesh) are engaged in this cultivation. The main cause of deforestation is concerned with shifting cultivation which causes environmental pollution.

Present paper highlights the causes of shifting cultivation amongst tribals and its effect on the environment. Paper also explores the means by which shifting cultivation may be stopped and quality of environment can be improved.

Keywords: Shifting cultivation, deforestation, air pollution, culture, tribal groups, plantation, awareness programme.

Shifting cultivation, also alluded as Jhum, Podu, Dahi, Bavada, Guhar, Marhan, Erka, Swidden and slash and burn; is considered to be one of the most ancient system of land based food producing tracing its emergence in the Neolithic times around 7000 B.C. when man began to plant, cultivate and improved species of edible plants by selection coupled with domestication of some animals. It is also taken as replica of pre agriculture stage; and the first step in transition from food gathering and hunting to food production. This archaic system has its continuance not only in India but also in several parts of the word. Magnitude of the practice may be minimal but anthropologically remains significant for its socio-cultural ramification and economic pursuit in the wake of meeting quest of hunger-food-a primary need of the people. It is widely established that food crops play an inevitable role in providing nutrition and thereby ensuring health of the people, but production of the same following traditional method, under constraints of geo-ecological conditions cast aspersions on adequacy or sufficiency in terms of yield to meet the requirement of the people in substantive manner.

10.1 INTRODUCTION : BEGINNING OF SHIFTING CULTIVATION

Neolithic period was a boon for *Homo sapiens*, because it was the time of agriculture. Nomadism disappeared and stability came in light. In real sense the Neolithic period was the foundation of culture, which made man different from the other creatures. In the beginning of his existence, man was totally supported by the nature and nature was established as goddess. So those people, who put their faith after the advent of pastoralism and agriculture in the nature were not interested to hurt the earth with plough. So they burnt the forest, which was present in abundance and with the help of its ash, they produced cereals two or three times in the same land. After that, they searched another place and repeated the same procedure for their cultivation. So due to the change of agricultural land or place time to time, a specific term took its shape for this unique cultivation *i.e.* shifting cultivation.

In the beginning of the Neolithic period, the maximum part of the earth was packed with dense forest. People were also limited in numbers. So no one was aware about the deforestation due to the effect of shifting cultivation. But when population increased and forest decreased an imbalance was created between man and nature. Human being felt the importance of forest for the conservation of environment. Those people who are thinking about environment are of the post Neolithic period *i.e.* urban areas who have adopted all modern facilities for a cozy life, but many groups of *Homo sapiens* are still living in a natural way and are totally dependent upon nature, these people are from simple societies, and form different tribal groups. They are very far from the modern facilities and technical, mechanical life. They are also not aware about the degradation of the forest and its ill effects.

Pollution is not only concerned with air, water, sound, soil etc. it is also associated with culture of a group, which affects the environment and ecology. In India 698 (2001 census) tribal groups are living and most of them are forest dwellers. They fulfill all their needs from the forest. Today 80 tribal groups are engaged in the shifting cultivation across the country, which is causing environmental pollution.

Shifting cultivation is associated with the different tribal groups of India. Presently 10% of the total tribal population of India is engaged in this tradition, nearly 90 lakhs tribal people are doing this cultivation. Shifting cultivation is mainly concerned with tree burning of tree wood. Its fumes emit carbon dioxide and this carbon is as hazardous similar to the carbon from the vehicles. This problem should be taken seriously, which has been from eyes of environmentalist.

10.2 METHODOLOGY

Present study is concerned with Adi (652), Apatani (16109), Galung (7) Khampti (47) and Nishi (5387) tribal groups living in lower Subansiri district, near to Itanagar, Arunanchal Pradesh, who are practicing shifting cultivation. The study was carried out during 15 November 2004 to 10 December 2004 and 1000 respondents were interviewed for the purpose, both male and female between the age group of 18 to 70 years. Respondents were chosen with the help of random sampling. In this study interview technique has been used along with secondary data to see the causes and effect of shifting cultivation on environmental pollution.

10.3 RESULTS AND DISCUSSION

Shifting cultivation is concerned with 80 tribal groups of the India. Among the Naga (eastern region) it is known as Jhum. In Orissa – Kaurai, veribaja, Dahi, Pod-chasa, Gidiaya, Dourgaum-chasa are the famous names of the shifting cultivation. In this cultivation, a tribal group selects a part of the forest and cuts it down with axes. After cleaning the forest, they leave the months trees for a few at the same place. After that, they burn these trees. As the result of burning, ashes occur on the surface of the field. They spread it in the field properly. This work is done before the starting of rainy season.

Though there has been spurt in the cereal production over the years through agricultural revolution and advancement in techniques and innovation of implements etc. hardly any effort seems to have been made to improve the yield in shifting cultivation method which poses threat to those depending upon the system and cannot rely on its produce which is just sufficient for not more than three to four months in a year. The north-eastern states of India, inhabited, predominantly by tribes, show poor yield of crops in Jhum as compared to production of the same under permanent cultivation based on modern agriculture system. As a result on an average the tribal people in northeastern states for want of food security from their production, require external support (under public distribution system) for about a period of eight months every year.

Sense of food security under Jhum can be achieved with the efforts of agricultural scientists by way of their involvement in related areas viz. Agricultural engineering and extension services besides land works as they are the pioneer of green revolution in the country. Also related aspects of land tenure, use and ownership pattern need a fresh look, if food security is to be given priority to Jhum- available method of cultivation in north-eastern states and its practitioner tribes (Plate I, Fig. A).

10.3.1 Increase in area of shifting cultivation:

Table 1: Comparison of census record regarding shifting cultivation

1991 census	2001 census	Result
1,51,295 hectares land engaged	1,52,703 hectares in use	1408 hectares land has increased in the last decade

The table 1 shows that shifting cultivation has increased in the last decade which is promoting environmental pollution.

10.3.2 Causes of shifting cultivation

Shifting cultivation is in practice due to the respective feeling among the tribals for they see earth as Maa. They do not want to dig the earth because it will hurt the earth. This cultivation is also in existence due to their poverty and lack of awareness. They are not in position to buy new and specialized implements.

Table 2: Why do they practice shifting cultivation?

	Due to culture		No alternative		Lack of awareness		Poverty		Total	
	No.	%	No.	%	No.	%	No.	%	No.	%
Male	188	42.3	117	26.3	78	17.5	62	13.9	445	44.5
Female	209	37.7	188	33.9	67	12.1	91	16.4	555	55.5
Total	397	39.7	305	30.5	145	14.5	153	15.3	1000	100.0

Table 2 Shows that maximum respondents 397 (39.7%) are engaged in shifting cultivation due to culture. It has been their cultural practice that they do not dig or plough as they do not want to hurt their Maa by digging it while, 305 (30.5%) respondents are engaged in this work as they have no other alternative for ensuring food security. 145 (14.5%) have no awareness

of modern implements of cultivation, whereas 153 (15.3%) respondents are engaged in shifting cultivation due to poverty.

10.3.3 Effects of shifting cultivation

Some government projects are increasing this cultivation – e.g., a project on modification of implements of shifting cultivator by North Eastern Regional Institute of Science & Technology (NERIST), Nivjuli, is promoting this cultivation which is causing deforestation and environmental pollution. Shifting cultivation is mainly concerned with tree burning which ultimately leads to carbon dioxide in the atmosphere (Plate I, Fig. B Hut made by tribals in the forest, also seen is maize crop raised by them).

Table 3: Showing the effect of shifting cultivation on environment.

	Increase in deforestation		Air pollution		Decrease in no. of Animals		Total	
	No.	%	No.	%	No.	%	No.	%
Male	201	45.2	97	21.8	147	33.0	445	44.5
Female	306	55.1	88	15.9	161	29.0	555	55.5
Total	507	50.7	185	18.5	308	30.8	1000	100.0

Table 3 Shows that maximum respondents 507 (50.7%) are of the view that shifting cultivation leads to deforestation. We have already noticed that in India forest covers only 19% of the land which is far below the standard parameter of 33% of the land. 308 (30.8%) respondents believe that obnoxious fumes emitted by burning of trees causes suffocation and death of animals. It has led to lessening of numbers of animals in the forest and tribal areas. 185 (18.5%) respondents feel that burning of trees increases heat in the environment and causes air pollution. Carbon dioxide in a green house gas and is responsible for global warming.

10.3.4 Means of environment protection

In order to protect the environment from the hazards of shifting cultivation literacy ratio should be increased and tribals should be shifted towards the other services, it will reduce pollution and deforestation This new alternative will give a new dimension to their lives. Government projects like improvement of implements of shifting cultivator should not be carried out so as to avoid environmental pollution. So far as respondents are concerned they gave following responses in favour of environmental protection.

Plate I. Fig. A. Part of a hill showing terrace farming and huts of tribals; **Fig. B.** Hut made by tribals in the forest, also seen is maize crop raised by them).

Table 4: What do you want for environmental protection ?

	Extensive Plantation		Domestic/Small scale industries		Literacy/awareness programme		Total	
	No.	%	No.	%	No.	%	No.	%
Male	106	23.8	56	12.6	283	63.6	445	44.5
Female	186	15.5	176	31.7	293	52.8	555	55.5
Total	292	29.2	232	23.2	576	57.6	1000	100.0

Table 4 Shows that 576 (57.6%) respondents are in favour of literacy and awareness programmes, so as to disseminate the knowledge about the hazards of shifting cultivation. With knowledge and awareness they can opt for other services and it would reduce the burden on the environment. 292 (29.2%) are in favour of extensive plantation in the area so that loss of forest cover may be compensated. Whereas, 232 (23.2%) are of the view that domestic and small scale industries should be promoted to provide alternative jobs and means of living. It will also ultimately reduce environmental pollution.

10.4 CONCLUSION

Thus we see that shifting cultivation which plays havoc with the environment and poses threat to posterity must be checked and stopped. For this awareness programmes should be conducted amongst tribal society so that they may adopt alternative means for food security instead of shifting cultivation. To bring them into main stream they should be given preference in certain jobs.

Acknowledgement

I am very thankful to Dr. Ramesh Chandra Deputy Director, Anthropological Survey of India, North Eastern Region who has provided the valuable information regarding Jhum/shifting cultivation.

References

Atal, Y. 1965 *Adivasi Bharat; Economic system*, Rajkamal Prakashan 30-45pp.

Chandra R. 2001 Shifting cultivation and food security: An experience of some north eastern tribes, ISCA, New Delhi session.

Dewangan K.N. 2001 Anthropometric Survey of Jhum Cultivator of Arunanchal Pradesh, IIT Mumbai, Seminar on Ergonomics.

Hasnain N. 1997 *Janjateya Bharat*, Jawahar Publishers and Distributors 20-30pp.

India Census 1991, New Delhi 142p

India Census 2001,

Vidharthi L.P. 1974 Bhartiya Adivasi, Economic system, R.P. Tandon Bhawan 113-150pp.

CHAPTER - 11

Optimizing dedusting system design with pleated filter cartridges

Satish Parekh

Solids Handling Expert, Parekh Global Services, Atlantic III, Ami Society No. 1, Opposite Yes Bank, Race course, Vadodara 390 007

ABSTRACT

According to estimates of World Health Organization (WHO), more than one lakh people in India die due to particulate pollution every year. Primary solid particulate matter that is emitted directly into the solid phase is significant category of fine particulates. It results largely from combustion of fossil fuels in automobiles and industrial processes, construction activities and wind erosion etc. Industrial process may produce different types of dust. For maintaining environment clean technologies are developed to make use of filter cartridges instead of filter bags or filter hoses to filter the dust from the air in various industrial processes. The paper addresses the need for the knowledge and can be taken for cartridge users for accurate design and efficient working of their pollution control systems.

Key words: Dedusting, Health, Pleated filters, Pollution control

11.1 INTRODUCTION: POLLUTION DUE TO DUST

In the contemporary global scenario, India has emerged as one of the very rapidly growing economy, competing with developed nations in terms of industrial growth. This rapid industrial growth has put forth along with it several challenges, amongst which effective maintenance of pollution free environment in an efficient manner is a foremost one. For efficient and effective maintenance of clean environment, it has been a practice in the technologically developed western countries to make use of filter cartridges instead of filter bags or filter hoses to filter the dust from the air in various industrial processes. Today, pleated filter cartridges are available to suit a wide range of industrial applications. The limited knowledge of the cartridge puts forth a lot of problems for industry as well as the users.

This paper addresses the need for the knowledge and care to be taken for cartridge users for accurate design and efficient working of their pollution control systems.

With advancement in the industrial processes, today, we come across more than 10,000 types of dust. Each of them has its own distinctive characteristics which affect the design and working of the filter cartridges. So, first of all, these characteristics must be understood.

Dust: Dust can be defined as visible and invisible (to naked eye) floating and settled particles of solid matter.

11.2 PHYSICAL PROPERTIES OF DUST AFFECTING FILTER SYSTEMS

11.2.1 Bulk density

Bulk density can be defined as the weight of the particles per unit volume of air. This property is most important for the design of the filter elements as it directly affects the surface area required for filtering the given quantity of dust (assuming similar particle size, distribution and shape). The lower the bulk density, the higher the amount of surface area of the cloth required and ultimately lower surface load is to be considered while designing the filter cartridge. Also, the dust with lower density remains suspended in air for longer time and hence while cleaning the cartridges with air jet, the dust particles tend to stick on other cartridges thus preventing effective cleaning procedure.

11.2.2 Particle size, shape and distribution

These three characteristics combine in different ways to affect other properties such as air retention, permeability, angle of repose, minimum conveying velocity etc. We usually design the filter system considering a mean size of a particle. The mean size is the one at which 50 of the particles lie on the particle size distribution diagram. Particle distribution is an analysis of the bulk material particles broken down by percentage into various size fractions. The transport gas velocity in the pipe depends upon the combination of particle size, shape and its distribution.

11.2.3 Solid ratio

It is defined as the weight of dust that can be carried per unit weight of the air.

Usually, for filtering systems using filter cartridges, the solid ratio is in the range of 0.1~0.2. i.e 100gms~200gms of dust can be carried per kg of air.

11.2.4 Moisture Content

Any moisture on the surface of the dust results in surface tension, which inhibit the movement of the dust particles relative to each other causing them to become more cohesive. So, the dust will tend to stick to the filter media surface thus adversely affecting the cleaning process of cartridges.

11.2.5 Temperature

This property is important for the selection of proper filter media as the filter media must withstand, without getting damaged, the temperature of the incoming dust. Some types of dust (for eg. 5 mol Borax) agglomerate at a high temperature and become sticky, thus sticking to the filter media and posing problems for effective cleaning of the filter media.

Dust like wood saw, pulverized coal etc., ignite if exposed to high temperature. Such dust will damage the filter media if directly exposed to it. The air temperature and air volume are interconnected by the General Gas Equation. The air volume is generally at ambient temperature of 32~35°C. The air volume has to be recalculated using general gas equation. It is important to have real actual volume of air for calculation of the specific load on filter cartridges.

11.2.6 Abrasiveness

Dust with higher hardness (like corundum, carborundum) etc are more abrasive in nature.

Abrasive dust affects the filter systems in two ways:

1. It will cause erosion of the pipeline components,
2. It will cause damage to the filter media.

So, while selection of suitable material of construction of components as well as filter media for abrasive dusts, these two things should be taken into basic consideration.

11.2.7 Cohesiveness and Adhesiveness

Cohesive force is the force with which a body attracts another body of same material. Dust particles with high cohesiveness will attract each other and form clinkers which stick in to the pleats of the cartridges and block them. In such cases it is recommended to use lesser pleat depth and less number of pleats.

Adhesive force is the force with which a body attracts another body of different material. This is important consideration as dust particles with higher adhesiveness will tend to stick to the surfaces inside the filter system and on the surface of filter cartridges. Also, if the cartridges in the filter housing are closely spaced, there are chances that while cleaning the cartridges alternatively with jet of air, dust with high adhesiveness will stick to the cartridges which are not being cleaned at that moment.

11.3 CHEMICAL PROPERTIES OF DUST AFFECTING FILTER SYSTEM

(i) *Hygroscopicity*: Hygroscopic dust (for example salt, sugar etc)will absorb moisture from the surroundings. This property will affect permeability and cohesiveness properties. With increase in hygroscopicity, the permeability of the material decreases and its cohesiveness increases. This leads to the material becoming sticky and thus causing difficulties in filtration. For this, the proper selection of filter media needs to be done.

(ii) *Corrosiveness*: Corrosive dust will react chemically with material of construction (MOC) of the filtration systems as well as cartridge and causing their deterioration. The proper selection of corrosion resistant MOC has to be considered.

(iii) *Explosiveness*: Dust with this tendency (for example carbon based dust like sugar, coal dust etc) tend to explode when exposed to spark, flame or heat. The types and characteristics of explosive dust are discussed in detail in ATEX 100a-RL 94/9/EC which is effective from 1st July 2003.

11.4 SELECTION CRITERIA FOR FILTER ELEMENTS USED IN FILTER CARTRIDGES

A wide variety of cleaning media are available today to suit various applications. The selection criteria of these filter media comprises of the following. :

(i) Plants and projects

(ii) Characteristics of dust.

Usually, the filter media used are of chiefly made up of polyester, polypropylene or cellulose fibers. For special applications, different types of coating are applied on the surface of the cloth.

For example:

(i) For dust having static charge, coating of aluminum is applied on polyester fiber cloth to enhance its antistatic properties.

(ii) For the dust in high quality food industry, a special filter media with combination of polyester base carbon fiber and protective PTFE film is used for water and oil resistance.

11.5 MECHANISM OF DUST FILTRATION

The diagram below shows the general setup of the dust filter system.

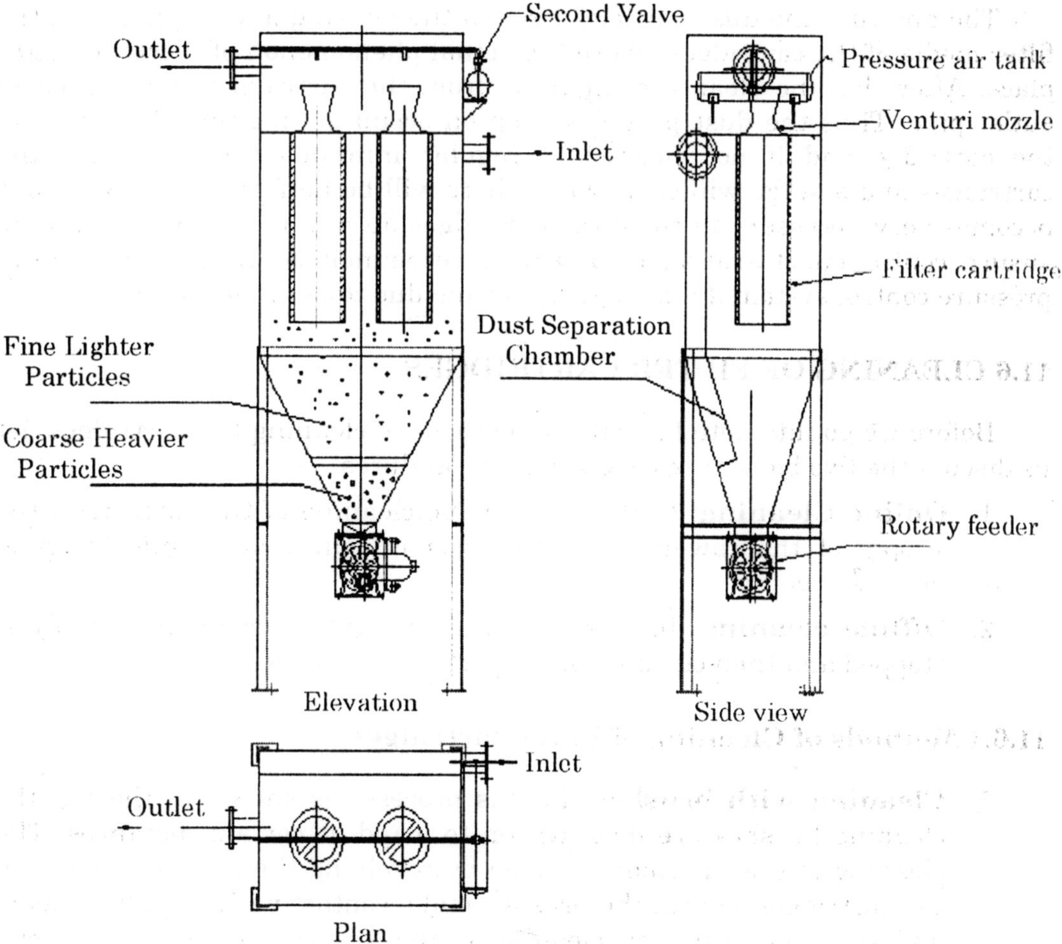

Air with a high velocity enters through the inlet pipe into the dust separation chamber. If the air at such a high velocity is made to impinge on the cartridges directly, it will damage the filter element. So, to reduce the velocity and hence the momentum and kinetic energy of the dust laden air, it is made to enter the air inlet chamber where the velocity comes down and is just above the floating velocity. The principle behind the working of dust separation chamber is that when a dust particle is made to change its direction, it loses considerable amount of its velocity (and ultimately kinetic energy).

This is most important as a considerable amount of coarse particles will be separated by their own weight and get collected at the bottom of the housing and only a certain fraction of the total dust contained in the air impinges on the cartridge. Thus, the dust load on the filter cartridge is greatly reduced. The ultimate benefit being less number of cartridges is required for filtering the same amount of air.

The air with fine dust particles rises in the housing and impinges on the filter media of the cartridges where the actual phenomenon of filtration takes place. After this the clean air emerging from the cartridges is conveyed to outlet pipe. Thus the dust particles keep on accumulating on the surface of the cartridge and if not cleaned at regular intervals they will clog the cartridges and a stage will come when there will be no filtration at all. So it becomes very necessary to clean the cartridges at certain interval of time to ensure continuous filtration of air. It is recommended to use differential pressure control system if dust load is varying due to plant operation.

11.6 CLEANING OF FILTER CARTRIDGES

Before we go into detail about the methods of cleaning the cartridges, let us discuss the two basic types of cleaning procedures

1. **Online Cleaning**: In this type, while cleaning of the cartridges, the supply of the raw gas or crude gas is continuous while cleaning process is going.
2. **Offline cleaning**: In this type, the raw gas or crude gas supply is stopped and then cleaning of the cartridges is done.

11.6.1 Methods of Cleaning of Filter cartridges

1. **Cleaning with brushes**: In this process, as shown in the fig, the cleaning brushes are mounted on to a rod at certain distances. The cleaning of the cartridges is done by revolving the rod from inside of the cartridges so that the brushes make contact with the filter media and clean the dust. This cleaning system can be used only for special type of cartridges. This is done offline only.
2. **Cleaning with Pulsed Air Jet:** This is the most widely used method of cleaning. In this method, a pulsed jet of air is made to impinge on the inner side of the cartridge. Due to this the dust sticking on the filter media gets separated. This method can be and is generally used online. Devices are available on market, where a pipe

with horizontal arms is mounted inside the cartridges and which turns, once the air blow is given (refer fig "B"). Only one device can work at a time as it needs controlled blow of air with pre-defined volume of air. The average value for rotating the arm will be 6 Rpm. The minimum pressure for good cleaning will be 7 barabs.

Another method of cleaning involves use of special accessory called venturi nozzle. Here the air is made to impinge on to the inner surface of the cartridge via nozzle (refer fig "C"). The important factors determining this process are pulse interval, pulse time and volume of air.

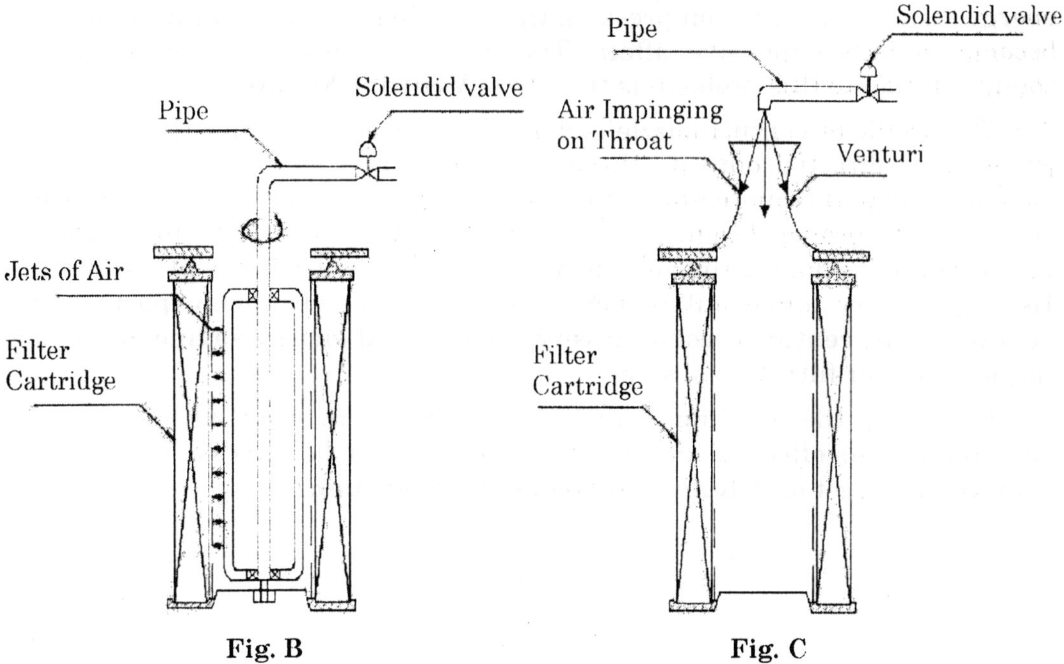

Fig. B **Fig. C**

(i) Pulse interval: The interval at which the cleaning air jet is to be supplied can be in two ways:

(a) **Jet at fixed time interval:** In this method, the solenoid valve is activated by a timer unit at a certain fixed interval of time. This interval is selected based on filtering load on the cartridge.

(b) **Jet at selected pressure drop:** In this process, the solenoid valve is activated by a pressure sensing device at a certain value of pressure drop. With the accumulation of dust on the filter media, the pressure drop in the cartridge is reduced; the solenoid valve is then activated by the pressure sensing device which senses lower pressure drop. When the cartridges are cleaned and pressure drop is restored the solenoid valve is deactivated and supply of cleaning air stops.

This is the best method of cleaning as its working is in line with dust load on cartridges. Thus, it takes care of the dust load fluctuations in the plant operation.

(ii) Pulse Time: The pulse time is generally in the range of 0.1~0.2 seconds.

(iii) Volume: Research shows that the cleaning process is directly related to the volume of the air being made to impinge on the cartridge. The higher the volume, the better is the cleaning efficiency. Hence, it is necessary to have high volume of air for effective cleaning. But, looking to the high energy costs involved in compressing the air, a high volume of compressed air becomes a very expensive affair. The most cost effective and technically sound solution to this problem is the use of **Venturi Nozzle.**

The profile of venturi nozzle is designed in such a manner that when the jet of air is directed onto its throat, a temporary vacuum is created in its vicinity. Due to this more amount of air flows from the surrounding inside the venturi. With proper design of venturi, the volume of air that can be induced in the venturi is double the volume of compressed air from the pressure tank. Here, it is very important to note that there are no moving parts in the venturi. Thus, venturi becomes a very low cost and energy saving equipment in the whole air filtration system.

Following the guiding principles discussed in this article, a most resourceful and effectual dedusting procedure can be carried out. We can achieve the target of "**Clean and Green Environment**".

CHAPTER - 12

Monitoring and control of urban air pollution in India

Hardik Bhatt and Hitesh V. Dhamecha

Civil Engineering Department, Polytechnic,
The Maharaja Sayajirao University of Baroda, Vadodara – 390 002

ABSTRACT

Urbanization and industrialization has provided humanity with many material and social benefits. At the same time, it has brought in its wake many material and social problems. One of these is the Environmental Pollution. Industrialization and urbanization have resulted in a profound deterioration of India's air quality. Sources of air pollution, India's most severe environmental problem, come in several forms, including vehicular emissions and untreated industrial smoke. The industries like thermal power plants, petrochemical industries and crude oil refineries, pharmaceutical and chemical industries are the major culprits of air pollution. However, in large metropolitan cities, the exhaust gases from the traffic (fuel driven vehicles), also contribute to a great extent to the air pollution problems.

This chapter deals with the sources of air pollution, concentration of air pollutants in the ambient atmosphere and monitoring and control measures for the same in the urban areas.

Keywords : Monitoring, Control, Urban Air pollution

Ecologists recognize pollution as a serious threat to the quality of life and possibly even to its existence. Interests in environmental concerns have become intense in recent years, with much pressure being exerted on public and corporate sectors to alter policies and practices, which in the past have resulted in the degradation of the environment. This study is a triumph to explore the sources, issues and possible solutions to the problem of the urban air pollution. Of the 3 million premature deaths in the world that occur each year due to outdoor and indoor air pollution, the highest number are assessed to occur in India. According to the World Health Organization, the capital city of New Delhi is one of the top ten most polluted cities in the world. Surveys indicate that in New Delhi the incidence of respiratory diseases due to air pollution is about 12 times the national average. According to another study, while India's gross domestic product has increased 2.5 times over the past two decades, vehicular pollution has increased eight times, while pollution from industries has quadrupled.

Sources of air pollution, India's most severe environmental problem, come in several forms, including vehicular emissions and untreated industrial smoke. Apart from rapid industrialization, urbanization has resulted in the emergence of industrial centers without a corresponding growth in civic amenities and pollution control mechanisms.

12.1 SOURCES OF AIR POLLUTION

The sources of air pollution can be broadly classified as natural and man-made. The natural sources include the air pollution problems as caused by the biogeochemical cycles ever-existing in the nature. The man-made sources of air pollution can further be classified as the pollution problems in rural and urban areas.

Major Air Pollutants

Table 1. Major Air pollutants their sources and effects

Pollutant	Sources	Effects
Ozone		
A gas that can be found in two places. Near the ground (the troposphere), it is a major part of smog. Higher in the air (the stratosphere), it helps block radiation from the sun.	Ozone is not created directly, but is formed when nitrogen oxides and volatile organic compounds mix in sunlight. That is why ozone is mostly found in the summer. Nitrogen oxides come from burning gasoline, coal or other fossil fuels. These are many types of volatile organic compounds, and they come from sources ranging from factories to trees.	Ozone near the ground can cause a number of health problems. Ozone can lead to more frequent asthma attacks in people who have asthma and can cause sore throats, coughs, and breathing difficulty. It may even lead to premature death. Ozone can also hurt plants and crops.

Carbon monoxide		
A gas that comes from the burning of fossil fuels, mostly in cars. It cannot be seen or smelled.	Carbon monoxide is released when engines burn fossil fuels. Emissions are higher when engines are not tuned properly, and when fuel is not completely burned. Cars emit a lot of the carbon monoxide found outdoors. Furnaces and heaters in the home can emit high concentrations of carbon monoxide, too, if they are not properly maintained.	Carbon monoxide makes it hard for body parts to get the oxygen they need to run correctly. Exposure to carbon monoxide makes people feel dizzy and tired and gives them headaches. Elderly people with heart diseases are hospitalized more often when they are exposed to higher amounts of carbon monoxide.
Nitrogen dioxide		
A reddish-brown gas that comes from the burning of fossil fuels. It has a strong smell at high levels	Nitrogen dioxide mostly comes from power plants and cars. Nitrogen dioxide is formed in two ways – when nitrogen in the fuel is burned, or when nitrogen in the air reacts with oxygen at very high temperatures. Nitrogen dioxide can also react in the atmosphere to form ozone, acid rain and particles.	High levels of nitrogen dioxide exposure can give people coughs and can make them feel short of breath. People who are exposed to nitrogen dioxide for a long time have a higher chance of getting reparatory infections. Acid rain can hurt plants and animals and can make lakes dangerous to swim or fish in.
Particulate matter		
Solid or liquid matter that is suspended in the air. To remain in the air, particles are usually less than 0.1 mm wide and can be as small as 0.00005 mm.	Particulate matter can be divided into two types–coarse particles and fine particles. Coarse particles are bigger than 0.002 mm and are formed from sources like road dust, sea spray and construction. Fine particles are smaller than 0.002 mm and are formed when fuel is burned in automobiles and power plants.	Particulate matter that is small enough can enter the lungs and cause health problems. Some of these problems include more frequent asthma attacks, respiratory problems, and premature death. Particulate matter can also make clothes and other materials dirty.
Sulfur dioxide		
A corrosive gas that cannot be seen or smelled at low levels but can have a "rotten egg" smell at high levels.	Sulfur dioxide mostly comes from the burning of coal or oil in power plants. It also comes from factories that make chemicals, paper, or fuel. Like nitrogen dioxide, sulfur dioxide also reacts in the	Sulfur dioxide exposure can affect people who have asthma or emphysema by making it more difficult for them to breathe. It can also irritate people's eyes, noses, and throats. Sulfur dioxide

	atmosphere to form acid rain and particles.	can harm trees and crops, damage buildings and make it harder for people to see long distances.
Lead		
A blue-gray metal that is very toxic and is found in a number of forms and locations.	Outside, lead comes from cars in areas where unleaded gasoline is not used. Lead can also come from power plants and other industrial sources. Inside, lead paint is an important source of lead, especially in houses where paint is peeling. Lead in old pipes can also be a source of lead in drinking water.	High amounts of lead can be dangerous for small children and can lead to lower IQs and kidney problems. For adults, exposure to lead can increase the chance of having heart attacks or strokes.
Toxic air pollutants		
A large number of chemicals that are known or suspected to cause cancer. Some important pollutants in this category include arsenic, asbestos, benzene and dioxin.	Each toxic air pollutant comes from a slightly different source, but many are created in chemical plants or are emitted when fossil fuels are burned. Some toxic air pollutants, like asbestos and formaldehyde, can be found in building materials and can lead to indoor air problems. Many toxic air pollutants can also enter the food and water supply, and people can be exposed when they eat or drink.	Toxic air pollutants can cause cancer. Some toxic air pollutants can also cause birth defects. Other effects depend on the pollutant, but can include skin and eye irritation and breathing problems.
Stratospheric ozone depleters		
Chemicals that can destroy the ozone in the stratosphere. These chemicals include chlorofluorocarbons (CFCs), halogens and other compounds that include chlorine or bromine.	CFCs are used in air conditioners and refrigerators, since they work well as coolants. They can also be found in aerosol cans and fire extinguishers. Other stratospheric ozone depleters are used as solvents in industry.	If the ozone in the stratosphere is destroyed, people are exposed to more radiation from the sun (ultraviolet radiation). This can lead to skin cancer and eye problems. Higher ultraviolet radiation can also harm plants and animals.

Greenhouse gases

Gases that stay in the air for a long time and warm up the planet by trapping sunlight. This is called the "greenhouse effect" because the gases act like the glass in a greenhouse. Some of the important greenhouse gases are carbon dioxide, methane, and nitrous oxide.

Carbon dioxide is the most important greenhouse gas, and it comes from the burning of fossil fuels in cars, power plants, houses, and industry. Methane is released during the processing of fossil fuels, and also comes from natural sources like cows and rice paddies. Nitrous oxide comes from industrial sources and decaying plants.

The greenhouse effect can lead to changes in the climate of the planet. Some of these changes might include more temperature extremes, higher sea levels, changes in forest composition, and damage to land near the coast. Human health might be affected by diseases that are related to temperature or by damage to land and water.

Source: Jonathan Levy, Harvard School of Public Health. Based on information provided by the Environmental Protection Agency

Pollution problems in urban area

Urbanization is said to be a quiet revolution in the third world that has escaped the attention of most policy makers. Concerning India, with only 30% of India's population living in urban areas, it is often said that India lives in it's villages, but according to an estimate over 50% of India's population will be living in urban areas by the year 2016 and there will be more than 50 cities with a population of one million and above.

The industries like thermal power plants, petrochemical industries and crude oil refineries, pharmaceutical and chemical industries are the major culprits of air pollution. However, in large metropolitan cities, the exhaust gases from the traffic (fuel driven vehicles), also contribute to a great extent to the air pollution problems. The overall impact of the above is the increased levels of air pollutants like smoke, gases, fumes, particulate matters etc. which are not only nuisance but also dangerous for living creatures. (*e.g.* In Delhi 865 tonnes of air pollutants were released daily in the year 1986-87).

12.2 SOURCES OF URBAN AIR POLLUTION

Industrial air pollution

The industrial air pollution mainly consists of exhaust gases as discharged from various industrial belts in urban area involved in various industrial activities like metallurgical plants, chemical plants, refineries, mineral production, pharmaceutical production, textiles and related operations paper and pulp production, sugar mills etc.

Table 2 shows the air pollutant emissions in New Delhi for the year 1997

Table 2. Trends in annual vehicular emissions in Delhi (thousand tonnes)

Pollutants	1990/91	1995/96	2000/01	2005/06	2010/11
CO	139	178	202	199	206
HC	55	71	83	82	83
NOx	33	44	54	62	74
SO_2	4	5	7	9	11
TSP	7	9	12	13	14
Pb	0.078	0.102	0.111	0.104	0.098

*Estimated values

The following graph shows various pollutant emissions in various counties.

The following table shows relative carbon emission in various Asian countries for the year 1990.

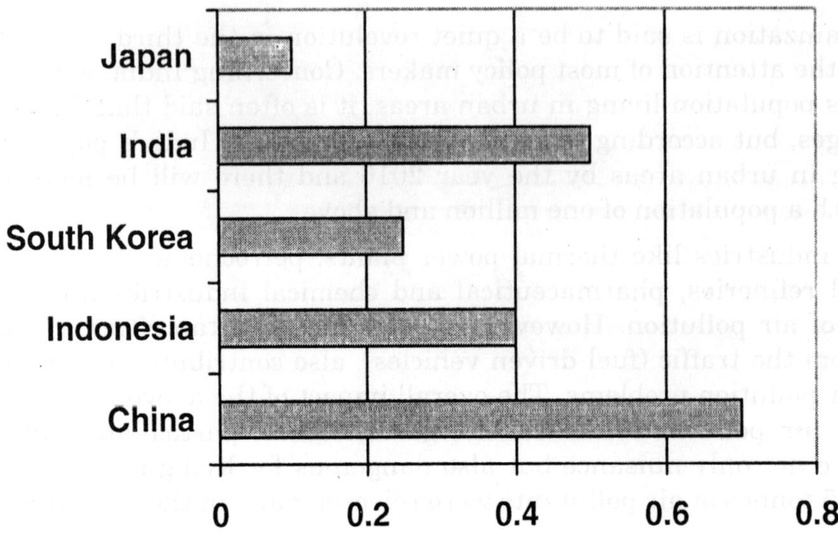

Pollution due to vehicles

The main three types of automotive vehicles used in our country are

- Passenger cars powered by four stroke gasoline engine.
- Motorcycles, scooters and auto rickshaws powered mostly by two stroke engines and
- Large buses and trucks powered by four stroke diesel engines.

Of the above types of vehicles the pollution is mainly due to,

- Exhaust emissions
- Crank-case emissions and
- Evaporative losses.

Considering the type of engines powering the vehicle, the diesel powered vehicles are less polluting ones, as the diesel engine uses a closed injection fuel system as diesel by property is less volatile than gasoline. It is estimated that a diesel engine exhausts only one-tenth of Carbon Monoxide as generated by a gasoline engine. However the major problems of the diesel engine are smoke and odour problems.

The following table 3 indicates the trend of fuel consumption in India from the year 1990 to 2000.

12.3 MONITORING OF AIR POLLUTION

The monitoring comprises of two important components (1) sampling and (2) analysis. Both of the sampling and analysis are equally important as if true representative samples are not collected, the results of analysis will not reflect the true picture of the ambient air quality.

The sampling will need the proper planning of sampling station network, interval of sampling, the quantity of samples to be collected and handling the samples during their transportation from sampling point to the laboratory.

Some of the instrumental methods used for analytical procedures for the measurement of gaseous pollutants are:

- Conductometric analysis working on the principle of the electrical conductivity of the solutions.
- Coulombometric analysis working on the principle of quantity of current flowing.
- Infrared absorption working on the principle of infrared light (longwave radiation absorbed by the pollutant).
- Flame ionization working on the principle of amount of electrons given off by ionization of pollutants.
- Flame photometry working on the principle of spectral intensity and frequency of light given off by spraying sample in to a flame.
- Colorimetry working on the principle of absorption of short wave 5600A°, or other wavelength by various pollutants and comparing with the calibration curves.

In addition to the above methods, wet chemical analysis can also be done.

For particulate matter the monitoring instruments used as:

- Dust fall jar

- High volume sampler
- Cascade impinger
- Electrostatic precipitators and
- Thermal receptors.

Now a days, equipments using laser technology to determine the particulate size and even size distribution of the particles are available.

Table 3. Petroleum, oil and lubricants consumption (thousand tonnes) in transport: 1990/91 to 1999/2000 [TERI Energy Data Directory and Yearbook 2001/2002]

Sector	1990/ 91	1992/ 93	1994/ 95	1996/ 97	1997 / 98	1998/ 99	1999/ 2000[a]
				High Speed diesel			
Road Transport	1864	2057	2156	2224	2433	2350	2409
Activation	11	12	6	6	8	19	33
Shipping	193	183	207	240	211	219	262
Railways	1453	1417	1398	1498	1401	1419	1380
Other transport[b]	15292	18057	20975	26389	27053	28091	29615
				Light diesel oil (regular)			
Road Transport	1	1	1	2	2	2	3
Activation	1	1	1	1	1	1	4
Shipping	67	66	62	46	41	45	43
Railways	4	2	2	2	4	2	2
Other transport[b]	13	15	17	15	16	13	7
				Fuel oil (regular)			
Road Transport	45	7	8	8	9	0.1	6
Activation	4	5	3	3	2	0.1	22
Shipping	300	235	323	249	214	4.4	209
Railways	22	15	13	12	12	0.3	11
Other transport[b]	76	77	93	79	78	1.5	103

[a] Provisional

[b] including agricultural retail trade

Note: This table excludes consumption of motor spirit, which is, by and large, used for personalized transport.

Source: MoPNG.2001. *Indian Petroleum and Natural Gas Statistics* (Various issues). New Delhi: Economic and Statistics Division, Ministry of Petroleum and Natural Gas.

Table 4. Air Quality in Seven Major Cities During 2001

Pollution level	Annual Mean Concentration Range (μg/m³)				
	Industrial (I)			Residential (R)	
	SO_2 & NO_2	RSPM	SPM	SO_2, NO_2, RSPM	SPM
Low (L)	0-40	0-60	0-180	0-30	0-70
Moderate (M)	40-80	60-120	180-360	30-60	70-140
High (H)	80-120	120-180	360-540	60-90	140-210
Critical (C)	>120	>180	>540	>90	>210

State / City	SO_2		NO_2		RSPM		SPM	
Area Class	I	R	I	R	I	R	I	R
Andhra Pradesh – Hyderabad	L	L	M	M	M	H	M	H
Delhi	l	l	l	M	C	C	H	C
Gujarat – Ahmedabad	L	L	M	M	C	C	H	C
Karnataka – Bangalore	L	L	L	M	H	C	L	H
Maharashtra – Mumbai	L	L	L	L	M	H	M	C
Tamilnadu – Chennai	L	L	L	L	M	H	L	M
West Bengal – Kolkata	L	L	H	H	H	C	M	C

Table 5. Industrial emissions inventory for Bombay, 1973

Industry	Sulphur dioxide					
	(t/a)	%	(t/a)	%	(t/a)	%
Chemical	48,180	45	2,920	19	7,045	29
Textile	27,010	24	1,351	9	4,453	18
Power Plants	22,630	21	10,366	67	10,950	46
Others	10,950	10	803	5	1,643	7
Total	108,770	100	15,440	100	24,091	100

After NEERI, 1991a

Table 6. Monthly mean lead concentrations, 1971-73

Monitoring Site	Monthly Mean concentration (Microgram/m³)
Bandara	2.20
Bhandup	1.80
Chembur	1.30

Dadar		1.70
Dongri		1.00
Ghatkopar		0.40
Goregaon		2.40
Kaliwada		0.40
Laibaug		1.10
Sion		2.00
Bombay	Mean	1.40

After NEERI, 1991a

12.4 CONTROL OF GASEOUS POLLUTANTS

Controlling the emission of gaseous pollutants include the following techniques.

- Adsorption by liquids wherein mostly packed or spray towers are used to reduce the exhaust gases.
- Adsorption by solids where beds of common adsorbents like activated carbon, activated Alumina, silica gel etc. are used.
- Combustion where the ultimate goal is the oxidation of the contaminants to harmless gases like CO_2, H_2O, Sulfur Dioxide and Nitrogen Dioxide.
- Incineration or afterburning, which is a combustion process to remove combustible air pollutants. This process is important from several aspects as indicated below.
- Almost all highly odorous pollutants are combustible or are changed chemically to less odorous substances when heated sufficiently in presence of oxygen.
- Organic aerosols that cause visible plumes are effectively destroyed by afterburning. Coffee rosters, meat smoke houses and enamel baking ovens produce such organic volatile compounds.
- Certain organic gases and vapours, if released to the atmosphere, become involved in smog reactions, flame afterburning is very effective in reducing such gases and vapours.
- Some industries such as refineries produce large quantities of highly combustible waste in the form of gases and volatile materials. The safest method of disposal of such wastes is by burning them in stake flares or specially designed furnaces.

Table 7. Emission standards with effective date for new petrol vehicles in India under type approval test (g/km) [Towards cleaner auto and fuel technologies in India: issues, recommendations and manifesto for road map]

Vehicle type	Carbon monoxide				Hydro-carbon	Hydrocarbon + nitrogen oxides		
	1Apr-91	1 Apr-96	1 Apr-00 (Euro I)	1 Apr-05 (Euro II) a	1 April 1991	1 Apr-96	1 Apr-00 (Euro I)	1 Apr-05a (Euro II) a
Two wheelers	30-Dec	4.5	2	-	12-Aug	3.6	2	-
Three-wheelers	30-Dec	6.75	4	-	12-Aug	5.4	2	-
Cars with CCb	-	4.34-6.20c	2.72	2.2	-	1.50-2.18b	0.97	0.5
Cars without CC	14.3-27.1	8.68-12.40c	2.72	2.2	2.0-2.9	3.00-4.36b	0.97	0.5

a. proposed;

b. catalytic converter;

c. the range corresponds to the engine capacity stratified into three classes: up to 1400 cm^3, 1400-200 cm^3, and over

Notes

1. Euro I is effective from 1 June 1999 and Euro II will become effective from 1 April 2000 for private (non-commercial) vehicles in the National Capital Region.
1. The test will be as per Indian driving cycle (which simulates average driving pattern such as acceleration, cruising, deaceleration, etc.) with warm start. However, with effect from 1 April 1998, the test will be as per Indian driving cycle procedure, till 1 April 2000.
1. Durability
2. A 20% relaxation is given in the conformity of production standards for carbon monoxide and combined hydrocarbon + nitrogen oxides for all categories of vehicles. However, the relaxation limit is reduced to 16.5% for cars.
1. Evaporative emission should not be more than 2 g. per test.
1. There should be no crankcase emission.

12.5 CONTROL OF THE PARTICULATE POLLUTANTS

The following are the techniques adopted for the control of particulate pollutants from the industrial air into the atmosphere.

- Gravitational settling by settling chambers
- Inertial impaction using cyclone separators
- Diffusion and direct interception using fabric bag filters

- Wet scrubbers
- Venturi scrubbers
- Cyclone scrubbers etc.

12.6 CONTROL OF THE VEHICULAR EMISSIONS

An emission control programme aims a reducing the concentration of CO, HC and NOx in the exhaust. The main approaches to achieve this purpose are:

Engine modifications

- Engine design modification which includes variations in
    - Air-fuel ratio
    - Load on the engine
    - Speed of engine operation
    - Spark retardation
    - Compression ratio of engine
    - Air injection pattern of engine
    - Fuel injection pattern of engine etc.
- Modification in the Carburetor design to use leaner air-fuel ratio may be used during idling and cruise operations.

Exhaust gas oxidation

This includes

- An addition of **after-burners** to the exhaust system that can completely burn the partially burnt HC in the emission gas
- Providing **Exhaust manifold reactor** that allows high temperature mixing of air-fuel so as to facilitate the complete burning of HC.
- Use of **Catalytic converters** to reduce the emissions from an engine.

Exhaust emission control by fuel variation

So it can be seen that both the methane and steam reformed fuels are very attractive from air pollution control point of view but there are technical hindrances to project them as conventional vehicle fuel alternatives.

Table 8. Table represents a comparison of exhaust gas emission for different fuels

Parameter	Propane Versus Gasoline	Methane Versus Gasoline	Stream reformed Hexane
Power reduction	6%	11%	20%
CO reduction	99%	99%	98%
HC reduction	87%	99%	99%
NO reduction	60%	59%	87%

Source: A course in I. C. Engines by Mathur and Sharma

12.7 CONCLUSIONS

- Though there have been considerable advances in dealing with problems of air pollution, with many legislative measures adopted in many countries, excessive and unnecessary pollution is prevalent in industrial areas, those happen to be a part of urban areas.
- Up till now complete elimination of the pollutant emission in to the atmosphere is not possible, but by proper implementation of the available pollution abatement techniques and legislative interference, the pollution level could be brought down to safer levels.
- Also the modification in the industrial processes and effective designing of the automobile engines, can figure out the remedies for the air pollution problems in a better manner.

CHAPTER - 13

Indoor air pollution and problems of Biodeterioration in museums of Gujarat

Arun Arya and V.K. Muliya[1]

Botany Department, Faculty of Science,
The Maharaja Sayajirao University of Baroda, Vadodara 390 002
e mail: aryaarunarya@rediffmail.com
[1]Christ College, Rajkot - 360 005

ABSTRACT

Museum collections comprising of materials of cultural, artistic, historical and scientific significance from different sources serve as an index of civilization between the present and past and constitute our bequest to the future. India is one of those few nations which have a very rich ancient culture. The problem of deterioration caused by biological agencies assumes a serious threat not only in India but also in many countries. The Baroda Museum and Picture Gallery is famous for its collections of European oil paintings all over Asia. It was founded in 1887 by Maharaja Sayajirao Gaekwad III. The picture gallery was completed in 1914. Oriental Institute has collection of old paper and palm leaf manuscripts. Maharaja Fatehsinh museum has good collection of paintings made by Raja Ravi Varma, a prince of the Travancore state. He made many outstanding paintings on mythological subjects and portraits of Baroda Royal family. It was started in 1961. Museums of departments of Museology, Archeology, and Botany have large number of objects made of paper cloth and organic materials.

It is the purpose of **Conservation Science** for cultural property to contribute to their preservation and restoration through scientific research on the materials, traditional techniques and the causes and mechanisms of deterioration. It is possible to preserve the cultural properties effectively only when we can assess clearly the causes and mechanisms of their deterioration. Aerobiological studies has resulted in assessing air borne spores and fungal hyphae, which can settle on organic objects and can cause irreparable damage. The present study reports indoor aeromycoflora of **Baroda Museum and Picture gallery**, Department of Botany and

Museology. 14 different species of fungi were recorded in picture gallery as compared to 12 and 10 in Dept. of Museology and Botany respectively. Gravity fall method was used to sample indoor aeromycoflora in culture plates. The damage caused to miniature paintings, pastel colour painting, wooden screen, palm leaf manuscripts, books, charts, botanical objects like herbarium specimens and ethnobotanical objects is described in details. A rare object **Egyptian mummy** harbouring the growth of ***Emericella nivea*** on the fingers of feet is described. It is one of the 7 mummies present in Indian museums. Use of sterile brush and inoculation needle was done to get some fungal inoculum to initiate the growth in petriplates. Cello tape technique was also used to isolate fungi from valuable objects. *Aspergillus terreus* and *Cladosporium herbarum* were isolated from herbarium specimens of *Lycopodium*. These two herbarium sheets are 98 years old. *Aspergillus niger* the most common airborne fungus was frequently encountered causing damage to books in library.

Key words : Aeromycoflora / Bio-deterioration / Baroda Museum and Picture gallery / Fatesingh Museum / Calico Museum / Lakhota Museum / Manuscripts

Museums are not merely storehouses but are centers for education in service of Society. They provide non-formal education. Museums have collection of cultural, artistic, ethnological archeological, historical and scientific significance. Foster (1986) quoted" A good museum attracts, entertain, arouses curiosity, leads to questioning and this promotes learning. It is an educational institution that is set up and kept in motion that it may help the members of the community to become happier, wiser and more effective human beings" (Dana, 1909).

The word **Museum** is derived from the Greek **Mouseion**, which was a place of contemplation, a philosophical institution or a temple of Muses. ICOM (1990) defines museum as non profit making, permanent institution in the service of the society and of its developments, communicates and exhibits for the purpose of study, education and enjoyments, material evidences of people and their environments.

The Baroda museum and picture gallery (Plate I Fig. A) is famous for its collection of European oil paintings all over Asia. The buildings of Baroda museum are situated in pastoral environment of Sayaji Park. This Museum is built in the Indo Sarascenic style of architecture in 1894. Department of Museology has also a good museum. This is also a pioneering institution in the field of Museum training at the postgraduate level associated with the M. S. University. The department was established fifty years back in the Baroda Museum and picture gallery. The department is presently housed in a new three storied building in the campus of Faculty of Fine Arts opposite Sayajibaug near river Vishwamitri. Although the department started in its new building in 1976, the Museum was thrown open to public in the year 1989 only. Museum of the Department possesses different types of objects such as tribal musical instruments, woods carving, objects made of paper mache, paintings and paper manuscripts etc.

Museology Department and Baroda Museum are situated with in a radius of 1km in Sayajigunj area of Baroda City. Two other spots Oriental Institute and Fatesinh Museum are situated 7 km apart. Oriental Institute is a part of the M.S. University of Baroda. It has collection of large number of books and old Journals. Paper and palm leaf manuscripts are placed here. It is situated on Palace road. Sadhana Press occupies the ground floor and Oriental Institute on 1st floor. Earlier it was housed in Smt. Hansa Mehta Library.

Biodeterioration can be defined as any undesirable change in properties of material of economic importance brought about by the activities of living organisms. Biodeteriogens may be plants, microbes and animals including human beings. Biodeteriogens like fungi, bacteria and actinomycetes pose severe threat to museum objects. Microbes get entry into indoor atmosphere through wind currents and settle on various objects by impaction (Gregory, 1973). Aerospora forms a normal component of the earth's atmosphere. Aerobiologists have correlated such data with clinical allergy. These studies are helpful in assessing the deterioration and management of environment. The aerobiological data may be used for the formation of disease fore casting system of plants (Mishra 1992) and human beings (Davis and Smith 1973; Spieksma 1980).

Paper is a very complex hygroscopic system with its homogenous network of pores and capillaries of different shapes. Persians and Greeks used parchment and tender skin, in the beginning of Christian era palm leaves were used as writing material, in India Bhojpatra (*Betula utilis*) was used. Actual paper is derived from the plant *Cyperus papyrus* used by Egyptians in 3000 B.C. Chemically paper is cellulose, which is formed from 300-1000 glucose units. Archival purpose paper is made up of cellulose fibre, rosin/gelatin and Kaolin as filler. Paper cellulose can be hydrolysed by endo and exo Beta 1,4 glucanase, cellobiohydrolase and betaglucosidase into glucose which is then converted into ethanol. Fungi can be categorized into two groups: (1) Fungi which start the degradation process like *Penicillium* and *Aspergillus*, (2) Fungi which can utilize the cellulose or its products as *Trichoderma viride* and *Chaetomium* spp. Role of certain fungi present in indoor air and biodeterioration caused by them is discussed in present paper.

13.1 MATERIALS AND METHODS

Various methods have been suggested to isolate indoor aeromycoflora. In the present study use of culture plate exposure method has been used as gravity slide exposure method is found to have certain drawbacks (Nigam and Pathak 1991) like:

(i) Similar spores are not identified upto generic level (e.g *Aspergillus, Penicillium* and *Mucor* etc.)

(ii) Some spores are too small or too transparent or do not have sufficient individual characteristics e.g. yeast and *Phoma* etc.

(iii) Fungal hyphae are not identified.

(iv) Many spores can not be trapped due to strong wind.

During isolation from Egyptian mummy cellotape technique was adopted. This technique is also suitable to study surface fungi or from dust etc. Care should be taken to use new cellotape and cut small pieces before plating them on PDA plates.

13.2 STUDY OF INDOOR AEROMYCOFLORA

It is noteworthy that 24 spp. of fungi belonging to 15 genera were isolated from Baroda Museum in 1999 and 22 spp. belonging to 17 genera were isolated in 2000 (Table 1). Presence of viable spores of *Aspergillus japonicus*, *A. wentii*, *Chaetomium atrobrunneum*, *C. ganglegarum*, *Lewia infectoria*, *Monascus rubur* and *Phoma nebulosa* is noteworthy in indoor air. These biodeteriogens have the potential to cause plant as well animal diseases at the same time can biodeteriorate cultural property. According to Nair (1972), Fungi that can grow on any suitable substrata play a prominent role in deterioration of Museum Materials. Tilak and Kulkarni (1972) and Tilak *et al.* (1970-1972) failed to detect such fungi during their surveys in Ajanta caves. Conidia of Aspergilli and Fungi imperfecti dominated in their slides. They concluded that the excreta of bats and birds residing in these caves serves as feeding source for the microorganisms and this act as a secondary source of aerospora. Air in the Mummy chamber harboured different mycoflora (Table 2). Species of *Aspergillus* dominated, in all 28 colonies were produced in 5 petriplates.

13.3 PROBLEMS OF BIODETERIORATION

Microorganisms produce different types of symptoms on objects they cause discoloration, changes in shape or disfigurement, loss in firmness and if not cared will spoil the whole aesthetic look with in a short time. This change may be further enhanced if environmental conditions are favourable.

Table 1. Indoor aeromycoflora of Egypto-Babylonian Gallery in Baroda and other museums

Fungi isolated	EG	PG	FM	LM	SM
Zygomycotina					
Absidia repens	5	-	1	-	-
Choenephora sp.	02	-	-	-	-
Mucor hiemalis	10	7	5	2	10
Rhizopus stolonifer	10	5	8	2	7
Syncephalastrum racemosum	-	2	2	-	-
Ascomycotina					
Aspergillus awamori	-	-	5	-	-
A. flavus	7	5	5	3	5
A. fumigatus	-	2	8	2	-
A. japonicus	7	-	-	-	-
A. niger	20	15	15	3	8
A. parasiticus	-	-	1	-	-
A. ustus	5	-	-	-	-
A. wentii	7	-	-	-	-
Chaetomium atrobrunneum	6	-	-	-	-
C. gangligerum	5	5	4	-	-
C. globosum	2	-	-	-	-
Emericella nidulans	5	-	5	-	-
E. variecolor	2	-	2	-	-
Lewia infectoria	5	-	5	-	-
Monascus rubur	2	2	2	-	-
Penicillium citrinum	2	2	2	-	-
P. nigricans	5				
Fungi imperfecti					
Alternaria alternata	4	4	2	2	5
Cladosporium herbarum	8	1	-	2	8
C. cladosporioides	1	2	15	-	3
Curvularia lunata	5	5	5	3	3
C. verruculosa	2	-	-	-	-
Fusarium oxysporum	5	2	2	1	3
F. pallidoroseum	-	2	2	-	-

Monilia sitophila	4	5	2	0	2
Phoma multirostrata	5	4	4	-	-
P. nebulosa	8	-	2	-	-
Trichoderma lignorum	2	5	6	-	-
T. viride	5	-	-	-	-
Sterile white mycelium	10	4	4	1	5
Sterile black mycelium	4	2	2	0	3
Total number of fungal colonies	133	86	109	18	72

*Data represent the total number of colonies that appeared on 5 P D A plates after 7 days of incubation; -, absence of colony

EG – Egypto-Babylonian Gallery, PG – Picture Gallery, Baroda museum, FM- Fatehsinh museum, LM – Lakhota museum, Jamanagar, SM – Shreyas museum, Ahmedabad

Table 2. Percentage occurrence of different fungi in air and dust of mummy chamber on 16 April 1999

	Mummy Chamber			
	Air		Dust	
Fungi isolated	Colonies*	Percentage occurrence	Colonies*	Percentage occurrence
Aspergillus flavus	10	35.7	8	29.6
A. fumigatus	5	17.8	2	7.4
A. niger	5	17.8	5	18.5
Emericella nivea	2	7.1	5	18.5
Penicillium citrinum	2	7.1	2	7.4
Mucor hiemalis	2	7.1	5	18.5
Unidentified white mycelium	2	7.1	-	-
Total no. of colonies	28	-	27	-

*Data represent the total number of colonies that appeared on 5 P D A plates after 7 days of incubation; -, absence of colony

Table 3. Potential allergens found in aeromycoflora of Baroda Museum and Department of Museology

Sr. No.	Baroda Museum	Department of Museology
1.	*Aspergillus flavus*	*Aspergillus fumigatus*
2.	*A. sydowii*	*A. niger*
3.	*A. versicolor*	*Aspergillus sp.*
4.	*Cladosporium herbarum*	*Penicillium sp.*

5.	*Emericella nidulans*	*Rhizopus stolonifer*
6.	*Monilia sitophila*	
7.	*Penicillium* sp.	
8	*Rhizopus stolonifer*	

Table 4. Symptoms of biodeterioration of different kinds of museum materials (Nair 1980)

Sl No	Symptoms	Example
1.	Stains and Discoloration	Caused by surface growth of Fungi on paper, leather, textile etc.
2.	Disfigurement	Positive changes in the surface of a painting due to fungus
3.	Fibrilization	Loss of coherence in fibrous material as in wool attacked by microorganisms
4.	Impenetrations	Invasion of the lumen of certain natural fibers by fungi.
5.	Changes in mechanical properties	Loss of weight and strength of wood on reduction or breaking
6.	Changes in chemical properties	Break down of cellulose by microorganisms, foxing on paper etc.
7.	Development of odour	Mildewing on paper, cloth etc.

Table 5 shows list of various fungi casing damage to museum objects in Baroda during 1988-2001. Different types of objects are studied like miniature paintings, manuscripts and herbarium sheets on paper. Musical instruments made of bamboo and bottle gourd, leather strings of Tambura, panther skin, even human skin in Egyptian Mummy kept in Baroda Museum and wooden plaque with decorative work (Plate I Fig. B)

Isolations made during March - April 1987 from two post-medieval period miniature paintings (Rajasthani) revealed the presence of *Chaetomium gangligerum* L. Ames. Here the painting surface turned black, obliterating the figures and the colourful aspect of painting. Greenish black velvety coloured growth was observed. From another miniature painting kept in Museology Department two more fungi *Aspergillus tamarii* Kita and *Cladosporium herbarum* Link. were observed. Isolation made from pastel colour painting kept in Fatesinh museum showed presence of *A. niger* and *A. flavus* on both the sides (Plate III; Fig. B & C).

Wood has been used by man since palaeolithic times for varied purposes. Wood is very stable ligno-cellulosic polymer, most wood species deteriorate due to biodegradation. It is being of organic origin is metabolized by many fungi for food. Spores and growing mycelia are mostly responsible for altering aesthetic appearance of an art object by creating residues. On surface of the object, mycelia can form dry and brittle crusts, wrinkled and corrugated colonies and some times loose elastic, fluffy structures. Walston (1976) has worked on the problems of deterioration of ethnographical objects of Australian museum. In Tibetan Gallery of Baroda Museum a Chinese wooden screen is placed. It's one of the panels developed deterioration problem. The panel was placed where there was some seepage in the wall due to fault in outlet of a wash basin. The lacquer got deformed and due to loose nature precious stones got separated from screen (Plate I; fig. B). Another observation was flaking of paint due to loss of adhesion in binding medium. On isolation *Aspergillus niger* (Plate I; fig. C) and *A. nidulans* were found.

Table 5. Fungi isolated from different museum objects.

Objects	Place	Organisms
Rajasthani miniature painting	Baroda Museum and picture gallery	*Chaetomium gangligerum* *Aspergillus versicolor* *A. terreus, Penicillium sp.*
Miniature Painting	Museology Dept.	*Aspergillus tamarii* *Cladosporium herbarum*
Pavri A. (Musical instrument)	Museology Dept.	*Aspergillus japonicus* *A. flavus* *Penicillium citrinum*
Pavri B (Musical Instrument)	Museology Dept.	*Cochliobolus pallescens* *Phoma sp.* *Fusarium semitectum*
Parsi Manuscript	Museology Dept.	*Cochleobolus lunatus* *Chaetomium cymbiforme* *Aspergillus niger*
Chinese wooden screen	Museum and picture gallery Baroda	*Aspergillus ustus* *A. nidulans*
Pastel colour Painting	Fatehsinh Museum	*Aspergillus niger* *A. flavus*
Palm leaf Manuscript	Oriental Institute	*A. niger* *Nigrospora sphaerica*
Egyptian Mummy	Baroda Museum and picture gallery	*Aspergillus state of Emericella nivea*
Paper Manuscript	Oriental Institute	*Chaetomium globosum* *Chaetomium fusisporale* *Aspergillus niger*

Herbarium sheet	Museum of Botany Department	*Cladosporium herbarum* *Alternaria alternata*
Herbarium sheet	Museum of Botany Department	*Curvularia clavata*

A wooden object found affected by fungi was Pavri, a musical instrument used by Kunkuna tribes of Dangs area in Gujarat. It is an air bellowing instrument. On isolation from joints growth of *Aspergillus japonicus* and *A. flavus* was observed along with *P. citrinum* Thom. (Plate II; Fig. D). From another Pavri instrument presence of *Curvularia* state of *Cochliobolus pallescence* (Tsuda & Ueyama) Sivan along with *Fusarium semitectum* Berk. & Rav. and *Phoma* sp. was observed. To prevent these instruments Shah and Arya (2001) suggested some curative treatments like cleaning with absolute alcohol and then fumigation with 5% solution of Thymol. A regular cleaning of show cases and periodical spray of 0.25% Pentachlorophenol is suggested.

Coffin box (as seen in Plate II, Fig. A) is at present removed for treatment against insect borers. White coloured powdery growth was observed in fingers and toe of right leg (Plate II Fig A). After identification the fungus has been confirmed as *Emericella nivea* Wiley & Simmons (Arya *et al.* 2001, Plate II, Fig. C). Its conidial stage *Aspergillus* was also observed in culture along with perfect stage. Fig. B shows cellotape technique to isolated fungi from artefacts.

Two spp. of *Chaetomium i.e., C. globosum* and *C. fusisporale* (Plate IV, Fig. D) were present on paper manuscript (Plate IV, Fig. A, B, C). Presence of *Phoma multirostrata* was found in one of the manuscripts (Plate IV, Fig. E) Ricelli (1999) studied fungal growth on paper samples. Fungi like *Aspergillus terreus, Chaetomium eslatum, Stacybotrys atra* and *Penicillium chrysogenum* grew on paper stored in high relative humidity and the addition of some new synthesized compounds to the different samples of paper inhibits the fungal growth. Books present in library of Botany department showed the presence of *Aspergillus niger* other fungi isolated included *Trichoderma viride, Chaetomium cymbiforme* and *Rhizopus stolonifer*. Consolidation of books was also observed.

Maharaja Fatesinh Museum was opened to public in April 1961. It houses world famous oil paintings by Raja Ravi Varma. Many objects of bronze, marbles and oils by Venetian Artist, Felici are displayed here. On first floor in a ceremonial room of an 18th century French Palace, four pastel paintings are displayed. The pastel painting in coloured chalk depicts the crown princes of France. The engravings are after paintings by Angelica Kauffmann. During rainy season in 2004 a severe fungal attack was observed in these paintings (Plate III Fig. B & C). On isolation *Aspergillus flavus* and *A. niger* were identified.

Plate I. A. Front view of Baroda Museum and Picture gallery; **B.** Wooden screen showing attachment of precious stones; **C.** Conidiophore and conidia of *Aspergillus niger* X 940.

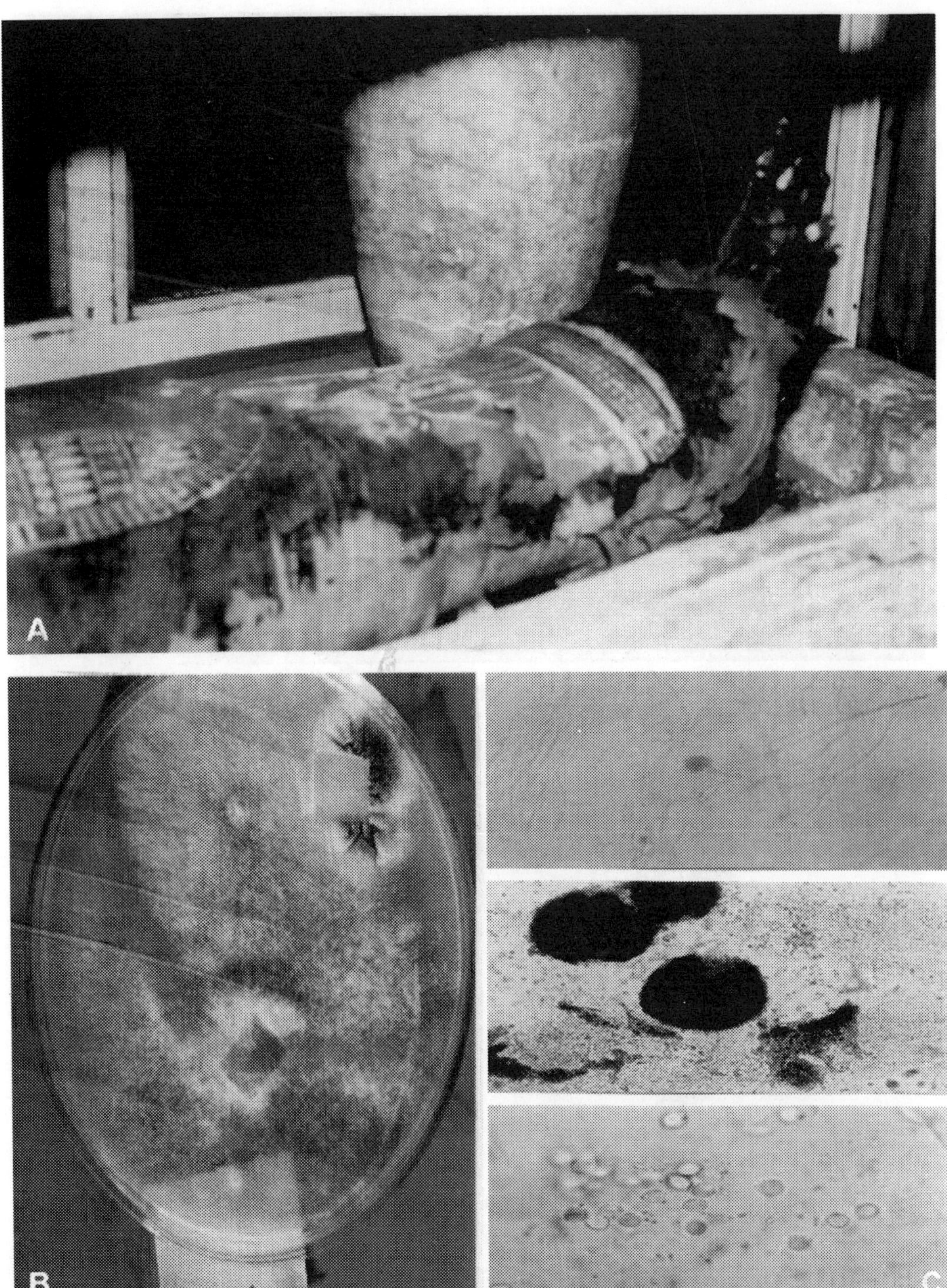

Plate II A. part of the Egyptian mummy showing white coloured fungal growth on toes. Also visible is the big white canopic jar. **B.** Cellotape method to isolate fungi from Mummy. **C.** Cleistothecia and ascospores of *Emericella nivea* X 126

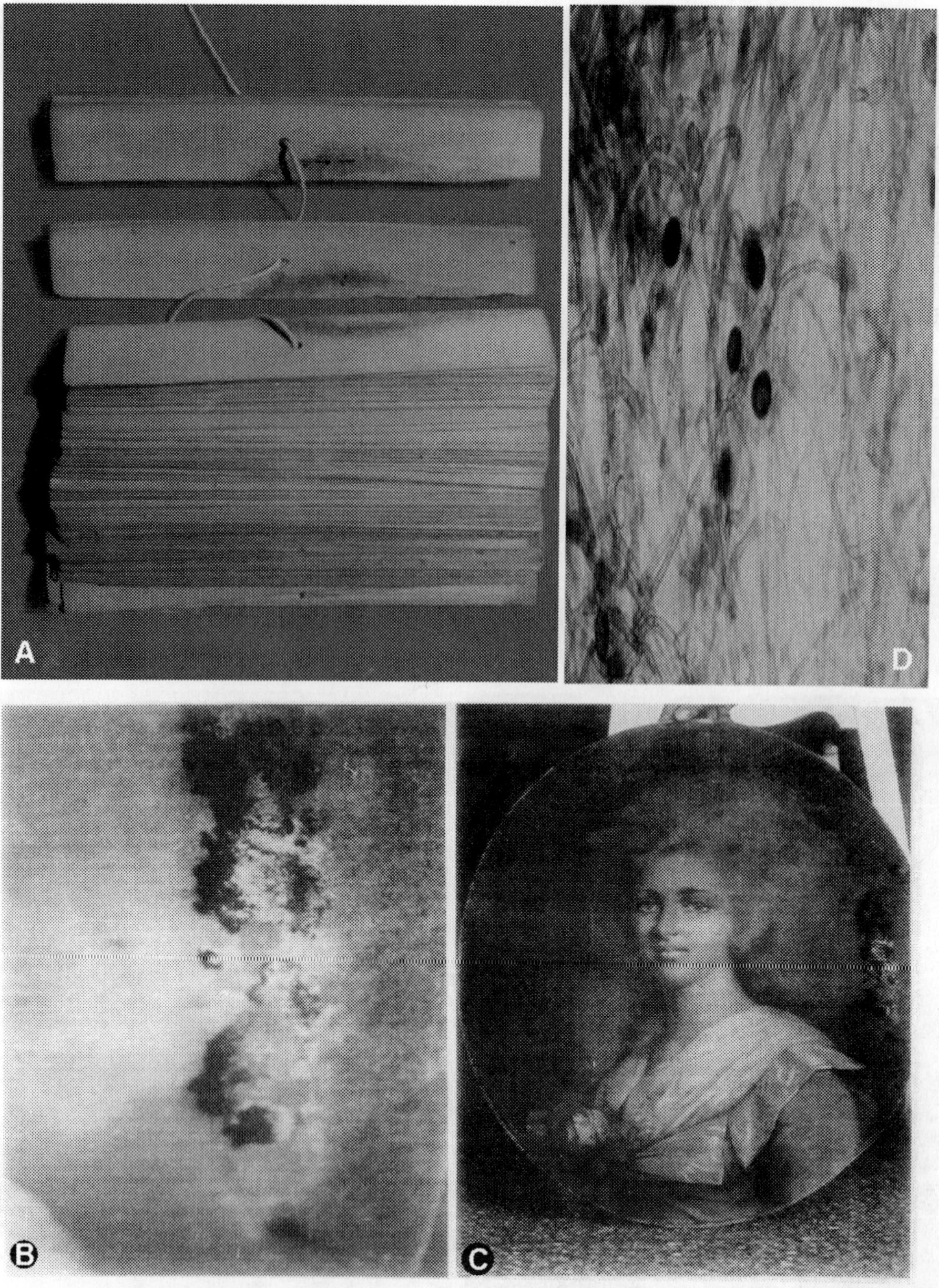

Plate III. A. Orrian palm leaf manuscript showing deterioration; **B.** Luxuriant growth of fungal biodeteriogens on back side of Pastel colour painting. **C.** Pastel colour painting showing growth of *Aspergillus niger* and *A. flavus* on right hand side of painted lady. **D.** Spores and mycelium of *Nigrospora sphaerica*.

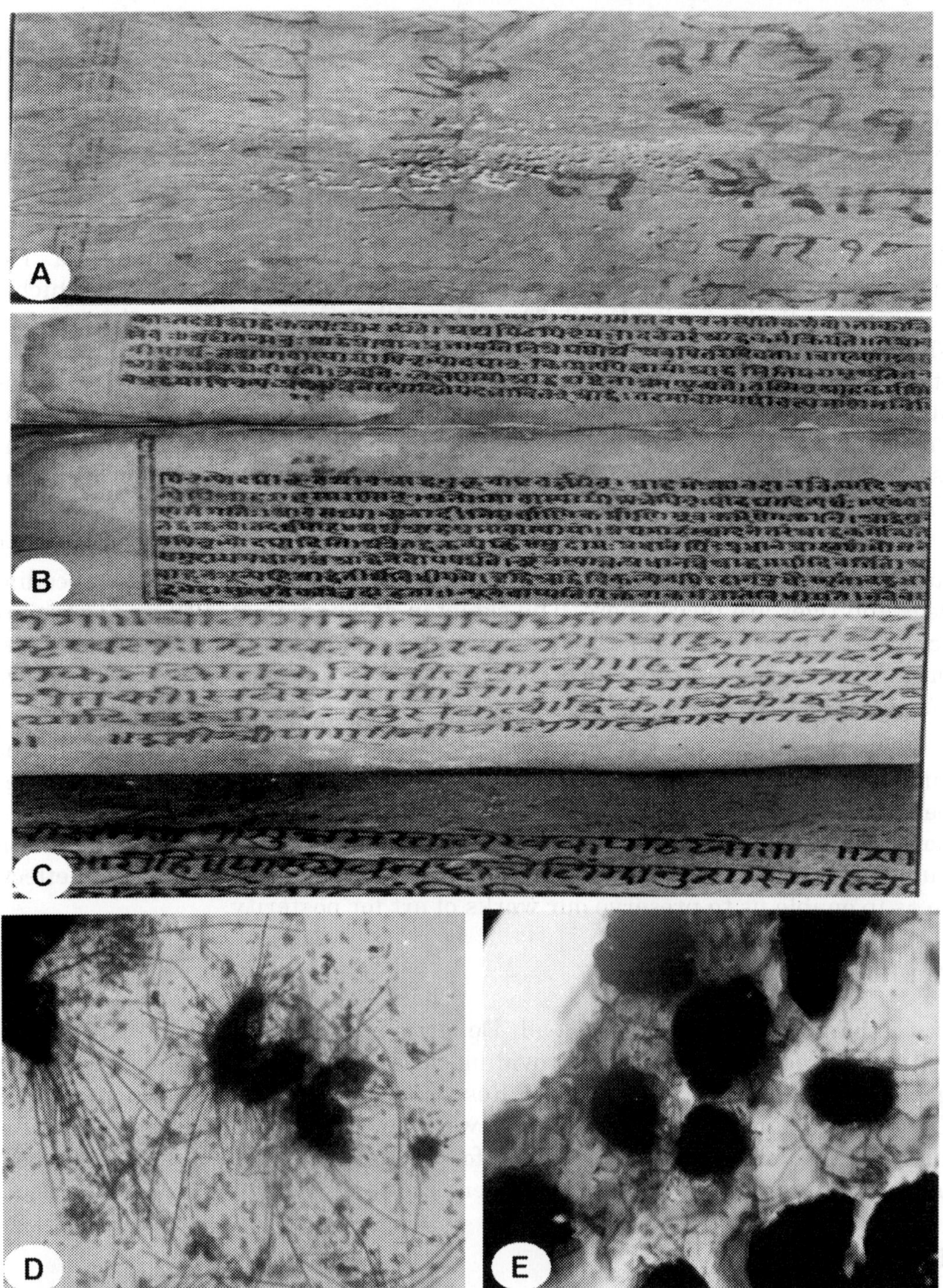

Plate IV. A-C Damage of Paper Manuscript by fungi; **B.** Perithecium of *Chaetomium sp.* with straight hairs X 126 **C.** Pycnidia of *Phoma multirostrata* X126.

From an Orrian Palm Leaf manuscript (Plate III, Fig. A) isolations were made in Jan.-Feb. 2004. Fungus *Nigrospora spharica* Penz. (Plate III, Fig.D) was isolated from black coloured spots of manuscript. The conidia and mycelium of fungus are shown in, conidia are dark black in colour, this is a leaf infecting fungi and produces dark pigment in culture thus damaging the beauty of museum objects.

Calico museum of textiles is sponsored by Sarabhai Foundation, Ahmedabad. Calico was one of the earliest mills started in Ahmedabad. The museum exhibits included a big display room showing a big tent in which a decorated bullock cart is placed. The sides are decorated with wall hangings. There is a table on which dresses are displayed. In different rooms sarees and other cloth materials is displayed. A room houses the cloths used during Mughul period. Decorative *Patola* and *Madhubani* of Bihar, Zari sarees of Varanasi and Kashmiri shawls are preserved here in good condition. Isolations from certain cloth objects resulted in species of Aspergillii and *Chaetomium globosum*.

Lakhota Museum of Jamnagar was started by King of Jamnagar H.H. Ranjitsinh ji. There is one gallery for animals, stuffed animals like crocodile, deer and tiger are placed here. The skeleton of whale is also there. There are sculptures, good collection of ornaments made up of bead, paintings are also present there. *Aspergillus niger*, *Curvularia lunata*, *Fusarium* sp., *Alternaria alternata*.

The need of the hour is to make periodic survey of museum objects to find out any damage caused by microbes, to isolate the fungi, study the detailed behaviour of fungi and suggest suitable control methods. These methods need not be just effective but meet the other criteria like preserve aesthetic quality, economically viable, eco-friendly, broad spectrum, with no side effects so as to enable us to preserve our works of art for posterity.

Acknowledgements

I thank Prof. M. Daniel, Head, Department of Botany, The M.S. University of Baroda, Vadodara for providing necessary laboratory facilities. I am grateful to the Director, Agharkar Research Institute, Pune, and the Head, Division of Plant Pathology, IARI, New Delhi, for confirming the identity of the fungi. Thanks are due to U.G.C. for awarding a Project on 'Eco-friendly approach to control fungal biodeteriogens in Museums of Gujarat'.

REFERENCES

Arya, A., Shah, A. R., and Sadasivan S. 2001. Indoor aeromycoflora of Baroda Museum and Deterioration of Egyptian Mummy. *Curr. Sci.* 81(7) : 793-799.

Chile, S. 1993. Study of airborne fungal spores at Jabalpur with special reference to allergenicity . Ph. D. Thesis, R. D. University, Jabalpur.

Davis, R.R. and Smith, L. P.1973. Aspects of human allergy and fungal spores. *Clinical Allergy* 3:23-267.

Foster, R.A. 1986. The larger provincial museums and art galleries. In: *Manual of curatorship*, Butterworths, London 66-75.

Gregory, P.H. 1973. *The Microbiology of the Atmosphere*, Leonord Hill Books, Aylesbury, England 1973 2nd edn. pp. 377.

ICOM 1990. ICOM Statutes code of Professional ethics, Paris pp. 3.

Lakshmikant and Mathur, S.N. 1991. Biodeterioration of cultural property by cellulolytic fungi. In: *Biodeterioration of Cultural Property* (eds. Agrawal O.P. and Dhawan S.) MacMillan India Ltd. New Delhi 168-172.

Mishra, G. and Mishra, K.B. 1992. Aeromycoflora of a mica mine. In : *National symp. on Management of microbes in Service of mankind*, Allahabad Univ., 19-21.

Nair, S.M. 1972. In: *Conservation in Tropics*. Proc. of AST-Pacific Seminars on con. of cul. Properties (Ed. Agrawal, O.P.) Rome 150-158.

Nair, S.M. 1980. Insect pesticides of textile materials in museum. In: *Conservation in tropics* ed. O.P. Agrawal pp. 150.

Nigam, R.K. and Pathak, N.C. 1991. Environmental mycology of a tannery and textile unit at Kanpur: Allergical aspects. In: *Biodet. of Cultural Prop.* (Eds. Agrawal O.P. and Dhawan S.) Macmillan India Ltd. New Delhi: 375-386.

Ricelli, A. 1999. Fungal growth on samples of paper: Inhibition by new antifungals. *Restaurator*, 20 (2): 97-107.

Shah, N. R., and Arya, A. 2001. *Studies on some fungal biodeteriogens.* Bhartiya Kala Prakashan, New Delhi, 212.

Spieksma, F.T.M. 1980. Importance of aerobiological studies for the prediction of human disease. In: *Aerobiological Studies for the prediction of human diseases* Aerobiol. Symp. Munich : 307-315.

Tilak, S.T., Sharma, B.R.N., Sengupta, S.R., and Kulkarni, R.L. 1970. The deterioration of Ajanta and Ellora paintings. In *Con. of Cul. Prop.* In India Proceedings of V. Seminar, New Delhi: 77-82. ,

Verma, K.S. and George, A.M. 1997. Aeromycoflora of Jabalpur. *Res. J. Chem. Environ.* 1: 29-31.

Wadhwani, K., Srivastava, A.K., Jamil, Z., Rizvi, H.I., Mishra, A.K. and Negi, H. 1986. *Jour. Rec. Adv. Appl. Sci.*, 1: 8-12.

Walston, S. 1976. Proceedings of the ICCM National conference, Canberra pp. 145.

CHAPTER - 14

Reduction in protein and chlorophyll contents in *Calotropis gigantea* (L.) R.Br. growing near stone crusher

B.R. Pandit and P.G. Prasannakumar

Ecology division, Department of life Sciences, Bhavnagar University, Bhavnagar - 364 002 India
E-mail: bharatrpandit@yahoo.com

ABSTRACT

Air pollution problems in India show a highly irregular pattern due to a non-uniform pattern of industrialization. Air pollution adversely affects vegetation, materials, structures, animals and human health. The present study was carried out for the determination of protein and chlorophyll contents in *Calotropis gigantea* [L] R. Br. growing dominantly near stone crushing unit during summer, monsoon and winter seasons. Various operations in stone crushers produce fugitive dust emissions in the form of heavy siliceous cloud.

The present study revealed that the reduction of protein and chlorophyll contents is due to high amount dust from stone crusher. Many workers have reported reduction of protein, fat and trace elements due to the high emission of dust from stone crusher. Again it was reported that the dust from the stone crushers may produce several diseases. Silicosis is an important form of disease observed in animals caused by inhalation of dust containing crystalline silica.

In the present study it was observed that protein and chlorophyll contents were reduced significantly from control to 50m distance from stone crusher. High amount of protein and chlorophyll were found during monsoon and low during summer. It was due to high dusting during summer and low dusting during monsoon. The stone crusher in the middle of human settlements was without any pollution control device and may be harmful for local population at a longer run.

Key words: *Calotropis gigantea*, Protein, Chlorophyll, Stone crusher, Pollution

Ever-since India became independent, considerable progress has been made and today it is among the first ten industrialized countries of the world. As a result of rapidly growing industrialization and urbanization, it is being recognized that substances harmful to plant and animal kingdom are being released into the environment as air pollutants. Air pollution is an increasingly important environmental issue in India. Rapid growth of population, with a large proportion of the population living in urban areas, heavily over-loaded transport vehicles and badly maintained automobiles, all contribute to urban air pollution problems.

The air pollution do not remain confined near the source of emission but spread into far of rural environs depending upon topography and meteorological conditions, especially wind direction and speed and vertical and horizontal thermal gradient prevailing in the area.

14.1 ANTHROPOGENIC ACTIVITIES : SOURCE OF POLLUTION

Air pollutants of major concern are particulate matter and gaseous pollutants. Particulate matter may be cement dust, stone crusher fugitive dust, coal dust, fly ash, lead particles, oxides of magnesium and iron etc. Sources of particulate pollutants are a variety of natural and anthropogenic activities. Air pollution problems in India show a highly irregular pattern due to a non-uniform pattern of industrialization.

Air pollution adversely affects vegetation, materials, structures, animals and human health. During the last two decades, there has been a lot of interest in the study of biological response of air pollutants such as particulate matter, oxides of carbon, sulphur and nitrogen and photochemical oxidants. Stoker and Seager (1972) have estimated the major air pollutants due to natural and anthropogenic emission globally. Dave (1976) estimated the extent of pollution of dry air by anthropogenic sources of pollutants. The effect of air pollution on the morphology, physiology, and biochemistry of plants have long been recognized (Koziol and Whatley, 1984; Treshow, 1985; Ahmed *et al.,* 1988). The gaseous pollutants as well as fly ash particulates cause, various injury symptoms on the leaves. These injuries adversely affect the physiological functions especially the photosynthetic activity of the plants (Trivedy and Singh 1995). There are only a few reports of reduction of proteins in plants due to air pollution (Rube and Kreeb, 1985; Banerjee *et al.* 1983; Singh *et al.* 1985; Trivedy and Singh 1995).

The present study was carried out for the determination of protein and chlorophyll contents in *Calotropis gigantea* growing near a stone crushing unit (Figs. 3-5) during summer, monsoon and winter seasons. Various operations in stone crushers produce fugitive dust emission in the form heavy siliceous cloud (Suess *et al.*,1985, Fasset and Irish, 1963)

14.2 MATERIALS AND METHODS

The plant studied for the present investigation is *Calotropis gigantea* (L.) R. Br. (Fig. 1 & 2). The fresh leaves of the same size were collected from in and around the stone crusher near Bhavnagar University, Bhavnagar. The area was divided into five sites. All these sites were kept at a particular distance from stone crusher i.e. 50m, 100m, 300m, 500m and control. Leaves of *C. gigantea* were collected from University campus kept as control. The method of Lowry *et al.* (1951) was adopted to determine the total protein contents and Chlorophyll contents were estimated by using spectrophotometric method (Arnon 1949).

14.3 RESULTS AND DISCUSSION

Results are given in Table 1 and 2 the amount of protein decreased from 500 m distance (from stone crusher) to 50 m. Amount of protein was found maximum at control and minimum at 50 m distance from stone crusher during all seasons. At all the sites the protein concentration was found to be lowest during summer and highest during monsoon. This may be due to the deposition of dust on leaves and hence reducing the metabolic activity in plants during summer than in monsoon. Trivedy and Singh (1995) reported reduction in protein contents due to the air pollution. Such reductions also have been reported by Singh and Rao, (1980); Banerjee *et al.* (1983) and Singh *et al.* (1985).

The dust holding capacity of leaves varies according to leaf structure, its surface geometry, epidermal and cuticular features of leaves and height and canopy of trees. Similar observations have also made by Agrawal and Tiwari (1997), and Singh (2000). Perhaps phyllotaxy, leaf orientation and sessile or semi sessile nature of leaves also play important role in dust deposition which caused less movement of leaves and keep leaves in horizontal direction, thus provide maximum surface for dust deposition. The metallic dust deposition and their effects are as follows,

Mercury: Necrotic spot on the inter venial areas on lower surface of leaves

Lead: Poisonous to human being & damages to vegetation

Copper: Badly damages vegetation

Cadmium: In plant chlorosis, foliar injury, root injury and in man cardiovascular problems and hypertension

Nickel: Damages vegetation

Zinc: Foliar injury and harmful to human beings

Asbestos: Cancer of lungs

Changes in amount of chlorophyll at various distance from stone crusher is given in Table 2. The results show that a reduction in chlorophyll contents from control to 50 m distance from stone crusher. The maximum amount of chlorophyll content (chlorophyll 'a', chlorophyll 'b' and total chlorophyll) found at control and minimum at 50 m from control to stone crusher, a gradual reduction of chlorophyll content was found. In all the cases maximum chlorophyll content was found during monsoon because during monsoon minimum dusting has been noticed; minimum chlorophyll content was found during summer, as maximum amount of dusting has been observed. The dusting of leaves causes the reduction of light absorption and retardation of chlorophyll content might be due to this reason. The variation of dusting was inversely proportional to the amount of protein and chlorophyll contents during different seasons.

Table 1. Change in concentration of protein (mg/g dry wt./d.m.) at various distances from stone crusher

Distance from Stone crusher	Seasons		
	Summer	Monsoon	Winter
Control	1.5	1.6	1.5
500 m	1.2	1.4	1.25
300 m	0.55	1.0	0.75
100 m	0.45	0.95	0.72
50 m	0.42	0.95	0.70

The dust particulates are also responsible for change in various biological activities in plants. Decrease in total amount of chlorophyll in the polluted areas than control may be attributed to the alkaline condition caused by solubilization of dust particles in cell sap and its responsible for chlorophyll degradation and reduced photosynthetic activity. Dust deposition on leaf surface reduces the synthesis of chl-a due to shedding effect (Bohme, 1963). Reduction in chlorophyll due to dust deposition have been observed by Singh and Singh (1978, 1981), and Singh (2000). Thus increase in dust deposition and subsequent decrease in chlorophyll may be positively correlated.

Fig. 1. *Calotropis gigantea* [L.] R.Br.

Fig. 2. *Calotropis gigantea* [L.] R.Br. white variety

Fig. 3. Stone crusher

Fig. 4. Stone crusher

Table 2. Changes in concentration of Chlorophyll (mg/g fresh wt.) at various distance from stone crusher

Distance from stone crusher	Summer		
	Chlorophyll 'a'	Chlorophyll 'b'	Total Chlorophyll
Control	0.017	0.027	0.044
500 m	0.011	0.025	0.036

300 m	0.007	0.017	0.024
100 m	0.007	0.016	0.023
50 m	0.004	0.008	0.012
		Monsoon	
Control	0.073	0.125	0.198
500 m	0.071	0.110	0.181
300 m	0.056	0.087	0.143
100 m	0.041	0.069	0.110
50 m	0.037	0.061	0.098
		Winter	
Control	0.054	0.089	0.143
500 m	0.044	0.082	0.126
300 m	0.042	0.078	0.120
100 m	0.037	0.062	0.099
50 m	0.029	0.056	0.085

Fig. 5. Smoke release from stone crusher

The present study revealed that the reduction of protein and chlorophyll content is due to high amount of dust from stone crusher. Pandey and Rastogi

(1983) reported that, the reduction of protein, fat and trace elements is due to the high emission of dust from stone crushers. Aslam *et al.* (1992) reported that the dust from stone crushers may produce several diseases. Silicosis is an important form of this disease observed men and in animal caused by inhalation of dust containing crystalline silica. A variety of dust control devices have been used for stone crushers in developed countries. But in India none of the stone crushers are using any pollution control device. The present study clearly demonstrate that the dust from stone crusher adversely affect plant metabolism.

REFERENCES

Agrawal, S. and Tiwari, S. 1997. Observation on Kiln dust pollution in Patrapli, region : Raigarh (M.P.) *Flora and Fauna* **3**(3) : 131-133.

Ahmed, K.J., Yunus, M., Singh, S.N., Srivastava, K., Singh, N. and Kulshreshtha, K. 1988. *Survey of Indian plants in relation to atmosphere pollution*. p. 283-306.

Arnon, D. I. 1949. Copper enzymes in isolated chloroplast polyphenol oxidase in *Beta vulgaris*. *Plant Physiol*. 24: 1-15

Aslam, M., Minocha, A.K. and Karla, P.D. 1992. Fugitive dust emission from stone crushers. *Indian J Environ, Hlth*. 34 : 186-191.

Banerjee, A.K., Sukar, R.K. and Mukherjee, S. 1983. Reduction in soluble proteins and chlorophyll content in a few plants as indicators of automobile exhaust pollution. *Internat. J. Environ. Studies*. 20: 239-243.

Bohme, H. 1963. Schadlichkeit Von Staub aus Zement worken fur wwaldbestande. Allg. Forstig. **18** : 107-111.

Dave, J.M. 1976. *Problems of Air pollution control in India*. Academic Press, New York. U.S.A.

Fasset, D.M. and Irish, D.D. (Eds.) 1963. *Industrial Hygiene and toxicology*, Vol. 11 Toxicology, Interscience Publishers, New York, p 2303-23-4.

Koziol, M.J. and Whately, F.R. (Eds.) 1984. *Gaseous air pollution and plant metabolism*. Butterworths, London.

Lowry, O.H., Rosebrough, N.J., Farr, A.L. and Randall, R.J. 1951. Protein measurement with the folin phenol reagent *J. Bio. Chem*. 193: 265-277.

Pandey, D.D. and Rastogi, S. 1983. Response of grain characteristic of Maize to stone crusher dust pollution. Proc. 80th Ind. Sc. Cong. Part- III (Abst.) p. 220.

Rube, R. and Kreeb, K.H. 1979. Enzyme activities and chlorophyll and protein contents in plants as indicators of air pollution. *Environ. Pollut*. 19: 19-137.

Singh, R.B. 2000. Monitoring of dust pollution by higher groups of plants around dust polluted habitats in Sonbhadra, U.P. *Ind. J. Environ. and Ecoplan*. **3**(1): 163-166.

Singh, S.N. and Rao D.N., 1978. Possibilities of using chlorophyll and potassium contents in plants to direct cement dust pollution. *Ind. J. Env.* **20** : 10-13.

Singh, N. and Rao, D.N. 1980. Growth of Wheat plants exposed to cement dust pollution. *Environ. Pollu*. 24: 75-81.

Singh S.N. and Rao D.N. 1981. Certain responses of of wheat plants to cement dust pollution. *Environ. Pollu.* (Ser. A) **24** : 73-81.

Singh, S.N., Yunus, M., Srivastava, K., Kulshreshtha, K. and Ahmed, K.J. 1985. Response of *Calendula officinalis* L. to long term germination with SO_2. *Environ. Pollu.* Ser. A., 39: 17-25.

Stoker, S.H. and Seager, S.L. 1972. *Environmental Chemistry* Scot, Foreman and Co. IIIionis.

Suess, M.J., Grefen, K. and Reinisch, D.W. 1985. *Ambient air pollutants from industrial sources*, WHO Regional Office for Europe, Elsevier, New York, p 213-214.

Treshow, M. 1985. *Air pollution and Plant life*, John Willey and Sons. New York.

Trivedy, M.L. and Singh, R.S. 1995. Reduction in protein contents in a few plants as indicators of air pollution. *Poll. Res.* 14: 269-273.

CHAPTER - 15

Effect of air pollution on plants

Arun Arya

Botany Department, Faculty of Science,
The Maharaja Sayajirao University of Baroda, Vadodara 390 002
e mail : aryaarunarya@rediffmail.com

ABSTRACT

Materials found in atmosphere in quantities in excess than normal amounts required, are called air pollutants that may be either solid, liquid or gaseous and result from both natural and human (anthropogenic) processes. Due to air pollutants, damage to the plants is a trite. During last four decades studies have been conducted under field conditions to see the effect of various air pollutants on plants. The physiology of plants changes and sometimes visible symptoms result in form of chlorosis or discolouration due to collapse of chlorophyll. Their epidermal micromorphological characters may change and reduction in growth and plant biomass is reported. Controlled fumigation experiments with pollutant gases are conducted and leaf injury or reduction in primary production is studied. Interaction between particulate dust and foliar damage are recorded. Most of the plants are adversely affected by acid rain in terms of reproduction, germination, growth, survival, yield and several biochemical and physiological traits. Additions of acid become harmful to plants particularly in soils which lack free calcium carbonate. Pollen viability is reduced at low pH. It was observed that initial injury occurred to adaxial leaf surface of *Phaseolus vulgaris* and *Helianthus annuus* near trichomes and stomata after exposure to simulated acid rain. Acid rain is believed to be harmful to vegetation both through direct effect on foliage and indirectly by making the soil impoverished of nutrients. The paper deals with problems of air pollution and their effects in plants.

Keywords : Air Pollutants, Suspended Particulate Matter, SO_2, O_3, Symptoms on plants

Rapid industrialization and urbanization in India has created problems of environmental pollution in various metropolitan cities. Various studies have brought to light the harmful effects of environmental pollution on plants under natural and controlled conditions. Salisbury (1927) published a comprehensive account of environmental influences on foliar epidermal features. Studies conducted by Yunus *et al.* (1979), Bhairava Murty and Kumar (1983) indicated the possibility of utilizing leaf cuticular and epidermal features as indicators of environmental pollution- normal as well as man made. SO_2 concentrations vary in a wide range of traces to 800 $\mu g/m^3$ (Rao *et al.* 1983). Location of thermal power plants and concentrations of industries mainly contribute to the high levels. The recommended permissible level for SO_2 is 60-80 $\mu g/m^3$ the effect of SO_2 on vegetation is harmful (Bell and Bedi,1983; Heagle *et al.* 1984). According to Dubey (2005) SO_2 appears to be the most important air pollutant in relation to crop damage.

Indian cities are among the most polluted in the world for their air pollution levels. There are too many cars and too many industries belching out fumes, cities like Ankleshwer and Surat in Gujarat, Delhi, Alwar in Rajasthan, Howrah in West Bengal and the Union Territory of Pondicherry are feared to have exceeded pollution limits. The air in these cities is harmful to plant and human health. Ambient air quality status derived from a network of 290 stations covering 90 towns/ cities in recent years indicates that while SPM is consistently critical in many cities, the concentration of NOx and SOx, is also increasing and is already transiting from moderate to high to critical levels.

The industrial units in India have largely been concentrated in the states of Gujarat, Maharashtra, U.P., Bihar, West Bengal and M.P. the high concentration of SOx and NOx is, therefore, often found in these cities. Obviously, new industrial states in Delhi, Punjab, Rajasthan and Andhra Pradesh are also manifesting a disturbing upward trend in air pollution. Air pollution is the sum total of pollutants contributed by transport, industry and power sectors.

Plants in general are subjected to attack by a large variety of diseases. Over 30,000 plant diseases have been recorded from various countries about 5,000 of which have been reported from India alone (Rangaswami 1993). These diseases handicap man's attempt to feed and clothe himself as soon as he started cultivating plants. Biopollutants in form of bacteria, fungi and viruses caused localized and systemic symptoms in plants. Ultimately diseases result in reduced yield.

Air pollutants interact with plants and alter their normal functioning. Characteristic morphological injury symptoms develop when plants are exposed to a high concentration of pollutants for a short duration or to chronic levels for relatively prolonged periods. Although exposure to lower concentrations does not produce any morphological injury symptoms, biochemical processes are altered significantly. Hence plants become stressed due to air pollutants. Increasing efforts are being made to use plants for detecting air pollutants (Manning and Feder 1980).

15.1 AIR POLLUTION : SOURCES

Air pollution is the presence of substances in atmosphere, generally resulting from the activity of man, in sufficient concentration, present for sufficient time and under circumstances which interfere significantly with the comfort, health or welfare of persons or with the full use or enjoyment of the property.

Air pollution identification is possible by-

1. Sensory Recognition
2. Physical measurement of pollution
3. Effects on plants animals and buildings

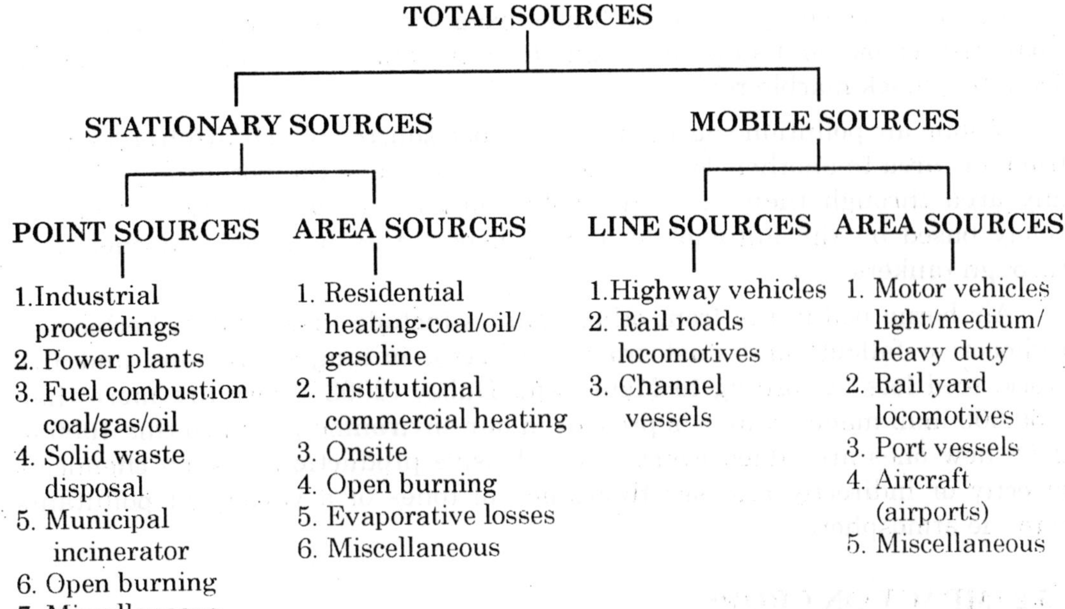

Gaseous air pollutants originate from fixed sources and mobile sources.

Fixed air pollution sources include those sources which are located in one place which include industries. Industries are major producers of carbon monoxide (CO), Carbon dioxide (CO_2) Sulphur dioxide (SO_2), NOx and other particulates. Industrial process exhibits its own pattern of air pollution. Petroleum refineries are responsible for extensive hydrocarbon and particulate pollution. Iron and steel mills, metal smelters, pulp and paper mills, cement and asphalt plants-all discharge vast amounts of various particulates.

NEERI's survey of five locations in Agra indicated that the Taj had highest concentration of SO_2, the daily or short-term average being 107 $\mu g/m^3$. In contrast, Agra fort had only 38 $\mu g/m^3$, and Itmad-ud-Daulah's tomb, Sikandra and Fatehpur Sikri each had 35 $\mu g/m^3$ or less. The theoretical studies of the IMD and Teeneco, however indicated that the long term SO_2 pollution at a distance of 40 km would be 1.3 and 1.7 $\mu g/m^3$ respectively. The highest concentration would be in December, of the order of about 7 $\mu g/m^3$. The results of these theoretical studies raised considerable pollution within Agra: two old thermal power units of 10 MW each, about 250 foundries were used for shunting.

It is not surprising that Central Pollution Control Board data showed the air near the Taj is becoming unhealthy. Sulphur dioxide has risen 66 per cent in little more than decade particulate matter already exceeds the MEF norms and continues to rise. Nitrogen oxides has rising steadily since 1989. The environmentalist felt that Taj and various other monuments may be irreparably affected by corrosive fumes from the refinery (Shivajirao, 1978). Coal dust, smoke and sulphur dioxide from the stack gases are likely to turn Taj into a black marble replica.

Mobile air pollution sources include those sources which are free to move from one area to another. Hence, mobile sources have the potential to pollute any area through their visit. Air pollution from mobile sources is almost solely based on the automobile and transportation of fuels and chemicals through tankers.

Air borne pollutants from other sources include insecticides, herbicides, radioactive fallout, and dust from fertilizers, mining operations and live stock, feed lots. According to Mehta and Dubey (2003) there are more than 7,50,000 man made chemicals present in our environment and to this 1000 to 2000 new ones are added every year. Massive production of such chemicals directly or indirectly releases thousands of tones of a variety of pollutants into the atmosphere.

15.2 IMPACT ON CROPS

According to Dubey (2005) a close review of the data analysis given in the ODA report which is an integration of 600 national and international papers and based on field studies shows that crop yield under air pollution stress is going to be reduced depending upon the geographical extent of the cultivated areas which is suffering from air pollution. Studies related to wheat, maize, rice and other crops indicate that crop damage is noticeable within 6 km radius of the location of air pollution on crop plants, which may extend over to 8-10 km radius from the point source. Dubey (2005) estimated loss of 1.22 million tones in wheat, maize and rice crops in India due to SO_2 pollution at the level of 6 mg m^{-3}.

Prasad (2004) while studying effect of pollutants from refineries in Assam oil fields observed that there was impact of industrial activities on chlorophyll content of rice, mango and betel crop up to 3-4 km. The total chlorophyll content was reduced, however, the reduction was less in case of banana and coconut leaves. Yield of vegetable, fruit, maize and wheat corps was reduced by 5-20%. According to Treshow (1985) ozone concentration in the range of 6 to 8 ppm for a few hours a day can cause crop yield losses in wheat, potato, corn and soybean.

Crop yield declined from 10-50% in response to the levels of SO_2 (28-52 ppb), NO_2 (25-58ppb) and particulates (306-762 μg m-3) in the Singrauli area (Agrawal, 2005). Maximum reduction in yield was observed for *Pisum sativum* (40-50%) and minimum in *Brassica campestris* and *Sesamum indicum* (10-15%). The lower biomass was observed in *Oryza sativa* at sites receiving higher dust load (Singh *et al.*, 1990).

15.3 ACID RAIN AND PLANTS

Gaseous sulphur and nitrogen compounds of atmosphere are presumed as the precursors of acid precipitation as natural or anthropogenic. The oxidation of SO_2 and NO_x leads to the formation of sulphuric and nitric acids. Analysis of more than 1500 precipitation samples with a median pH of 4.0 revealed that in 80-100% of the cases, low pH was attributed to sulphuric and nitric acids(Galloway *et al.,* 1976). Precipitation of hydrogen ion contents is known to be 60% due to sulphuric acids, 34% due to nitric acid and 6% due to various organic acids. Pollutant concentration in rain generally decrease as rainfall amount, intensity and duration increase. Amount of rain recorded, however, exerts the most significant influence on pollutant transfer.

It was 1952 when the phenomenon of acid rain was identified in Manchester city of England, but it was described more in 1972. The Northeast USA suffered the great damage from acid rains and its at least 10 states had rainfall within a pH of 4.0 to 4.5 and individual rainstorms with pH below 3.0. Canada today is the major source and major victim of acid rain. According to Ontario environment department 140 acidified lakes in the province got devoid of fish and every year the damage costs $180-250 millions. However, Japan and China have managed to reduce the oxide emission of sulphur and nitrogen to the acceptable levels.

Most of the plants are found to be adversely affected by acid rain in terms of reproduction, germination, growth, survival, yield and several biochemical and physiological traits (Kumar, 1997; Dhaka, 1999; Verma, 1999). Additions of acid become harmful to plants particularly in soils which lack free calcium carbonate. Pollen viability is reduced at low pH in *Zea mays* (Kumar, 1997). Evans *et al.* (1997) observed initial injury to adaxial leaf surface of *Phaseolus vulgaris* and *Helianthus annuus* occurred near trichomes and stomata after exposure to simulated sulphate acid rain. Lesion frequency was not correlated with density of either stomata or trichomes but was correlated with degree of leaf expansion. Acid rain is believed to be harmful to vegetation both through direct effect on foliage and indirectly by making the soil impoverished of nutrients (Rathier and Frink, 1984).

15.4 POLLUTION DUE TO CEMENT MANUFACTURING PROCESS

The first cement plant in India was commissioned in 1914 with installed capacity of 10,000 tone per year at Porbander, Gujarat by Indian Cement Company Limited. Since then industry has come a long way. At present there are 128 large cement plants in India having installed capacity of 152.14 million tones and actual production of 152.56 million tones (CMA 2004). India ranks world's second largest cement producing country after China. There are three distinct processes for manufacturing of cement, mainly wet process, semi dry process and dry process. Today with the advancement of technology, dry process has been adopted with calciner. Cement is manufactured in four basic stages (Saxena and Kumar, 2005) :

(a) Quarrying and crushing; (b) Grinding and blending of raw material

(c) Clinker production; (d) Finished grinding

Cement manufacturing due to dry process contributes significantly to air pollution. Cement plants produce particulate matter as a result of the continuous feeding of raw materials into cement kiln and rapid counter current flow of combustion gases over raw feed. Particles that become entrained in combustion gases and are removed from the kiln by exhaust gases are known as cement kiln dust.

Mehta and Dubey (2003) studied ambient air quality in the cement producing areas of M.P. Results revealed that SPM was maximum around old established plant like J.K. White Cement works, while in Vikram Cement which was established after the J.K. White is having the lesser SPM concentration. Of all these Aditya Cement was having minimum SPM. These observations directly corresponds to the equipment installed and their performance for dust control in these respective units. Particulate emission from cement factories results in the higher stomatal and trichome densities with smaller epidermal cells and trichomes as compared to non-polluted (Yunus and Ahmad, 1980). *Syzygium cumini* also shows conspicuous decrease in size of epidermal cells and an increase in number of epidermal cells and stomatal frequencies (Singh, 2005).

Yunus and Ahmad (1980) studied the effect of air pollution on leaf epidermis of *Psidium guajava* L. 25 samples of guava leaves were collected from plants growing in polluted environments around the cement factory at Churk, district Mirjapur in U.P. The major air pollutants at Churk were SO_2, CO, CO_2, cement dust and carbon particles. The upper epidermis was characterized by injury symptoms represented by pockets of dead or degenerated cells. The foliar injury was probably by cement dust as this was confined mostly on upper surface. The frequency of trichomes in leaves of polluted population was two times greater on the upper and three times greater on the lower epidermis. The increase in trichome frequency in response to air pollution has also been reported by Sharma (1977) and Yunus *et al.* (1979). Trichomes act as insulators, providing shade to the leaf surface (Treshow 1961). The size of trichomes in leaves of polluted population was smaller than those in leaves of healthy populations. Uma *et al.* (2003) observed decrease in number of epidermal cells and stomatal frequency of *Hibiscus cannabinus*. They observed reduction in total chlorophyll content. Indirabhai *et al.* (1988) opined that the reduction in growth might be due to reduced intensity of light energy available for photosynthesis in the dusted plants.

15.5 POLLUTION BY MINING ACTIVITY AND STONE CRUSHER

Particulate pollution is resultant of both natural as well as human activities. The human activities include building construction, stone crushing and open cast mining etc. the particulate pollution affects all the 3 major things *i.e.* air, water and soil. It in general affects the entire environment and adjoining ecosystem of the region. Kaumawat *et al.* (2004) reported eco degradation around dolomite grinding units at Tehsil Alirajpur District Jhabua, M.P. Pandit *et al.*, 2004 found reduction in protein and chlorophyll contents in *Calotropis gigantea* growing near stone crusher. Different workers have reported reduction of protein, fat and trace elements due to high emition of dust from stone crusher. Bose *et al.* (1983) reported air pollution studies conducted in Jharia coal fields in Bihar. They found high levels of SPM and gases like SO_2 and NO_2. The level of pollution was more in winter months.

Dust ascribed to industrial sources has been shown to reduce carbohydrate content in *Ficus religiosa, Ailanthus excelsa, Cordia mixa, Calotropis procera, Polyalthia longifolia, Ficus drupaceae, F. benghalensis, Terminalia catappa* and *Nyctanthes arbor-tristis* (Vora *et al.*, 1986). A reduction in the amount of reducing sugars in tree species from polluted sites may be due to clogging of stomatal pores which interferes with gaseous exchange and photosynthetic activities.

15.6 AIR POLLUTANTS : IMPACT ON VEGETATION

Air pollution effects on vegetation have been recognized for more than a hundred years (Magill *et al.,* 1956). Symptoms on various plants attributed to toxic atmospheric concentrations of gaseous pollutants in industrialized areas of Germany at the end of the 19th century are well documented (Schroeder and Renss, 1883). The methyl iso cyanate (MIC) gas leak in Bhopal during 1984 has been regarded as the worst industrial accident in India, which is related to air pollution. About 200,000 people were affected and severe loss of vegetation was observed the trees shed their leaves and blades of grass dried due to MIC exposure. On December 3rd 1984 the gas emitted from the factory spread over some 40 square km area and affected people seriously. MIC is invariably accompanied by Phosgene ($COCl_2$). The toxic effect of MIC is enhanced by $COCl_2$.

In Europe and North America active monitoring with bio indicators (especially vegetation) are widely used to monitor air quality (Manning and Feder, 1980). In Israel lichens and higher plants were exposed near industrial areas to detect the accumulation of heavy metals in plants (Naveh *et al.,* 1979; Gartey, 1987). Several indicator species have been used to map the extent of air pollution by Phytochemical, oxidants in Mexico (de Baur and Krupa 1990). In India Chaphekar (1972) studied visible injuries on plants through vegetational survey in Mumbai, Some plants were exposed and based on visible symptoms they were classified as sensitive or tolerant species (Chaphekar and Karbhari 1974).

The most important phytotoxic air contaminants are generally gaseous in nature. They may be characterized as either primary pollutants, causing effect in the state in which they emerge from industrial processes; or secondary pollutants, that is created in the atmosphere as a result of chemical reactions among precursors not especially toxic per se (Haagen-Smit *et al.* 1952). Sulphur dioxide is probably the chief offender among the primary pollutants (Leone 1985-86). Marginal chlorosis results in apple which extends down between the veins, with occasionally all markings tending to be intercostals. At very high pollution levels, the entire leaf is necrotic. According to Spierings (1967) apple trees loose their leaves early in high SO_2 concentration. Istas (1980) established the thresholds concentration for SO_2 injury to apple as 0.1-0.3 ppm. Brennan *et al.* 1967 observed an intercostal foliar necrosis of crab apple each spring in New Jersy when SO_2 level exceeded 0.10 ppm for a four hour period. The intercostal necrotic area became reddish brown in color. Sweet and sour oranges exposed to 2.6 and 6.5ppm SO_2 exhibited foliar symptoms only on young leaves. Necrosis also occurs in certain grape varieties (Fujiwara, 1970). Fruit injuries and cracks were also observed in apple exposed to SO_2 (Istas, 1980). Dry white straw coloured, marginal or interveinal blotches on leaves of broad leafed plants are common symptoms of this pollutant. Sensitive plants (exposed to 0.5 ppm conc.) are alfalfa, apple, barley, garden beans, *Bigonia,* black berry, cabbage, cotton, cucumber, mustard, *Petunia* (Plate II, Fig. B), white pine, pumpkin, raspberry, rose, soyabean, spinach, squash, strawberry, tomato, *Tulip*, Sulphur dioxide injuries may occur up to 30 km source and at concentration as low as 0.3-0.5 ppm. Conifers (Plate I, Fig. F) show brown to reddish brown necrosis at the tip of the needle, grassses develop lighten to white streaks. Black tip of mango is very common in few areas of the country. Resistant crops are corn, onion, pine, and potato (Dahiya 2000).

Sulphur dioxide enters through the stomata by diffusion where it reacts with water and gets converted into sulphite and bisulphate. These sulphites are highly toxic, as they bring about changes in leaf pH. The photoreduction of free sulphite utilizes sulphite oxidase and reduced ferredoxin from photosynthetic electron transport chain, which in turn, inhibits CO_2 fixation and photosynthesis. The reduction in photosynthesis appears in the form of reduced plant dry weight as shown in *Cassia fistula* (Raza *et al.* 1988), *Ficus religiosa* (Tiwari and Bansal, 1966) and *Mangifera indica* (Plate II, Fig. D) (Agarwal *et al.*, 1991).

The quantitative changes in leaf structure of evergreen plants (*Pinus sylvestris, Arctostaphylos uva-ursi, Vaccinium vitis-idaea*), deciduous (*Betula pubbescens, V. myrtilus, V. uliginosum*) and herbaceous (*Potentilla erecta* and *Solidago japponica*) under intense pollution due to sulphur dioxide and poly-metallic dust were investigated in Russia (Karvkina and Lyanguzova 2006). Plastidom (chondriom) content was calculated as a ratio of the plastidom area to the cell area; starch grains content in the plastidom as a ratio of the total starch grain area to the plastidom area. The plastidom is a totality of plastids in a tissue under study, while the chondriom is a totality of mitochondria in the same tissue. While studying plastidom no difference was found in the density of chloroplasts in the mesophyll cell sections of the needles of the current and the first two years in *Pinus sylvestris* from both the area under investigation. Only in the three year old needles the density of chloroplasts in the chlorenchyma cells in the intensely polluted area became 25% higher. Two year old leaves of *Arctostaphylos* from the area of intense damage showed 26% decrease of chloroplast density in a cell. No significant change of chloroplast density was found in *Vaccinium* under pollution during three years of vegetation. In the polluted area, 24% decrease of the chloroplast density was revealed in leaf chlorenchyma cells of *V. myrtillus*. At the same time, no change of chloroplast density was found in *V. uliginosum* leaves. Leaves of *Betula pubescens, Cornus suecica* and *Potentilla erecta* from the intensely damaged area showed constant chloroplast density per cell. The chloroplast density in *Solidago japponica* leaves was less by 15% in the polluted area.

Hydrogen fluoride (HF) has a more localized aspect since it is emitted from specific factories involved in such industries as PO_4 fertilizers, aluminium, glass, bricks or ceramics (Daines *et al.,* 1952). Fluorides may accumulate in plants to the extent that pose the danger of florosis to animals or human beings injusting contaminated plant parts. Fluoride injuries to apricot was first reported by De Ong (1946), in the vicinity of an aluminum plant in San Fransisco. Apricot leaf injury first appears 2-3 weeks after emergence, the earliest symptoms are a wilting of leaf edges, which becomes more pronounced with time. By the tenth day this area turns brown and necrotic. The necrotic area soon become detached from the leaf and drop up, leaving an irregularly shape margins. Narrow dark reddish brown bands delimit necrosis caused by successive fumigation, imparting a zonate appearance to the lesions (Jackobson and Hill, 1970). According to Istas (1980) the threshold HF concentration for apricot foliar injury is 0.1ppb. Some fruits may be more sensitive to fluoride than leaves. The best known example is **soft suture** disease of peach. Benson in 1959 described this physiological disturbance of fruit manifested by premiature ripening of the fruits along the suture. This is often accompanied by suture splitting from the flesh. Sensitive plants are citrus, barley, corn, *Gladiolus*, grapes, peach. Corn leaves develop yellow mottling prior to typical burning. Among the resistant plants are cotton, rose, squash, tobacco and tomato.

Ozone (O_3) is by far the most phytotoxic of the secondary or photo-chemical gasses. It is not emitted as such by any source, but has its origin in reactions between the nitrogen and oxygen of the atmosphere at high temperature to form nitrogen dioxide, which in turn, undergoes chemical transformation in the presence of automobile exhaust to produce (O_3) and other oxidants. O_3 damage, first recognized in California (Haagan-Smit *et al.* 1952) occurs through out the north eastern USA as well as the other countries of the world. Depending largely on light energy for its existence. O_3 causes damage primarily during the growing season and temperate climate. Ozone is formed at a higher rate during sunny days due to photochemical reactions. NO_2 is consumed in O_3 formation. Numerous studies have documented that roots showed proportionally greater growth reductions when exposed to O_3 than the shoots (Keutgen *et al.* 2005).

The guard cells loose water in contact with O_3 and collapse, partially closing stomatal pores and interfering with the photosynthetic process. Transpiration is also reduced. An. increase in reducing sugars in cells accelerates respiration. Cellulose synthesis and cell wall formation is affected adversely, due to the oxidizing effect of O_3 on the sulphydril groups of enzymes (Chaphekar 1982). Pigmented or un pigmented spots or bleaching on upper leaf surface, browning and death of conifer needle at the tip, fleaking, chlorosis or bronzing are main symptoms of injuries caused by ozone. Alfaalfa, Aster, garden beans, carnation, carrot, celery, cabbage, *Dahlia*, grapes, grasses, marigold, musk melon, onion, *Petunia*, potato, pumpkin, radish, spinach, squash, tomato and tobacco are most sensitive plants to ozone. Among resistant plants are citrus, *Gladiolus,* mint, pepper etc. Age of leaves is a factor to be considered in determining the relative resistance to air pollutants. Darley and Middleton 1961 demonstrated that oldest leaves of *Petunia* were severely injured by ozone while younger leaves were progressively less injured.

Crop yield losses due to O_3 within 80 km zone of four major cities of India, viz., Mumbai, Ahmedabad, Pune and Lucknow as estimated by Ashmore and Marshall were 40, 29, 19 and 12% respectively. Estimated total rice production loss due to O_3 in Kantoh district (Japan) ranged from 1.1 to 46% for 1981 to 1985 (Kobayashi 1999). The results of a field study in China showed that growth of vegetables grown around city (Chongqing) is not vigorous as in Beibei, relatively clean environment 50 m away from city resulting on average 24% reduction in final yield (Cheng and Chen 1991).

Ammonia (NH_3) is a gaseous pollutant only under accidental, unusual conditions, escaping from precooling system. In some places, it is known as phytotoxic air pollutant when it escapes from fertilizer factories, during manufacture of ammonia or nitric acid. Blackening or bleaching, spotting of leaves, brown lesions between veins, change of colour of fruits, etc. are the symptoms of ammonia injury. Chaphekar and Boralkar (1979) have reported reduction in seed germination in *Trigonella.*

Peroxy-acetyl nitrate (PAN) is one of the reaction products of nitrogen oxides with oxidants in air and is a major phytotoxic component of smog. On coming in contact with chloroplasts of mesophyll cells, it affects enzymes necessary to photo-phosphorylation and inhibits photosynthetic CO_2 fixation. Mudd (1963) showed that enzymes in sulphydril groups were sensitive to PAN. Unlike ozone younger leaves are more susceptible to PAN derived from erradiated auto exhaust.

15.7 PLANTS AS POLLUTION INDICATORS

Many plants produce visible symptoms when exposed to environmental pollution. They are now being acknowledged as effective alternatives to mechanical instruments for gauging pollution levels. Moreover, bioindicators facilitate effortless awareness generation among the masses, as they themselves are able to assess the damage. Bioindicators does not involve high cost as different plants are used.

Plate I. Fig. A *Holoptelia integrifolia* showing leaf tip damage; **Fig. B**. Chlorosis in *Datura stramonium* leaves; **Fig. C** Leaves showing tip burn in *Croton;* **Fig. D** Chlorosis in the leaf of *Garcinia mangostina;* **Fig. E** Twig of *Nerium* showing chlorosis symptom; **Fig. F** Small tree of *Pinus longifolia;* SO_2 pollution turns tip of the needles reddish brown.

Plate II. Fig. A Bonsai plant of *Citrus sinensis;* **Fig. B** Flowering twig of *Petunia;* this plant is sensitive against ozone; **Fig. C** Healthy plant of *Nerium;* **Fig. D** Flowering twig of *Mangifera indica;* reduction in photosynthesis occurs due to SO_2 pollution

Plate III. Fig. A A leafless tree of *Ficus religiosa* (Babool, neem and Peepal are recommended trees against pollution); **Fig. B** A twig of *Ficus religiosa* showing chordate leaves (inset).

Rapid industrialization has improved the standard of living but threatened the very existence of mankind. Man is going to conquer the universe with the help of technological developments but the unsustainable development has resulted in grave danger to the very existence of man. Our efforts in industrialization or development in the next millennium should go hand in hand with due concern for environment protection with concrete steps in this direction. Plants are good indicators of air pollution, and many of these do well to fix air pollutants also. In this context tree plantation, creation of green belts is one of the potential alternatives to mitigate air pollution. Plants act as sink for pollutants. They act as filters for SPM and their importance is well recognized to develop a clean and healthy environment.

It has been recognized since the mid 19th century that lichen are particularly susceptible to SO_2. However, there are large interspecies differences in degree of susceptibility which has been exploited as a tool for bio monitoring this pollutants. Various scales have been developed for predicting SO_2 concentrations based on lichen community composition on different substrates. The best known is that of Hawksworth and Rose (1970), who produced pollution / lichen map of England and Wales, which indicated that the more sensitive epiphytic species were confined largely to very clean location in the west and in the north. They used 18 point scale, ranging from a rich flora containing fruticose species in very low SO_2 sites to a complete absence of epiphytes in most polluted areas. The link between environment and air quality can be made in many ways. Lichens exist all the year round, they colonized almost every substrate and exhibit a wide range of shapes and colour, we can study these organisms and link them to potential sources of pollution such as roads or industry (Bell *et al.,* 2006).

15.7.1 Detecting Cadmium in Willow Leaves

Plants are extremely sensitive to environmental pollution, especially heavy metals. The phytoextraction of cadmium from terrestrial ecosystems result in direct and indirect toxic effects on leaf physiology affecting *inter alia*, nutrient uptake and the photosynthetic apparatus.

Pierre Vollenweider and colleagues from Federal Institute Forest, Snow and Landscape Research, Birmensdorf, Switzerland have investigated the distribution of Cd and associated structural changes in the tissues of the willow (*Salix viminalis*). Cadmium was cytochemically revealed using autometallography and the distribution of reaction products was studied with light and fluorescent microscopy. The advantages of physical development include very high sensitivity compared to electron microprobe analysis with a detection limit around 10 atoms and element stabilization against leaching by a sodium sulphide pre-treatment.

It was found that main Cadmium sink was in pectin rich layers of the collenchyma cell walls of the veins. Active storage was indicated by homogenous cell wall thickenings with cellulose and prothocyanidins. Cadmium micro localization and cell injury in the conducting phloem indicated metal cycling. In the leaf blade, oxidative stress and accelerated cell senescence increased in those areas of the mesophyll with a low cadmium content. Local cadmium accumulation in the veinlets near the leaf edges caused tannin plugging in xylem and necrosis in the surrounding mesophyll and upper epidermis.

15.8 POLLUTION PREVENTION

Problem of air pollution is attaining shape of monster for both agricultural crops as well as for forest trees. There are numerous species, which can be planted on agricultural fields, in orchard and surroundings of factories. In general, to control air pollution from thermal power plants, brick kiln and fertilizer plant following tree species can be planted.

Forest Trees: Babul, Sirish, Kadamb, neem, Shisham, Subabul, Khejri, *Eucalyptus*, peepal, *Artocarpus, Butea*

Fruit Trees: *Zizyphus jujuba, Aegle marmelos, Ficus carica, Sapindus* and *Psidium guajava*

REFERENCES

Agarwal M. 2005, Effects of air pollution on agriculture: An issue of national concern. *Natl. Acad. Sci. Lett.* 28(3-4): 93- 106

Agarwal M. Singh S.K. and Rao D. N. 1991, Biomonitoring of air pollution in urban and industrial sites. *J. Environ. Biol.* **5:** 211-222

Agarwal S.B. and Singh A. 2007, Effect of air pollution on growth and yield of three agriculturally important plants under varying levels of mineral nutrients. *Proc. Nat. Acad. Sci. India* 77 (B): II 158-171.

Azeem A. 1995, Response to some trees to urban air pollution, M. Phil. Dissertation submitted to SES, JNU, New Delhi

Bell D.C. and Bedi S.J. 1983, Effect of industrial air pollutants on dry weight and yield of tomato, *Proc. of Silver Jubilee symp, Int. Soc. Tropical Ecology,* Bhopal.

Bell J.N.B, Davies L. and Jarraud N.S. 2006, Detecting air pollution with plants in In: *Plant Response to environmental stress* (Tripathi R.D. *et al.*, Eds) International book distributing company, Lucknow, India pp 103 – 112

Bensone N. R. 1959, Fluoride injury on soft suture and splitting of peaches, *Amer. Soc. for Hort. Res. Proc.* 74: 184-188

Bhirava Murty P.V. and Kumar P.V. 1983, Air Pollution and epidermal traits of *Calotropis gigantea* (L.) R. Br., *Indian J. Air Pollution Control* 4(1 & 2): 23-26

Boralkar D.B. 1979, A contribution to the study of the effects of industrial air pollutants on plants. Ph. D. Thesis Univ. of Bombay, Bombay

Brennan E. Leone, I.A. and Daines R.H. 1967, Characterization of the plants damage problem by air pollutants in new Jersey, *Pl. Dis. Rep.* 51, 850-854

Brown D.H 1984, Mineral nutrition. In: *Bryophyte ecology*, A.J.E Smith (Ed), Chapman and Hall, pp 383 – 444

Chaphekar S. B. 1972, Effects of atmospheric pollutants on plants in Bombay, *J. of Biol. Sci.* **15:** 126

Chaphekar S. B. 1982, Air pollution and plants *Indian Rev. Life Sci.* 2: 41-66

Chaphekar S.B. and Karbhari M.R. 1974, Sulphur dioxide injury in plants- some experiments, *Geobios*, 1: 141-142

Cheng Y. and Chen S. 1991, Assessment of economic losses due to acid precipitation in Chongqing terrestrial ecosystems. *Atmos. Environ.* 6: 45-51

CMA, 2004, Basic data-Indian Cement industry, Cement Manufacturer's Association, Noida

Dahiya S. 2000, Air pollution and plant health, *Paryavaran Urja Times* 3(2) :141 – 142.

Daines R.H., Leone I.A. and Brennan E. 1952, The effect of fluorine in plants as determined by soil nutrition and fumigation studies, *Proc. of the U.S. Tech. Conf. on Air Pollution.* Mc Graw Hill Co., New York, 97-105

Darley E.F and Middleton J.T 1961, Carbon filtered protects plants from damage by air pollution. *Florists Rev.* 127 (3294, 1516)

de Baur L.I. and Krupa S.V. 1990, The valley of Mexico: Summary of observational studies on its air quality and effects on vegetation *Environ. Pollution.* 65: 109-118

Deong E.R. 1946, Injury to apricot leaves from fluoride deposit, *Phytopathology* 36: 469-471

Dhaka A. 1999 Impact of Air pollution on some plants, Ph.D. Thesis, CCS University, Meerut.

Dubey P. S. 2005, Urban air pollution and its management, In: *Urban Pollution-Issues and Solutions* (eds. Arya A., Bedi S.J., Jasrai Y.T. and Patel V.S.) Pub. by Nidhi Book Centre, New Delhi, 1-9

Evans L.S., Gmur N.F. and Kelsch J.J. 1977, Physiological and biochemical effects of atmospheric oxidants on plants. *J. Air Pollut. Contr. Assoc.* **16:** 467-471

Fujiwara T. 1970, Sensitivity of grape vine to injury by atmospheric Sulphur dioxide *J. Jap. Soc. Hort. Sci.* 39: 13-16

Galloway J.N., Linkens G.E. and Edgertor E.S. 1976, Acid rain precipitation in North Western United States *PH and Acidity,* **194**, 722-724

Garty J. 1987 Metal amounts in the lichen *Ramalina duriaci* (De Not.) Bagl. Trans planted at biomonitoring site around a new coal-fired power station after one year of operation. *Environ. Res.* **43**, 104-116

Haagen-Smit, A.J. Darely E.F., Zairlin, M. Hull H. and Noble W. 1952, Investigation of injury of plants from air pollution in the Los Angeles area. *Plant Physiol.* **37**, 18-34

Haegle A. A., Denis E. B. and Grady E. 1984, Injury to yield responses of soybean to chronic doses of ozone and sulphur dioxide in the field. *Phytopathology* **64(1)**: 132-136

Hawksworth D.L. and Rose F. 1970, Qualitative scale for estimating sulphur dioxide air pollution in England and Wales using epiphytic lichen. *Nature* 227: 145 – 148.

Indhirabhai K. Dhana Lakshmi S. and Lakhmanan K.K. 1988, A significant observation on the effect of cement dust, H_2S and NaCl on growth and development of *Dolichos biflorus* L. *Adv. in Pl. Sci.* **11**, 91-101 pp.

Istas J. R. 1980, Quelques notions au sujet de pollutants atmospheriques et leurs effects sur les plantes. Groupe de Travil pour la pollution on Agriculture. Ist. De Recharches chiniques, Terveuren, Boligique, 49pp

Jacobson J. S. and Hill A. C. eds. 1970, *Recognition of Air Pollution injury to vegetation*: A Pictorial Atlas, Air Pollut. Contr. Assoc.

Kobayashi K. 1999, Assessing the impacts of tropospheric ozone on agricultural production. *J. Japan Soc. Atmos. Environ.* 34: 162-175.

Keutgen A.J., Noga G. and Pawelzic E 2005, Cultivar specific impairment of strawberry growth, photosynthesis, carbohydrate and nitrogen accumulation by ozone. *Environ. Exp. Bot.* 2005 53: 271-280.

Kravkina I.M. and Lyanguzova 2006, Quantitative characteristics of the leaf structure under industrial pollution. In: Plant response to environmental stress (eds. R.D. Trpathi, K. Kulshreshtha, M. agarwal, K.J. Ahmad, C.K. Varshney, S.V. Krupa and P. Pushpangadan) International Book Distributing Co., Lucknow pp 303-315

Kumawat D. M., Ram H. K. and Haryani K. 2004, Ecodegradation around dolomite grinding units at tehsil Alirajpur district Jhabua M. P. *Nat. seminar on Ecodegradation due to Air Pollution,* IAAPC, Vadodara, 16th October 2004

Leone I. A. 1985-86, Air pollution diseases of fruit corps, Rev. Trop. Pl. Path. (eds. Ray Chandari S. P. and Verma J. P.) **2**: 263-286

Magill P. L., Holden F. R. and Ackley C. ed. 1956, *Air Pollution Hand book,* Mc Graw Hill Book Co., New York, 9-44.

Manning W. J. and Feder W. A. 1980, *Biomonitoring air pollutants with Plants,* Applied Sci. Pub., London

Mehta N. and Dubey P.S. 2003, Air quality in three different cement producing areas, In: *Air Pollution: development at what cost*? (eds. Jasrai Y.T. and Arya A.) Daya Pub. House, New Delhi 10-20 pp.

Mudd J. B. 1963, Enzyme inactivation by peroxyaectyl nitrate, *Arch. Biochem. Bio-Phys.* 102: 59-65

Nath V., Asthana A.K. and Sharma S. 2006, Environmental studies on bryophytes of Mussoorie (Western Himalayas) In Plant Response to environmental stress (Tripathi R.D *et al.* Eds) International book distributing company, Lucknow, India pp 173 – 177

Naveh Z., Steinberger E. H. and Chaini S. 1979, The use of biodindicators for monitoring of air pollutions by fluor ozone and Sulphur dioxide, In *Environmental biomonitoring assessment production and management-certain case studies and related qualitative issues* (eds Cairns J. Jr., Patil G. P. and Waters W. E.), Intl. Co-operative Publ. House, Fairland, Maryland, pp. 21-47

Pandit B. R., Prasanna Kumar P. G. and Prajapati S. 2004, Reduction in protein and chlorophyll contents in *Calotropis gigantea* (L.) R. Br. Growing near stone crusher. *Nat. seminar on Ecodegradation due to Air Pollution,* IAAPC, Vadodara, 16th October 2004

Pittsburgh, P.A. 1977, Effect of simulated acid rain on *Zea mays* L. Ph. D. Thesis, CCS university Meerut.

Prasad R. 2004, Some more observations in EIA studies of oil refineries *Indian J. of Air Poll. Control* IV (1): 61-66.

Rangaswami G. 1993, *Diseases of crop plants in India*. Prentice-Hall of India Pvt. Ltd., New Delhi 498 pp.

Rao C.S.G., Giti M., Basu A. R., Patra S. R., Gupta R. K. and Pampathvar V. C. 1983, Sulphur dioxide and sulphation rate levels in major areas of Calcutta metropolis, *Proc. Sym. Air Poll. Control* **1**: 73-90

Rathier T.M. and Frink C.R. 1984, Simulated acid rain: Effects of leaf quality and yield of broad leaf tobacco, *Water, Air, Soil Pollut.* **22**: 389-394.

Raza S. H., Murthy M.S.R. and Ahmed A. 1988, Air pollution tolerance index of certain plants of Nacharam industrial area, Hydrabad, India, *Ind. J. Bot.* **11(1)**: 91-95

Salisbury E.J. 1927, On the causes and ecological significance of stomatal frequencies with special reference to wood land flora *Phil. Trans. R. Soc. B.* **216:** 1-65

Saxena Y. K. and Kumar H. 2005, Vegetation for cement dust control and mine reclamation. *Indian J. of Air Pollution Control,* **V(II)**: 70-76

Schroder J. V. and Reuss. C. 1983, Beschadigung der vegetation durch rauch and oberharzer Juttenrauch Schaden. *Paul Parey,* Berlin

Sharma G. K. 1977, Cuticular features as indicators of environmental pollution *Water, Air and Soil Poll.* **8:** 15-19

Singh P. K. 2005, Plants as indicators of air pollution an Indian experience. *Indian Forester* : 71-80

Spierings F. 1967, Methods for determining the susceptibility of trees to air pollution by artificial fumigation. *Atmos. Environ. J.* 1: 205-210

Tiwari S. and Bansal S. 1996, Changes in morphological characters in *Ficus religiosa* and *Mimusops elengi* to exposure to Sulphur dioxide, *Acta Eco.* 18(1): 10-14

Treshow M. 1961, Environment and Plant response, Mc Graw Hill Book Co. New York.

Treshow M. 1985, Air Pollution trends in USA. *Indian J. of Air Pollution control,* 6(1):1-4

Uma Ch. Ramana Rao T. V. and Inamdar J. A. 2003, Effect of particulate (Cement kiln dust) pollution on *Hibiscus cannabinus* L. In: Air Pollution: Development at what cost? (eds. Jasrai Y. T. and Arya A.) Daya Publ. House, New Delhi 66-74

Verma S. P. 1999, Studies on Phototoxicity of acid rain. Ph. D. Thesis CCS University Merrut.

Yunus M. Ahmad K. J. and Gale R. 1979, Natural air pollutants and epidermal traits in *Ricinus communis* L., *Environ Poll.* 20(3): 189-198

Yunus M. and Ahmed K. J. 1980, Effect of air pollution on leaf epidermis of *Psidum guajava* L. *Indian J. Air Pollution Control*, 3: 62-67.

CHAPTER - 16

Eco-auditing as a strategic tool in managing the resource and energy in cement industries

K. Haryani, S. Maderia and A. Dangash

Institute of Environment Management and Plant Sciences, Vikram University, Ujjain, 456010 Madhya Pradesh, India

ABSTRACT

A continuous environmental auditing executed for several years in three cement plants conclusively prove that ecoauditing if done in detail on monthly basis for various components one can diagnose the point of strength and weakness of entire production process, resources and energy utilization trend. A careful review of the recommendation of the audit report can provide remedial measures for correction of the processes, for checking of the spoilage and waste of the raw material and resources at various points starting from mining to packing and transportation. From the same report one can visualize the area where recycling and reutilization is possible.

The case history of Vikram Cement, Aditya Cement and JK White Cement proved that the environmental audit can yield improvement in environmental quality as well as conservation and minimize economic losses. With the implementation of the recommendation and observation of the audit team the achievements in the conservation of water; in the production process is quite impressive. On the energy front there is a remarkable achievement in heat and thermal energy per ton of cement produced.

The achievements are more spectacular in Vikram Cement where auditing has been done for eight years continuously. In other two industries in four years itself, auditing has proved its utility in conservation of energy and resources and in turn economic benefit.

Keywords : Eco-auditing, Cement industries, Pollution, Energy

Eco-auditing is a significant tool in the problems of environment. The word acquired such a proportion of awareness that it has almost become synonymous to the word environment. The entire scenario of environmental concerns at village to global level is changing fast and attitude is growing towards the totality of protection. Obviously industries are no longer and will no longer be treated as culprits much because of the commitment generated, technological expertise implemented and the self consciousness emerged at every level. Because, the environment requires respect, dedication, devotion and designs for natural benefits :- means we need a management proposal which ultimately provides a regular guidance. The picture is getting clear and clear, the industrial environments are getting cleaner and cleaner, environmental management is gaining priorities over the profit oriented economical management.

16.1 ECO-AUDITING AND ENVIRONMENTAL MANAGEMENT

It is this reason that in India, even in the industrial sector, the fusion of technosphere-. biosphere has started yielding better Environmental Management Plans and Systems. True, that the environmental preservation is a task, no regulation, acts, laws, rules can make a sea change in our environmental scenario until people's participation and the industries, planners and authorities adopt voluntary exercises.

The objective of this paper is not to meet the standards and complete the consent conditions laid by Government agencies or Pollution Control Boards but the real objective is to obtain consent from our ecosystems and their components. No certificate is more valid than this. The environment management of Cement industry is not just a routine one but it is a fact sheet of a cohesive effort of people, executives, labours and other employees. The general concept that cement production is energy intensive and pollution generative process has been proved into misconceptions at these sites at least. The continuous auditing procedure on annual basis since last ten years at these sites and of much significance environmentally is, the improvement made as a consequence of the recommendations in these annual statements.

The over all scenario of comprehensive efforts of company in turn will have a long term impact on conservation of these resources - a policy quite compatible with the national policies.

Sites

The environmental auditing was done annually at different cement industries since last ten to fifteen years. The details about industries are mentioned in Table 1.

Table 1. General information about the industry

Details	Name of Industry		
	Vikram Cement	Aditya Cement	J.K. White Cement
Location	A unit of Grasim Industries P.O. KHOR, Distt. Neemuch (MP) 458470	A unit of Grasim Industries, P.O. Sawa-Shambhupura Dist. Chittorgarh (Raj.) - 312613	A unit of J.K. Industries Ltd. P.O. Gotan, Distt. Nagaur (Raj.) 342902
Longitude	75°58' E	74°37' E	73°45' E
Latitude	24°07' N	24°46' N	26°49' N
Annual Production (TPD)	Up to 1998 – 4000 From 1999- 7300	Up to 2001 - 3300 from 2001 - 4800	Up to 2000 - 530 From 2000 – 760
Year of establishment	1985	1994	1984
Soil Type	Black cotton soil with large patches of red laterite	Black/yellowish brown soil with red loam	Yellow sandy soil

16.2 RESULTS AND DISCUSSION

The energy and resource management is a key to the success for environmental quality. The very fact that cements production is highly energy intensive and resource exploitative, it is too critical for cement industries to manage these two aspects, which are rather key issues of environment management in general.

In nutshell, the strategy of resource saving yielded results with no compromise on the quality of the cement produced. This orientation in functional and management aspect of lime stone, other raw materials and water consumption is a significant achievement in it self. Since it corroborates with the national policy of natural resource conservation.

The situation of raw material utilization too is optimistic with a continuous reduction in water use pattern, mineral materials to be used and coal consumption for the per ton of cement or clinker production. Infact, there has been an over all attempt to cut the energy in all sectors whether it is grinding or packing or clinker formation.

The Tables 2, 3 & 4 gives the detailed information of the water use pattern and raw material consumption in manufacturing of the cement. The water conservation efforts shows, that a serious attempt has been made for water economy so as to bring it to the level of previous years. The good practices adopted in the domestic consumption of water have given positive results at Vikram Cement then other industries. The data exhibit that the company has taken water conservation as a high priority area: since the region is already a water scarce area. The industry has prepared a programme for rainwater harvest and this season will be the one, which will offer the result of it.

Raw material consumption pattern has been optimized, since the raw material use per ton of cement produced is decreased in last few years showing that processes are in good shape and are well stabilized. The mineral resource picture is more impressive with a reduction of per ton of cement, which is once again an achievement at these industries. The data of the few years and the current year exhibit a trend towards resource waste minimization through careful production and management.

In the raw material (limestone) sector the per ton clinker/cement is decreased by 3.55%, 15.64% and 1.45% at Vikram Cement, Aditya Cement and J.K. White Cement over 1995-96 respectively. Actually the conservation of limestone will be actually 0.049, 0.206 and 0.018 ton per ton of cement production. The maximum saving of limestone was carried out by Aditya Cement then other two industries. As far as raw material is concerned the situation is quite optimistic which can lead to the further saving of resources in next few years. It should be noted that this has been achieved through updating of technical know how and care to stop spillage.

No country, industry or a modern civilization can afford to be without energy. The industries require enough of it and cement industry in it self is in energy intensive process. In such a case it becomes more important rather obligatory for the industry to reduce the energy use per ton of clinker/cement, since ultimately it terminates to the saving of energy may be electrical may be thermal and may be even human one. Now cement is a key player product of market especially in developing countries as result whether it is thermal or electrical, it should be saved, which will mean a strategy based on field experience, keen observations and regular implementations.

The energy is the key of all functions of life, industries and systems of any kind and cement production requires large quantum of energy per unit of production which may be derived from coal, oil and electricity. The competition of cutting the price of cement and the energy for conservation of resources made the cement industry in general to evolve strategies for energy conservation. A comparative data of use of the coal, diesel and power has been calculated in terms of thermal or electrical energy per ton. The picture of conservation of energy resources has been continually positive since last four to five years have been continued to this year also with success.

There is a significant reduction in the thermal energy per ton of cement/ clinker i.e. A saving over 1995-96 is 8.07% at Vikram Cement, 21.19% at Aditya Cement and 17.13% at J.K. White Cement. A perusal of the data sheet indicates a continuous trend of energy saving from 1995-96 to 2002-2003. The similar trend figures for electric energy Kwh requirement per ton of cement produced was decreased at Vikram Cement and Aditya Cement i.e. 22.10% and 51.09% respectively over 1995-96 while there is a slight increased in electric energy consumption at J.K. White Cement i.e. 1.05% was observed. These data for achievement of the energy in the one decade in terms of resource and money as well as environment would amount to large saving of resources. The most significant attainment of the Aditya Cement is thus energy management.

The situation of raw material utilization too is optimistic with a continuous reduction in water use pattern, mineral materials to be used and coal consumption for the per ton of cement or clinker production. Infact, there has been an overall attempt to cut the energy in all sectors whether it is grinding or packing or clinker formation.

Finally the objective of ecoauditing is simple, if the pollution control can not yield results in natural resource conservation than these efforts and financial inputs are of less importance and poor significance because it will serve only one objective i.e. environmental quality improvement only but no resource economy. The following points will present a summary of these units on resource minimization and energy conservation.

1. Though the cement sector is in turmoil of recessive economy, these cement units managed financial allocation for the environmental management and pollution control efforts. The dust control and recycling of the trapped dust material, amounts to a definite saving of the mineral resources and an achievement in itself.
2. Water is becoming a precious commodity and is more so in a case of these cement areas, where water is in scarce. In such a situation cutting down of water consumption in domestic as well as cement production should be well appreciated since the quantity of water saved is enormous.
3. The water resource economy is further emphasized by the use of the entire quantity of treated sewage for the horticultural purpose that mean a straight reduction in the use of fresh water equivalent to $120m^3$/day.
4. The utilization of low-grade limestone for production of cement is another effort, which has a direct bearing on the saving of limestone from natural sources.
5. A critical assessment made after a continuous appraisal of the plant and machine operations indicate that energy saving efforts is efficiently put in all sectors of the plant. Also the spillage of oils, grease and other material is almost negligible and perhaps induction of TPM concepts has made it possible. Ultimately the resources are utilized and production cost is reduced.

6. One of the major achievement is in the energy use pattern per ton of clinker/cement, whether it is the electrical or thermal any type of energy. The net savings in last six/seven years in energy sectors has been significant. This achievement has not come suddenly; indeed it is because of recurrent efforts and modern technical inputs, which gave these results. On an economic scale the savings would be in terms of crores of rupees while in terms of resources saving i.e. coal, HSD, etc. which are priceless adventure.

At last one can justifiably conclude that the management of any industry if acts and behaves environmentally, it can become a symbol of industrial environmentalism.

Table 2. Water and Raw Material Consumption at Vikram Cement

I. Water Consumption (Kl/day)

	1995-1996	1996-1997	1997-1998	1998-1999	1999-2000	2000-2001	2001-2002	2002-2003
Industrial + Cooling	1145	933	1071	956	1509	1505	1670	1681
Domestic	1293	1269	1468	1338	634	649	506	463

Name of the product	Water consumption / unit of product (Kl/Ton)							
	1995-1996	1996-1997	1997-1998	1998-1999	1999-2000	2000-2001	2001-2002	2002-2003
Cement	0.17	0.174	0.173	0.163	0.2116	0.2230	0.2352	0.2350

II. Total Raw Material Consumption

Years	Consumption of raw material/ unit of clinker (per Ton)					
	Limestone	Laterite	Red Ochre	Fly ash	Total	*Gypsum
1995-1996	1.4288	0.0295	--	--	1.4583	0.0518
1996-1997	1.4241	0.0454	0.0093	--	1.4817	0.0529
1997-1998	1.4264	0.0439	0.0500	--	1.4807	0.0500
1998-1999	1.4200	0.0512	0.0115	--	1.4827	0.0500
1999-2000	1.4170	0.050	0.0115	--	1.4995	0.0518
2000-2001	1.4037	0.0650	0.0143	-	1.4830	0.0488
2001-2002	1.3980	0.0339	0.0459	0.0387	1.5165	0.0455
2002-2003	1.3781	0.0383	0.0592	0.0719	1.5475	0.0495

Table 3. Water and Raw Material Consumption at Aditya Cement

I. Water Consumption (Kl/day)

	1995-1996	1996-1997	1997-1998	1998-1999	1999-2000	2000-2001	2001-2002	2002-2003
Industrial + Cooling	512.78	451.55	653.53	678.01	1000	1065	1427	1445
Domestic	232.14	288.73	361.46	380.13	401.00	406.00	499.00	505

Name of the product	Water consumption / unit of product (Kl/Ton)							
	1995-1996	1996-1997	1997-1998	1998-1999	1999-2000	2000-2001	2001-2002	2002-2003
Cement	0.39	0.18	0.37	0.23	0.26	0.28	0.32	0.34

II. Total Raw Material Consumption

Years	Consumption of raw material/ unit of clinker (per Ton)					
	Limestone	Laterite	Red Ochre	Fly ash	Total	*Gypsum
1995-1996	1.5660	0.037	0.033	0.002	1.607	0.051
1996-1997	1.4345	0.0418	0.0096	--	1.5172	0.0492
1997-1998	1.4345	0.0481	0.0439	--	1.5366	0.052
1998-1999	1.4220	0.057	0.0439	--	1.4375	0.051
1999-2000	1.4190	0.0410	--	--	1.4356	0.0530
2000-2001	1.3366	0.0564	0.0177	--	1.4580	0.0473
2001-2002	1.3429	0.0246	0.0632	0.0003	1.4800	0.0505
2002-2003	1.3435	0.0215	0.0611	0.0005	1.4266	0.0515

Table 4. Water and Raw Material Consumption at JK White Cement

I. Water Consumption (Kl/day)

	1995-1996	1996-1997	1997-1998	1998-1999	1999-2000	2000-2001	2001-2002	2002-2003
Industrial + Cooling	455.4	487.58	460.00	494	591	497	422	410
Domestic	671.0	444.08	502	571	568	517	605	630

Name of the product	Water consumption / unit of product (Kl/Ton)							
	1995-1996	1996-1997	1997-1998	1998-1999	1999-2000	2000-2001	2001-2002	2002-2003
Cement	1.877	1.762	1.649	1.776	1.790	1.480	1.821	1.830

II. Total Raw Material Consumption

Years	Consumption of raw material/ unit of clinker (per Ton)				
	Limestone	China Clay	Fluorspar	Total	*Gypsum
1995-1996	1.2492	0.2843	0.0025	1.5760	0.0400
1996-1997	1.2210	0.3062	0.0040	1.5742	0.0430
1997-1998	1.2569	0.2637	0.0030	1.5713	0.0470
1998-1999	1.2709	0.2568	0.0029	1.5740	0.0434
1999-2000	1.2241	0.3251	0.0026	1.5380	0.0420
2000-2001	1.2261	0.3267	0.0037	1.5370	0.0400
2001-2002	1.2371	0.2651	0.0036	1.5393	0.0369
2002-2003	1.2375	0.2560	0.0034	1.4965	0.0375

*Gypsum consumption taken per unit (Ton) of cement

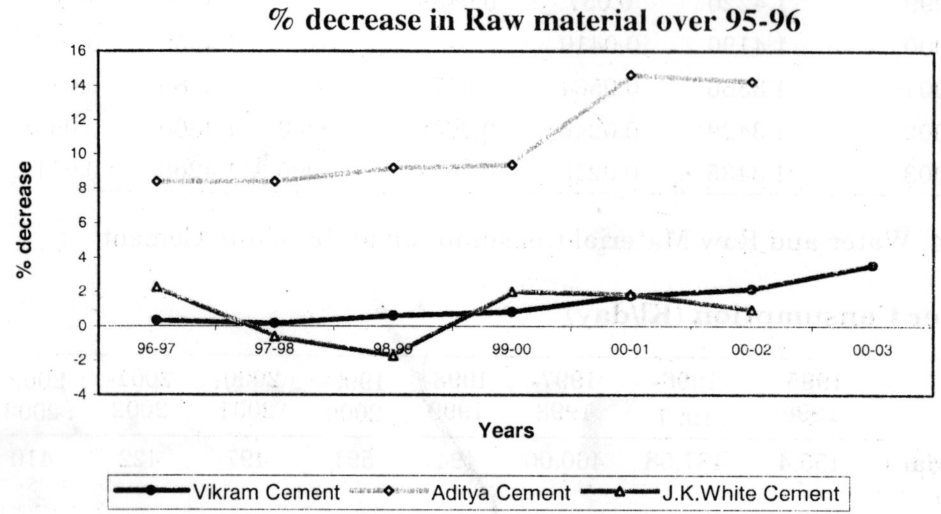

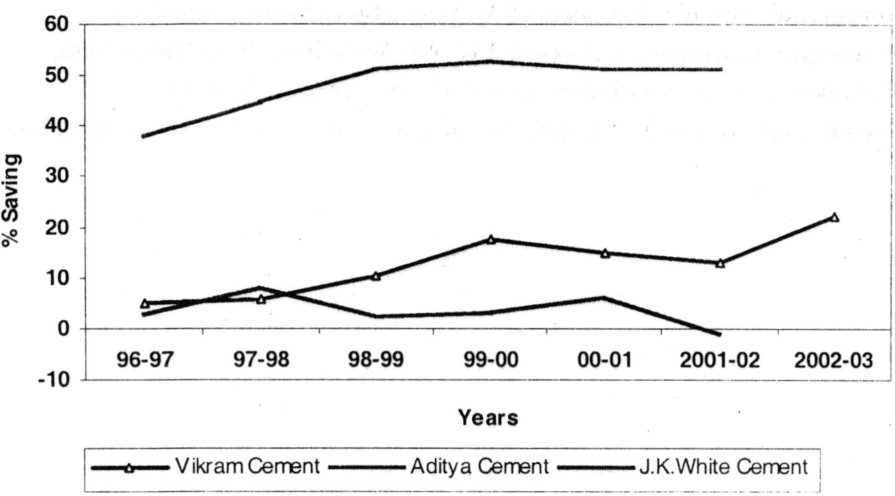

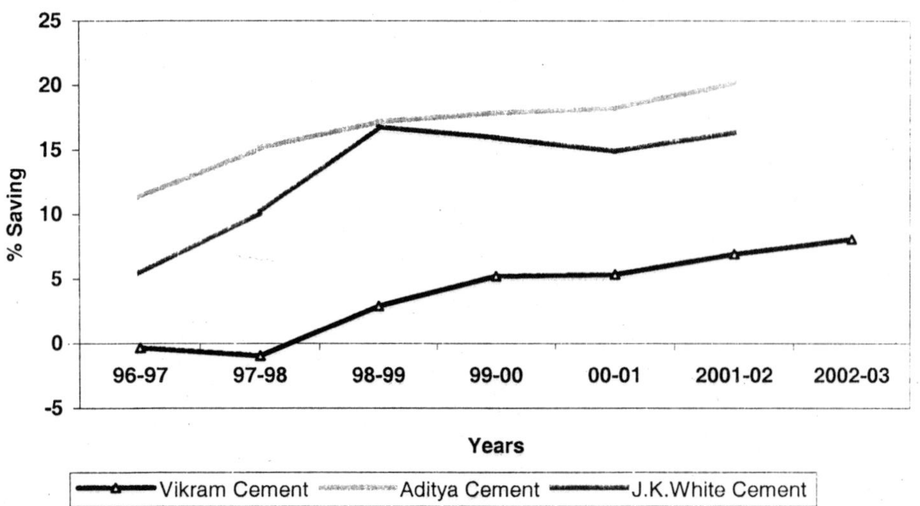

Acknowledgement

Authors extend thanks to Vikram Cement, Aditya Cement and J.K. White Cement for sponsoring the eco-auditing projects.

REFERENCES

[1]. Environmental statement reports of Vikram Cement from 1995-2003.

[2]. Environmental statement reports of Aditya Cement from 1995-2003.

[3]. Environmental statement reports of J.K. White Cement from 1995-2003.

[4]. http://classes.aces.uniuc.edu/ansci308/environmentalaudit.html.

[5]. Outline of environmental auditing. Ministry of Environment. Govt. of India.

CHAPTER - 17

Eco-degradation around Dolomite grinding units at Tehsil Alirajpur, district Jhabua (MP)

D.M. Kumawat, H.K. Ram and K. Haryani

Institute of Environment Management and Plant Sciences, Vikram University, Ujjain (MP) 456 010

ABSTRACT

Particulate pollution is a resultant of both natural as well as human activities. The human activities include industrialization, transportation, building construction, stone crushing and open cast mining etc. The particulate pollution affects all the three major sinks *i.e.* air, water and soil. It in general affects the entire environment and the adjoining ecosystems of the region. Various small and large dolomite extraction and crushing units are in operation at Tehsil Alirajpur District Jhabua (MP). Even though the dolomite is not chemically much harmful to the nature but a continuous particulate load in the region is disturbing various ecosystems to a great concern.

The disturbance caused by these industries is evident through the study made in the present investigation. The vegetation response and soil analysis studies suggest that a significant deterioration is being caused by these units.

Keywords : Eco-degradation, Vegetation, Dolomite units, Alirajpur, Jhabua (M.P.)

Particulate pollution is one of the major concern these days, may it be the particles from cement industries, thermal power plants, auto vehicular exhausts or even the road side dust. The stone quarry and processing of various ores and mineral rocks (crushing and grinding etc.) do cause a particulate load in and around the processing area. Even though it is taken to be a local problem but it may have adverse effects on the surrounding air, water and soil. In the present study an impact assessment of the particles generated through dolomite industries at, tehsil Alirajpur, District Jhabua (MP) on the surrounding soil and vegetation was made.

Dolomite is a double carbonate of calcium and magnesium [$CaMg(CO_3)_2$]. Theoretically, it contains 54.35% $CaCO_3$ and 45.65% $MgCO_3$. In other words, it contains 30.4% CaO; 21.7% MgO and 47.9% CO_2. Dolomite at Alirajpur is of sedimentary origin and is supposed to have been formed due to simultaneous chemical precipitation of calcium and magnesium carbonate during the drying of the sea lagoons and inland seas or by the chemical action of the enriched sea water magnesia on limestone. Dolomite contains usually impurities like silica, alumina and iron oxide. For commercial purpose percentage of combined impurities should not go beyond 7%, above which it becomes unsuitable for industrial use. It is then used only for road ballast, building stones, flooring chips etc. Metamorphosed dolomite forms commercial marble. Reports regarding effect of different particulates on soil and vegetation have been made by various workers in the past but specific studies on dolomite particulates are few.

17.1 MATERIALS AND METHODS

Study area: The present study was carried out at Alirajpur tehsil of Jhabua district in Madhya Pradesh State (India) (Fig. 1), where about 20 different small and large scale dolomite processing units are in operation. These units cover an area of about 10-15 km of Alirajpur tehsil. A few units have been in operation for more than 18 years and have affected the adjoining ecosystems to a great extent. The area has a dry sub humid climate with average rainfall 50-75 cm and possesses a medium black soil. It is situated at the southwest range of Vindhyachal hills with several wide and steep hillocks.

The latitude and longitude of the study area is 21°53' to 22°35' N and 73^0 30' to 75^0 01'E respectively.

Site Selection : Monitoring and sampling of the affected region was carried out at three different sites in down wind direction during winter season i.e. October to February. The sites were located at a distance of 0 to 5 km (Site I), 5 to 10 km (Site II) and 10 to 15 km (Site III) from the point source. A reference site was also selected about 20 km in up wind direction i.e. towards North East. The wind direction during the study period varied from West to South West.

Soil Sampling and Analysis: Soil samples at different selected sites around the industrial area were collected from 8-10 locations and were pooled for further analysis. Surface soil of about 1cm was removed and the soil samples were collected from a depth of 7-10cm randomly at each site and composite samples were prepared. Various parameters were analyzed following standard protocols.

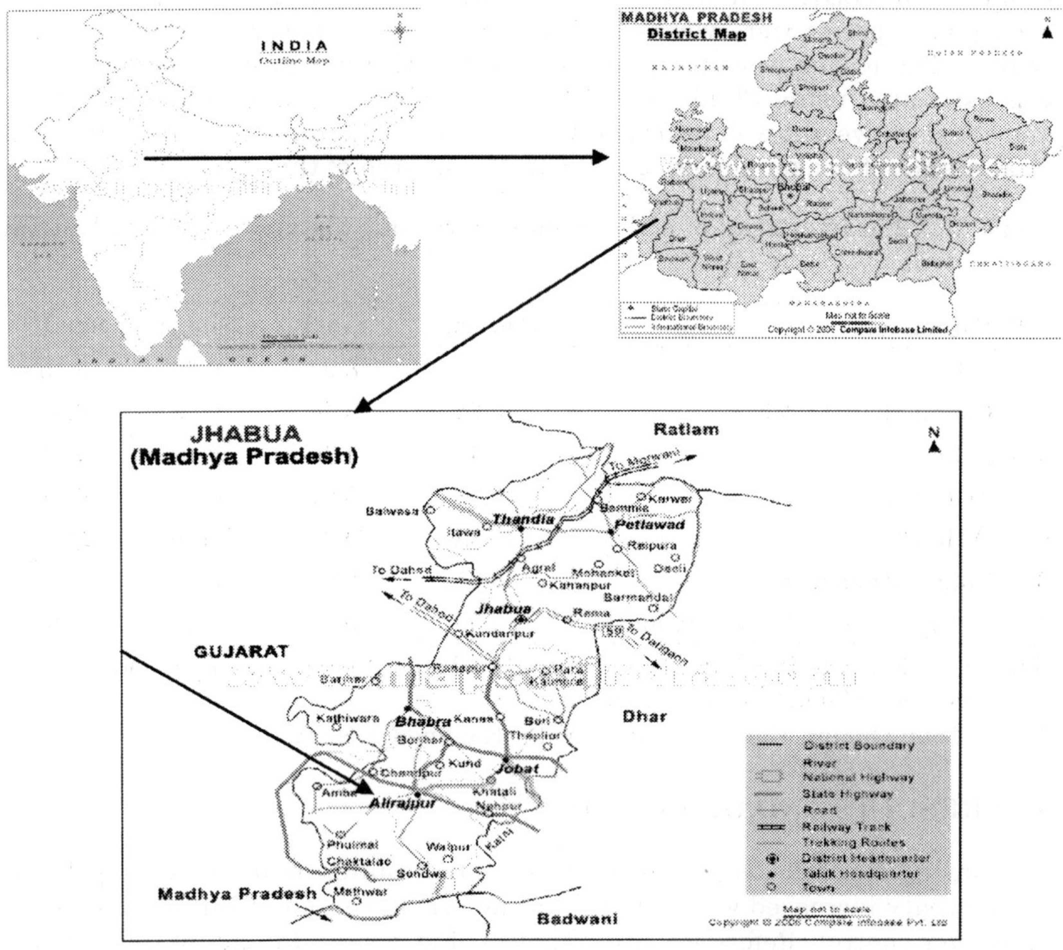

Figure 1. Location Map of Tehsil Alirajpur

Soil pH and Conductivity: The pH of the soil was measured by Systronics digital pH meter (1:5 soil suspensions) and the soil suspension was filtered for measuring conductivity of the soil using Conductivity meter. Total Organic Carbon (TOC) – Walkely and Blacks, (1934); Water Holding Capacity (WHC)-Thompson, (1957); Total Nitrogen-Jackson, (1962) and Dehydrogenase activity –Greaves *et al.* (1978).

Plant sampling and analysis : After a survey of the phyto-sociological status of different selected sites, a few plants which were common and had higher densities at all the three sites were selected (*Mangifera indica* L., *Madhuca indica* Gmel., *Ficus religiosa* L., *Butea monosperma* L. and *Calotropis procera* R.Br.) for certain physiological and biochemical studies.

The parameters analyzed were; Stomatal conductance (cm/s)- using Steady State Porometer LICOR-1600, USA; Total Chlorophyll content (mg/g)- Arnon (1949); Foliar Protein Content (mg/g) – Lowry *et al.* (1951); Leaf area dry weight ratio (LDR) – Rao (1987) and Total Dust Fall (g/m^2) – 10 leaves of each plant species were washed with 50ml distilled water in a pre weighed beaker. The leaves were then surface dried with filter paper and their area was estimated using Planimeter. The leaf wash was evaporated till dryness and the final weight was recorded. For total dust fall, difference in initial and final weight of beaker was divided by total leaf area of the species.

Table 1 : Characteristics of soil samples at various distances from industrial zone

S.N.	Parameters	Reference	Site I	Site II	Site III
1	pH	7.94	8.92	8.50	8.15
2	Conductivity (mMho)	0.24	0.44	0.38	0.32
3	Salinity (%)	2.40	4.40	3.80	3.20
4	TOC (%)	0.98	0.62	0.72	0.78
5	WHC (%)	25.00	19.00	21.00	23.00
6	Total Nitrogen (%)	0.05	0.04	0.05	0.05
7	Dehydrogenase Activity (μg/ml)	0.59	0.40	0.44	0.48

Site I = 0-5 km; Site II = 5-10 km and Site III = 10-15 km from the point source; Reference site = 20 km upwind direction.

TOC = Total Organic Carbon; WHC = Water Holding Capacity

17.2 RESULTS AND DISCUSSION

Among the selected parameters for soil, results, indicates that pH and conductivity increased while other parameters decreased at different selected sites over control (Reference site) (Table 1, Fig. 2 a-e). Impact of dust on soil is directly correlated with distances, dust level decreases with increase in distance from the point source. In case of dolomite dust particles being heavier caused more harm in the adjoining regions (upto 5km). Soil pH was slightly alkaline and it increased by 0.94 units at Site I, 0.56 units at Site II and 0.21 units at Site III respectively (Fig 2a). While in conductivity the increase was 0.20, 0.14 and 0.08 units at Site I, II and III respectively over control (Fig. 2b). The nutrient levels are also affected due to particulate load resulting in alteration of C and N content of the soil. The maximum decrease in organic carbon is 36.71% at site I while minimum at site III i.e. 20.28% (Fig. 2c). This reduction in organic carbon could be due to excessive quantities of dolomite dust in the area. Water Holding Capacity of all soil samples at various distances decreased when compared to reference site. The percent decrease at site I, II and III were 24%, 16% and 8% respectively (Fig. 2d). Total Nitrogen content of the soil samples at different selected sites also decreased. The change in total nitrogen content ranged between 4-19% at the three selected sites (Fig. 2e). Another important parameter i.e. dehydrogenase activity, which reflect the microbial activity status of the soil also changed due to particulate deposition. The decrease in dehydrogenase activity ranged between 0.11 to 0.19 units (Table 1).

17.2.1 Plant response

The foliar dust reveals the particulate load in the area. Dust load was more on plant species growing at site I as compared to other sites. The maximum dust content was observed in *Butea monosperma* followed by *Mangifera indica, Calotropis procera, Ficus religiosa* and *Madhuca indica* (Table 2; Fig. 3a). All other physiological and biochemical parameters showed an over all decrease when compared to their respective controls (Fig. 3b-e). The decrease as obvious was more in vegetation at Site I. Steem (1986) also reported about significance of dry deposition of dolomite particles on vegetation.

Table 2: Total Dust fall (g/m^2) of different plants at different site in industrial zone

S.N.	Name of Plant Species	Reference	Site I	Site II	Site III
1	*Madhuca indica*	5.24±0.21	8.47±0.18	6.47±0.23	4.12±0.18
2	*Mangifera indica*	3.98±0.43	10.78±2.16	8.96±0.94	6.43±0.87
3	*Ficus religiosa*	3.47±0.10	6.24±0.34	4.86±0.47	3.17±0.14
4	*Calotropis procera*	4.24±0.16	10.17±0.41	8.34±0.32	7.12±0.28
5	*Butea monosperma*	2.41±0.71	6.83±0.64	5.16±0.75	3.98±0.57

± Standard deviation

Table 3: Stomatal Conductance (cm/s) of different plants at different site in industrial zone

S.N.	Name of Plant Species	Reference	Site I	Site II	Site III
1	*Madhuca indica*	0.211 ±0.079	0.151 ±0.074	0.162±0.075	0.183 ± 0.068
2	*Mangifera indica*	0.382 ±0.067	0.266 ±0.079	0.284±0.098	0.349 ± 0.091
3	*Ficus religiosa*	0.296 ±0.064	0.221±0.062	0.229±0.059	0.259 ± 0.031
4	*Calotropis procera*	0.229 ±0.059	0.181±0.070	1.178±0.081	0.192 ± 0.072
5	*Butea monosperma*	0.248 ±0.054	0.208±0.075	0.207±0.072	0.229 ± 0.067

± Standard deviation

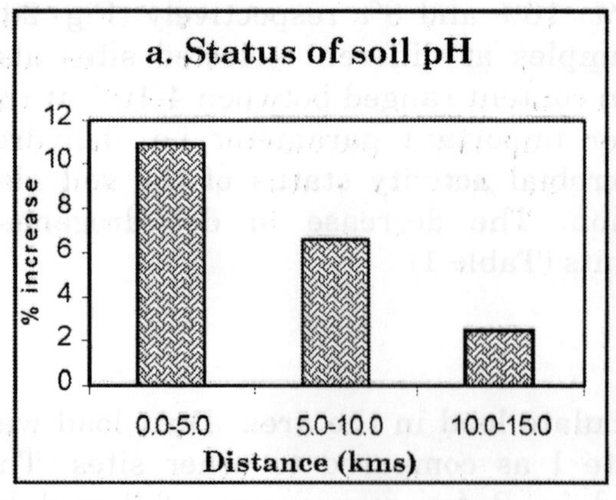

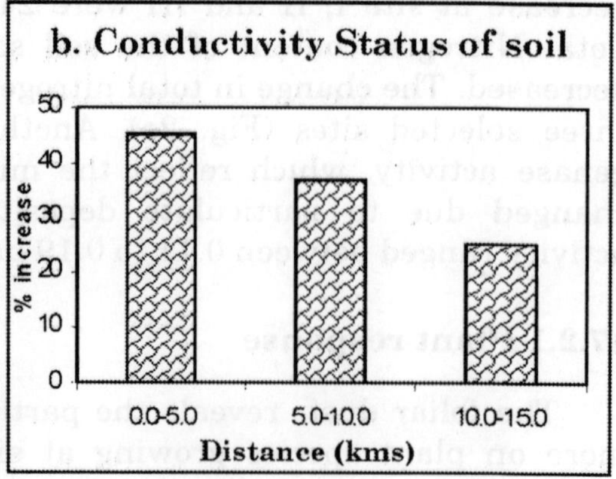

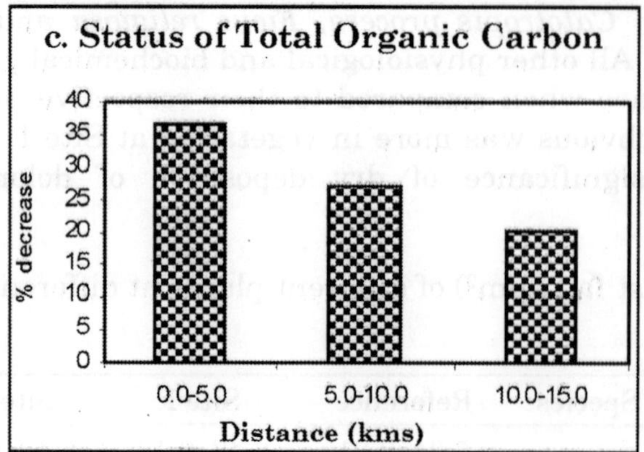

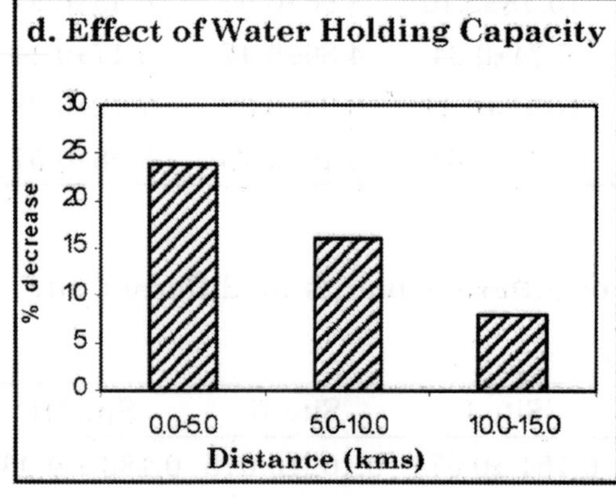

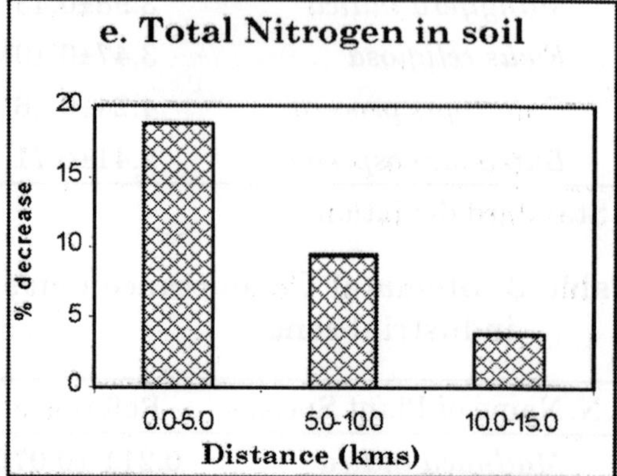

Fig. 2. Soil characteristics as affected by Dolomite Inudstry Pollution.

The stomatal conductance decreased in all plant species at all the selected sites. The decrease in stomatal conductance indicates the stress in the environment. The maximum decrease was in *Mangifera indica* followed by *Madhuca indica, Ficus religiosa, Calotropis procera* and *Butea monosperma* (Table 3, Fig. 3b).

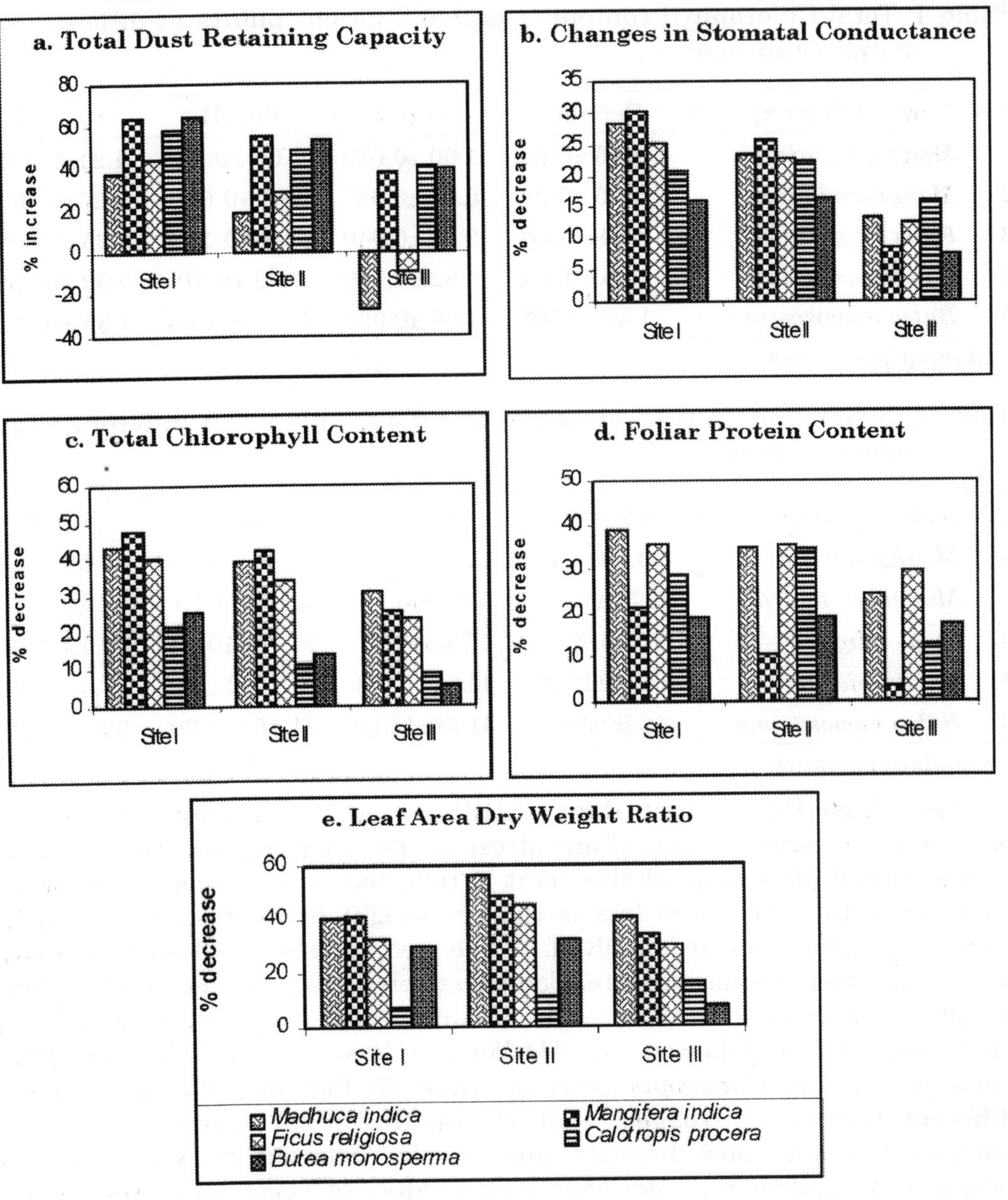

Fig. 3. Effect on vegetation by Dolomite Industry Pollution.

Total chlorophyll content study reveal that maximum pigment reduction was in *Mangifera indica* followed by *Madhuca indica, Ficus religiosa, Butea monosperma* and *Calotropis procera* (Table 4, Fig. 3c). Foliar protein content decreased maximum in *Madhuca indica,* followed by *Ficus religiosa, Calotropis procera, Mangifera indica* and *Butea monosperma* (Table 5, Fig. 3d).

Table 4. Total Chlorophyll content (mg/g) of different plants at different site in industrial zone

S.N.	Name of Plant Species	Reference	Site I	Site II	Site III
1	*Madhuca indica*	1.76 ±0.09	1.00 ±0.012	1.06 ±0.14	1.21±0.070
2	*Mangifera indica*	1.70 ±0.08	0.89 ±0.037	0.98 ±0.059	1.26±0.090
3	*Ficus religiosa*	1.64±0.092	0.98 ±0.040	1.08 ±0.140	1.25 ±0.080
4	*Calotropis procera*	1.05 ±0.170	0.82 ±0.110	0.93 ±0.10	0.96 ±0.038
5	*Butea monosperma*	1.43 ±0.065	1.06 ±0.025	1.23 ±0.070	1.35 ±0.075

± Standard deviation

Table 5. Foliar Protein content (mg/g) of different plants at different site in industrial zone

S.N.	Name of Plant species	Reference	Site I	Site II	Site III
1	*Madhuca indica*	48.20±7.75	29.35±8.60	31.40±10.10	36.55±9.70
2	*Mangifera indica*	54.30±12.60	42.79±6.75	48.44±7.75	52.40±5.48
3	*Ficus religiosa*	42.98±6.75	27.65±8.21	27.70±10.14	30.15±9.55
4	*Calotropis procera*	43.00±12.20	30.64±10.21	28.20±11.20	37.40±9.81
5	*Butea monosperma*	63.60±8.96	51.58±14.20	51.45±15.20	52.30±8.37

± Standard deviation

Leaf Area Dry Weight Ratio (LDR) is one of the important growth parameter for measurement of any stress in the environment. Any alteration in the normal physiology of the plant is reflected in dry weight production. Here, more the LDR ratio less is the dry weight production and therefore more is the plant affected. LDR of all the tree species at all sites increased when compared to reference site. Comparatively LD ratio was slightly higher at site I as compared to site II. Maximum LD ratio was observed in *Mangifera indica* followed by *Madhuca indica, Ficus religiosa, Butea monosperma* and *Calotropis procera* (Table 6, Fig. 3e). Overall study of different parameters suggest that of the five plant species studies, *B. monosperma* was most tolerant and sturdy, followed by *C. procera, F. religiosa, Mangifera* and *Madhuca indica*. More of vegetation study is still required to reach at some conclusive remark.

Dolomite mining at Bhutan is not only destroying its mountains but also the famous West Bengal-Darjeeling tea plantations in India (Majumdar, 2007). Air borne dolomite dust is choking the plants and thereby affecting the tea production. On the basis of the above results it is clear that particulate concentration/level of dolomite in the study area is sufficient enough to change/alter the vegetation as well as soil characteristics of the region. In the near vicinity of the point source a thick green belt with higher dust retaining capacity plants should be developed. Eco-degradation of the area can be minimized by sustainable utilization of grinding activity/practice. Also, there is a need to install various dust control devices / measures such as Bag filters, Cyclone, ESP etc. by the industries. Even though water and human health aspect was not included in the present study but still a regular health checkup of workers in the area should be carried out and environment awareness programs for the workers and local people should also be arranged. According to Mishra *et al.* (2004) mining operations not only disrupt the natural ecosystems in a significant way but also bring about a number of health hazards and socio-economic disturbances to the local people. Regular soil testing (especially in adjoining agricultural land) and proper soil amendments should also be carried out in the region to maintain better productivity.

Table 6. Leaf Area Dry Weight Ratio of different plants at different site in industrial zone

S.No.	Name of Plant species	Reference	Site I	Site II	Site III
1	*Madhuca indica*	78.17	176.90	130.99	131.91
2	*Mangifera indica*	67.66	131.31	115.79	103.30
3	*Ficus religiosa*	98.71	179.63	146.90	142.15
4	*Calotropis procera*	129.75	146.43	139.53	155.41
5	*Butea monosperma*	113.84	167.54	163.11	123.65

Acknowledgements

The authors extend their sincere thanks to Head of the Institute for providing necessary research facilities.

REFERENCES

Arnon, D.I. 1949. Copper enzymes in isolated chloroplasts. Polyphenol oxidase in *Beta vulgaris*. *Plant Physiol*., 24 : 1-15.

Greaves, M.P., Copper, S.L., Davies, H.A. Marsh, J.A.P. and Wingfield, G.I. 1978. Methods of analysis for determining the effects of herbicides on soil micro-organisms and their activities. Technical Report, Agricultural Research Council Weed research Organization, No. 45 pp.55.

Jackson, M.L. 1962. *Soil Chemical Analysis*. Asia Publishing House, Bombay.

Lowry, O.H., Rosenbrough, N.J., Farr, A.L. and Randall, R.J. 1951. Protein measurement with the folin phenol reagent.. *J. Biol Chem*. 193 : 265-275.

Majumdar, B. 2007. Mines destroy Bhutan's mountains, affect India. http:// www. Boston.com/news/world/asia/articles/2007/06/05 mines_destroy_bhutans_mou...

Mishra, P.C., Sahu, H.B. and Patel, R. K. 2004. Environmental pollution status as a result of limestone and dolomite mining – a case study. *Pollution Res.* 23(3) : 427-432.

Rao, M.V. 1987. Air pollution stress and plant response - A study at Dewas area. M.Phil. Thesis, Vikram University, Ujjain MP. India.

Steem, B. 1986. Dry deposition of 4-50Um dolomite particles on vegetation, flat surfaces and deposition gauges. *Atmos. Environ.* 20(8) : 1597-1604.

Thompson, L.M. 1957. *Soil and Soil Fertility - The Physical Properties of Soils*. Mc Graw Hill Book Co. Inc., New York.

Walkley, A. and Blacks, I.A. 1934. An examination of the Degtjareff method for determination of soil organic matter and a proposed modification of the chromic acid titration method. *Soil Science*, 37 : 29-38.

CHAPTER - 18

Changing atmosphere - the effects of global warming

Bindoo Bhatt and A.K. Gupta

Department of Geography, Faculty of Science,
The Maharaja Sayajirao University of Baroda, Vadodara-390 002
e mail: bindoobhatt@rediffmail.com; e mail: major_amitgupta21@yahoo.co.in

ABSTRACT

Energy from the Sun drives the earth's weather and climate and heats the earth's surface; in turn, the earth radiates energy back into space. Atmospheric green house gases (carbon dioxide, water vapour and other gases) trap some of the outgoing energy, retaining heat. Life known today would have not been possible without this green house effect. It is due thanks to the green house gases; the earth's average temperature is more hospitable; about 60°F. However, problems may arise when atmospheric concentration of green house gases increases.

According to the scientific evidences, the earth's temperature has risen by 1°F in the past century with accelerated warming during the past two decades There is new and stronger evidence that most of the warming over the last 50 years is attributed to human activities. Human activities have altered the chemical composition of atmosphere through buildup of green house gases. The heat trapping property of these gases is undisputed although uncertainties exist. What has changed in past few hundred years is additional release of carbon dioxide by human activities. Increased agriculture, deforestation, landfills, industrial production and mining also contribute a significant share of emission of green house gases.

Estimating future emission will depend on demographic, economic, technological, policy and institutional development and therefore is difficult. According one of such estimate, by 2100, in absence of control policies for emission, carbon dioxide concentration is projected to be 30-15% higher than today's levels. One of the most current and widely discussed threats that could lead to ultimate end of the existence of Earth and man, is global warming and the devastating effects. Scientists have been concerned about how fast the earth is heating up and how warming will affect crops and climatic condition. Several current trends clearly demonstrate that global warming will directly have impact on rising sea levels, melting of ice caps and world wide climatic changes. Thus, it can be concluded that global-warming represents a fundamental threat to all living things on earth. Global warming is regarded as the greatest challenge of 21st century. The study is conceptual in nature and deals with the degree of destruction caused by global warming, factors contributing to warming and finally the suggestions to decrease the rate of global warming.

Key Words: Atmosphere, Green house, Global warming, Temperature, Human activities.

The prediction of climate change due to human activities began with Svante Arrhenius, a Swedish Chemist in 1896. He realized that the industrial revolution that had then started and carbon dioxide released was increasing as world's consumption of fossil fuels increased more rapidly. He then had predicted that if atmospheric carbon dioxide doubled, earth would become several degrees warmer. He was referring to potential modification, a phenomenon called green house effect and global warming was understood, but little attention was paid to the prediction that had no apparent consequences for people living at that time.

Post World War II industrialization has caused dramatic jumps in amount of carbon dioxide in the atmosphere and prospects of considerable change becomes more real and threatening. An estimate of carbon dioxide in the atmosphere in the next century shows a substantial increase and as IPCC projects global warming of 1.4° to 5.8° by 2100. Several current trends clearly demonstrate that global warming will directly have impact on rising sea levels, melting of ice caps and world wide climatic changes.

Scientists have identified that our health, agriculture, water resources forests, wildlife and coastal areas are all vulnerable to the changes that global warming may bring. There are possibilities that warmer world could lead to more frequent and intense storms.

An attempt has been made to reveal some of the facts with the degree of destruction caused by global warming, factors contributing to warming and finally the suggestions to decrease the rate of global warming.

18.1 THE CONCEPT: "GREEN HOUSE"

A shield of atmospheric gases, primarily composed of nitrogen (N_2) 78%, and Oxygen (O_2) 21% and remaining are trace gases (CO_2) that surrounds the Earth. Many of these trace gases are called green house gases. These gases allow direct sunlight (relative short wave energy) to reach unobstructed and heat the earth's surface. Long wave radiation emitted by the warm surface of the earth is absorbed partially and then re-radiated by a number of trace gases-particularly CO_2 and water vapour-in the cooler atmosphere above. Atmospheric green house gases (carbon dioxide, water vapour and other gases) trap some of the outgoing energy, retaining heat.

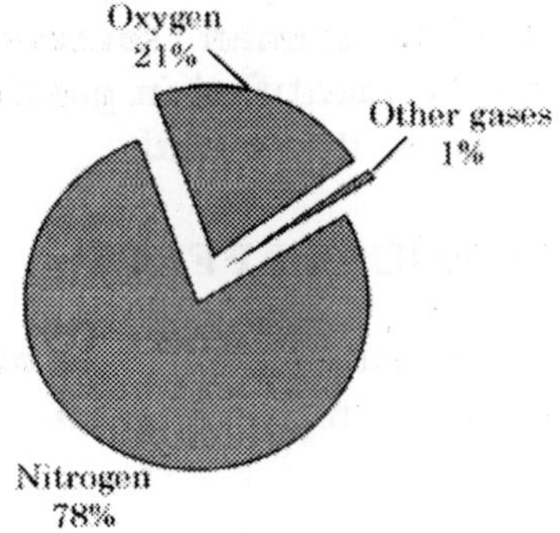

Fig. 1. Showing Composition of the Atmosphere

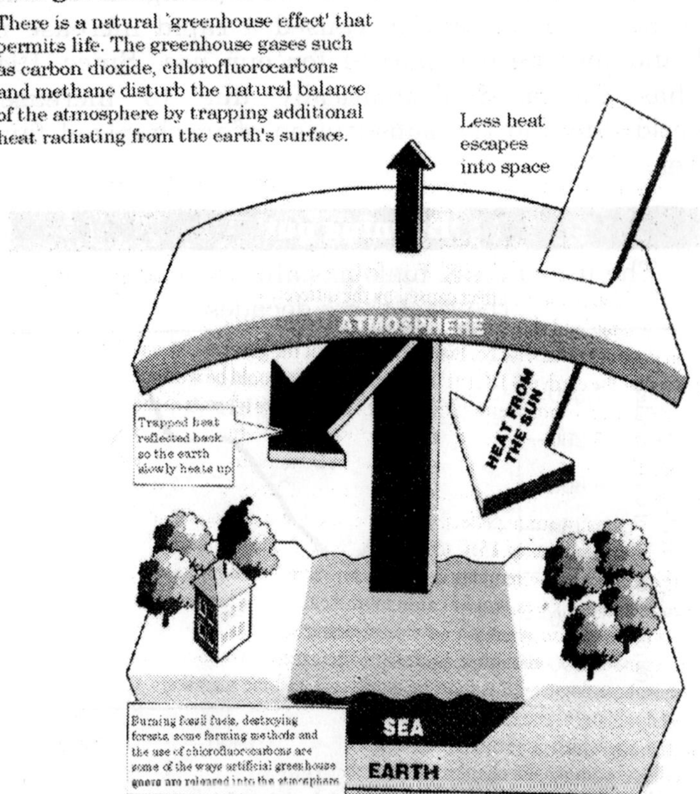

Fig. 2. Atmosphere green house gases (Source : Global Warming, Paul Brown)

Life known today would have not been possible without this green house effect. It is due thanks to the green house gases; the earth's average temperature is more hospitable; about 60°F. The green house effect is the term used to describe the retention of heat in the earth's lower atmosphere. In colloquial usage it often refers to the ***"enhanced"*** global warming which is considered likely to occur because of increasing concentrations of certain trace gases (green house gases) in the atmosphere. A large part of increased concentration is attributed to human sources. Although most of the anthropogenic emissions can be identified in particular countries, their effect is not confined to those countries,-it is global.

18.2 CAUSES OF GREEN HOUSE EFFECTS

Fossil fuels and the forests cause about half of the green house warming. Other gases (CFC, methane, nitrous oxide, tropospheric ozone) are responsible for the rest.

The intellectual power and effective use of technology and industrial revolution thereby changed the environment. Technology is partly responsible for explosive population growth and resulting in damage to earth's resources. The industrial revolution caused a rapid increase in population growth as oil and gas were exploited for our use. Since 1945 petroleum consumption has increased dramatically due to increased usage of automobiles worldwide and the substitution of mechanized farm machinery for animal power.

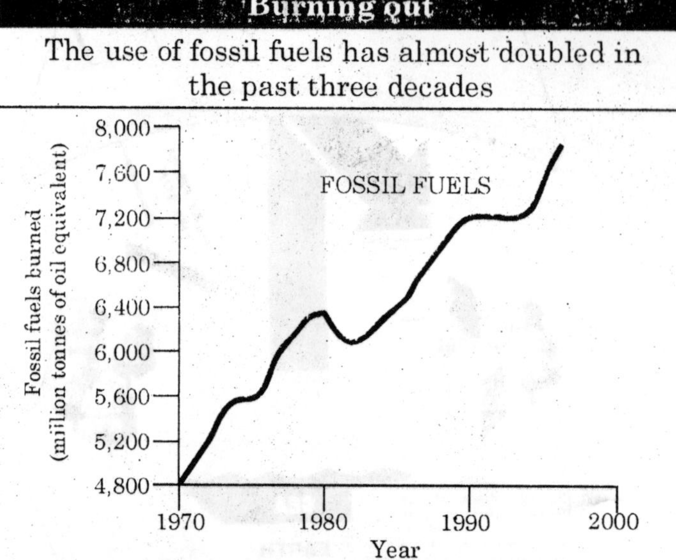

Fig. 3. Graph showing almost doubled in the past three decades (Source: Earth Day 2000,World Watch)

Trees play a unique role in the global carbon cycle; they are the largest land-based natural mechanism for removing CO_2 from the air. An acre of forest will absorb about 2.8 tons of CO_2 and 10 times more than an acre of crop land or grass land. Conversely, when trees are burned or dead/ decomposed, the locked carbon is released. Every day 5500 acres and over 50 million acres every year of rain forest are destroyed. The shrinking world forests are unable to absorb all CO_2 created by human while burning fossil fuels. According to FAO, the total net rate of forest loss was of 9.4 million hectares per year for 1999-2000. Recent calculations suggest that natural forest can store approximately 430 billion metric tones of carbon, more than that released by fossil fuel burning over the next seventy years.

Table 1. Annual Deforestation

REGIONS	ANNUAL DEFORESTATION (Million hectare per year)
Africa	2.3
Asia	2.5
Latin America	4.4
Pan-tropical	9.2

Global CO_2 levels have approximately increased by 0.4% each year, levels not experienced on the planets for millions of year.

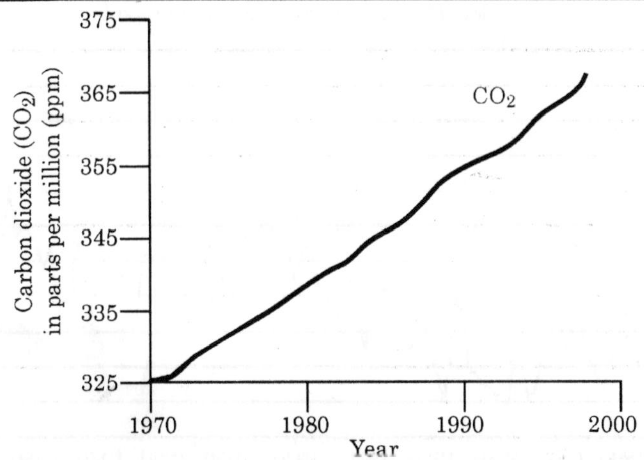

Fig. 4. Graph showing increased CO_2 12% in the last thirty years (Source: Earth Day 2000, World Watch, Vol. 13, No. 2)

18.3 THE GREEN HOUSE EFFECT: THE CURRENT TRENDS

The global effect of the green house cannot be directly predicted. However, there can be direct connection between certain natural phenomenon that supports the idea that some thing is changing. A number of indicators suggest that warming due to increased levels of green house gases is indeed seen since 1980. Ice core samples shows that carbon dioxide and methane levels are higher than any time in the past 160,000 years (*Nuclear Issues Briefing Paper 24 January, 2003).*

18.3.1 Weather and Climate

Weather disasters are increasing Instability in the atmosphere may also increase rapid changes and severe weather. Throughout the past century, global surface temperature have increased at a rate of 0.6°C/century but the trend dramatically increased to rate approaching 2.0°C/century in last 25 years. Global temperatures in 2001 were 0.51°C above the long term (1880-2000), thus making 2001 the second warmest year, next to 1998 (0.75°C). The Northern Hemisphere records 0.60°C and the Southern Hemisphere of 0.43°C. Notable temperature extremes during 2001 included harshest winters in decades across Siberia and Mongolia. Extreme minimum temperatures as low as -50°C were recorded in Russia in mid-January and maximum of 50°C in parts of Pakistan during early May, killing dozens. And a study of long-term European temperature published (Science, March 2004) showed that 2003 was the hottest summer in last 500 years. The climatic models predict faster heating of northern continental areas will have drier summers soils, extensive evaporation of ground moisture, hotter and cloudless summers.

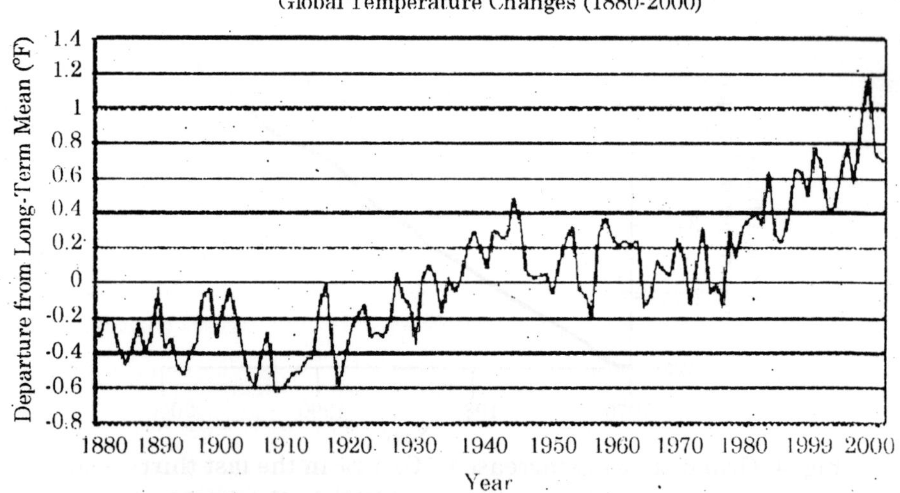

Fig. 5. Graph showing Global temperature Change

According to the scientist at School of Oceanography, Jadavpur, Sea Surface Temperature (SST) of Bay of Bengal increases by 0.02°C, which can have impact on intensity of cyclones. An increase in SST by 1°C leads to ten fold increased intensity of cyclones. The recent disaster of Hurricanes, Katrina, Rita, Wilma, El Nino, all directs a serious question to the meteorologists. Meteorologists have already seen an increase in severity of storms, rainfall and floods. Torrential rains and swollen rivers have caused chaos across eastern Europe, erratic rains in India, Pakistan and neighboring countries and the severe droughts as National center for Atmospheric Research, Colorado, USA, have increased by 30% in 2000s from 12% in 1970s.

Annual Global Precipitation Anomalies
January - December, 1900-2000

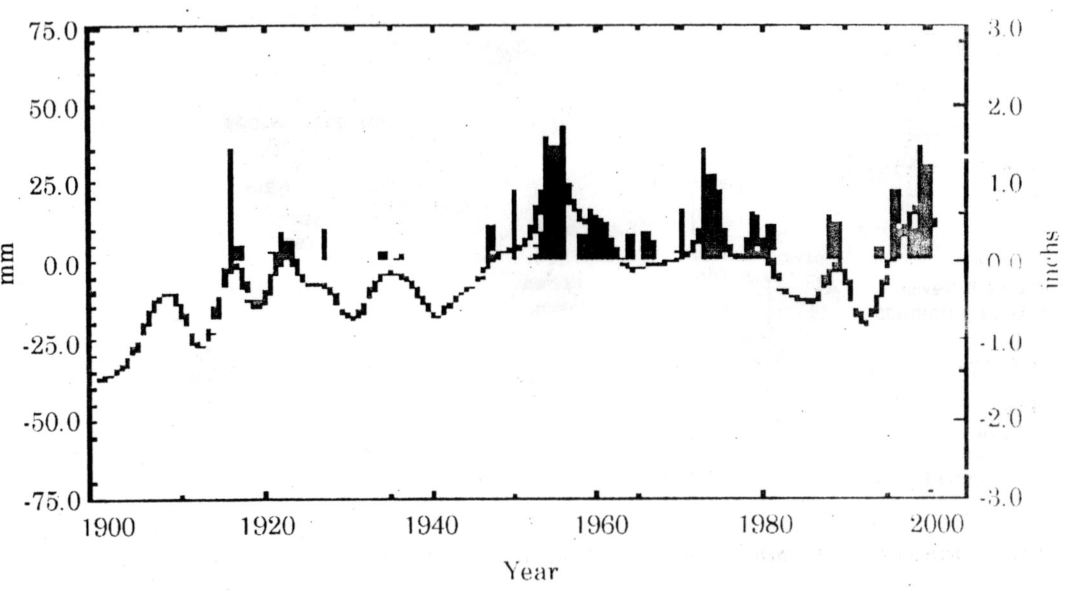

Fig. 6. Graph showing Annual Global Precipitation Anomalies

18.3.2 Sea level rise and melting glaciers

The evidence is that sea level is currently rising quite fast by historic standards. The problem of sea level change is still more complex science. There are two main reasons for rising oceans: the thermal expansion of water as the world is getting warmer and the melting of ice above sea level in mountain glaciers and at the poles. The range of ice around both poles continues to shrink, as it melts. Professor Pelto of the North Cascade Glacier Project reports that west Antarctica Ice Sheets have shrunk by 25% over past

25 years Antarctica Sea Ice is reported to be 33% thinner and 500 km further north than usual.

The melting of ice in Arctic especially release of vast store of methane hydrate now locked under permafrost and in sediments of ocean floor. Together with oxidation of high latitudes release of these carbon stores could add as much CO_2 to the atmosphere. Another factor would be snow and ice has a high albedo, they reflect energy when reduced it will increase energy absorption and atmospheric warming. It is observed that since the year 1979 snow cover is decreasing by 9% every decade and if this continues by the year 2100 Arctic will have no snow cover at all.

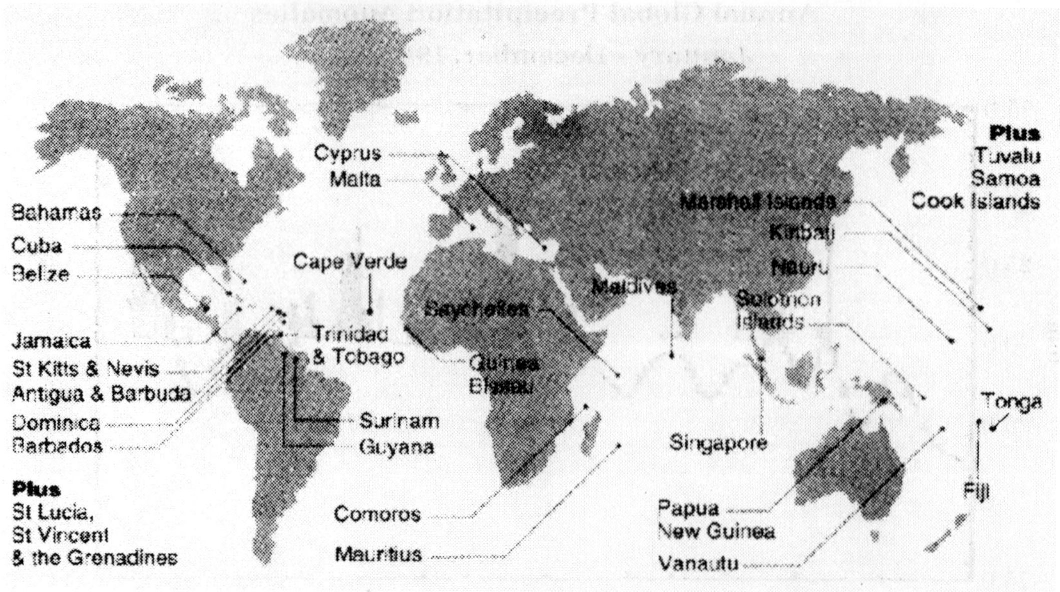

Fig. 7. Map showing Sinking of Small Islands (Source: Global Warming, Paul Brown)

The two giant ice sheets of Antarctica and Greenland have most of the planet's fresh water locked in them. Thus, a loss of only small fraction could have a noteworthy effect on sea level. The current estimates for sea level rise are of 15 cm by 2030 and 50 cm by the end of 21st century.

Global warming and consequent sea level rise will bring many problems. Small islands and low level coastal areas are predominantly at risk. Parts of the Maldives, Egypt and Bangladesh would be submerged and made uninhabitable by a rise of 50 cm. Extensive loss of land will occur and many million of people likely to be displaced. Some small islands in Pacific will experience disaster even if sea level rises by a few centimeters.

Global warming is a grave threat that has already caused substantial damage to the world's coral reefs (Science, 2003, 301). Further, it is causing the world's most biologically diverse aquatic system to fail.

The extra heating of water can cause extensive bleaching of corals killing the tiny creatures. The other effects include wave erosion of the coast line, partial inundation during storms and sea water intrusion into fresh ground water which is often the main supply making life more difficult. The economy of many islands is dependent on tourism, loss of which can spell financial disaster.

18.3.3 Health of Human beings

The health effects are divided into two: the direct effects, that is being killed by heat stroke or drowned and the indirect effects: getting nasty dose of cholera, malaria or plague which could be avoided otherwise. Extra heat means that tropical diseases can move north and south to trouble the immune population of the temperate regions.

Table 2. Major Tropical diseases likely to spread with global warming.

Diseases	Vector	Population at risk	Prevalence of infection	Present distribution	Likelihood of altered distribution with warming
Malaria	Mosquito	2,100 million	270 million	(Sub) tropical	+++
Schistrosomiasis	Water snail	600 million	200 million	(Sub) tropical	+
Filariasis	Mosquito	900 million	90 million	(Sub) tropical	+
Onchocerciasis (liver damage)	Black Fly	90 million	15 million	Africa/Latin America	+
African Trypanosomiasis (Sleeping sickness)	Tsetse Fly	50 million	25,000 new case/year	Tropical Africa	+
Dengue Fever	Mosquito	Estimates unavailable	-	Tropics	+
Yellow Fever	Mosquito	Estimates unavailable	-	Tropical south America &Africa	+

+++Higher +Lower (Source: Environmental Science, S.C. Santra)

The most susceptible groups are the old and the very young, who, find it difficult to adjust to heat and those with lung illness. The more obvious direct effect is extreme weather actions are the damage by storms, sea level rise, drought and/or erratic rains. The IPCC scientists say that climate change will have significant effects on the distribution and quality of surface water, including increases in both flooding and water shortage. Many of the world's major infectious diseases are caused by parasites and viruses transmitted to human via vectors. Diseases thought likely to increase in range and numbers of cases are malaria, dengue fever, African sleeping sickness, river blindness, elephantiasis, Chagas' diseases and many others.

The WHO estimates that 300-500 million people currently infected with malaria every year, of which two million die annually. This pattern of incidence is set to change dramatically, with malaria spreading widely with the predicted increase of 3°-5°C over most of land surface. Cases of plague have been reported in Russia, North Africa and USA. There are many other less known but are very unpleasant and kill or disable a number of people.

Malnutrition due to decrease in food production will be added threat to health. It is predicted that by year 2060 an extra 40-300 million will be at risk of hunger because of climate change.

18.3.4 Flora and Fauna

Plants reflect the possible response to the climate change to increasing temperature. For instance, a study of plants in Alps shows that some cold loving plants are starting to move to higher and cooler altitudes.

As with humans there are a large number of parasites that affect animals; due to temperature increase are probable to spread. There are evidences that the animals particularly young ones suffer from heat stress, which may be fatal. The threat to fisheries because of loss of the breeding habitat of coastal ecosystem is felt. If climate change induces the movements of ocean currents it could shift the fish population and change their feeding patterns. Scientific observation shows changes in distribution and abundance of zooplanktons and inter-tidal life. Warm water species increase even extended range while cold-water species decline or retreat. American ornithologist reports changing migration and range of hundreds of North American bird species.

18.3.5 Economics

Experts warn that economy can suffer as environmental changes caused by global warming begin to affect coastal properties, tourism and agriculture. Expected changes including sea level rise, hotter temperature, and increased variability in the weather-meaning both droughts and heavy rains would become more frequent.

The impact of climate change on agriculture could result in problems of food security and many threatened livelihood activities upon which the population depends. Changes in certain crops can affect imports and exports. Consumers would suffer from high food prices, producers would have little to share, and agricultural exports dwindle. Within agricultural system, the development of new pest and/or expanded range and greater resistance to control methods are certainly possible but difficult to foresee.

Finally, very little about how a regional and local economy responds to multiple changes. The affects on agriculture, water, forests, recreation etc. can be experience in chorus. History has seen many cases of regions and communities declining and depopulating when a crucial resource is exhausted, a industry on which a community is based fails or fails to keep pace with competitors or other areas are deemed more livable or more fashionable.

In addition, the costs to implement a worldwide plan to cut the production of CO_2 and other gases would cost approximately 3% of the world's total GDP.

18.4 INDIAN PERSPECTIVE

According to Dr. G.B. Pant (Director, Indian Institute of Tropical Meteorology), the fact that concentration of green house gases is increasing remains but what is more serous is the rate at which they are increasing. He also predicts that western coast of country will receive more rains while parts of eastern, central India will receive less rain. The countrywide temperature records suggest an increase in 3°C to 4°C, which is more evident in northern states. Scientists predict that it is expected that extreme heat and cold both will be experienced, with increased night temperature than day. With an increase in the Sea Surface Temperature (SST), the intensity, frequency and magnitude of cyclones in Bay of Bengal have been observed with height of sea level rising to 1.5 to 9 meters. The result will be that the lower coastal areas of Orissa, Andhra Pradesh and West Bengal will submerge. It is believed that some 18,000 years back sea level of India was about 120 meters lower than today.

According to a study done by International Commission for Snow and Ice (ICSI), "The melting of Himalayan glaciers is a burning issue". The Himalayan glaciers are extremely sensitive to changes in climate because they accumulate snow during monsoon and shed it in summers. This is unlike European and other higher latitude glaciers that accumulate snow in winter and recede in summers. This gives them time to build up their snow cover (Hasnain, 2002). Among all Himalayan glaciers Gangotri has been a matter of concern.

Data from Geological Survey of India shows that glaciers have been retreating at an average rate of 19 meters per year. The consequence of such recession is that the flooding of the Himalayan Rivers and mountain lakes are being filled faster, threatening thousands of lives and colossal floods. This was revealed during a study conducted by scientists from United Nations Environment Programme (UNEP) and International Center for Integrated Mountain Development (ICIMOD). Bursting of such lakes and a fast recession rate would cause large scale flooding, and mudslides and eventually drying up of the rivers. This would indirectly have important consequences for the water supplies, hydro-electricity generation and riparian habitats. Many spring in Kashmir have dried as the sources feeding spring are blocked or dried. Source dry up when glaciers retreat with increased temperature over a period due to global warming (Chandrasekharam, 2005).

18.5 OTHER VIEWS: UNCERTAINTIES ABOUT GLOBAL WARMING

Michel Christane in his book, "State of Fear" appeal to mass to take a note that issue of changing climate is just misleading and Russian scholars, Solar Physicists Galena Meshniv and Vladimir Bashkertserv bets on the issue of global warming. According to them in the years to come the earth will be cooling.

The idea is that climate scenario is so uncertain and thus the quantitative analysis of climate change impacts faces many difficult challenges. This leads another school of thought to believe the risk and danger of global warming just as an imaginary guess. Certain scientists believe that global warming is no threat and the planet is essentially cooling off. They argue that the factors causing the phenomenon and the measurements are not fully understood, and that it is impossible to draw any conclusion whether the warming of the earth is purely natural. They also believe that the trend is a false alarm and that it is not a sign of a fore coming global disaster. Human beings can adapt to changes caused.

In the opinion of the others, most activity is due to fluctuating sun's energy due to large sunspots, which is thought to be partially responsible for the "Little Ice Age". Also, some researchers believe that smoke from the burning tropical forests and grasslands causes strong cooling force on the climate. This cooling effect could nearly equal the warming power created by the fires. The issues regarding the rising sea level, the elevations of various coastal land areas are rising and sinking due to geological factors. Thus, the ocean levels may not rise as the continents may be sinking.

The researchers also assume that global warming is predicting a coming of an Ice Age. The last ice age occurred as the Earth's climate was warming.

In Arctic region more water would evaporate in summer, and fall onto land as snow in winter. The winters would not be so warm as to melt all of this snow, thus the glaciers would grow. Some carbon compounds released in the atmosphere may check global warming.

Following the patterns of many assessments to many optimistic, it appears to go against common sense. Crop yields are likely to be higher in future, agricultural prices will be different, land use pattern will change, the global trade picture will change and the entire set of technological options available will change. The historical response and even of today's agriculture system is irrelevant as agriculture is changing fast.

18.6 POLICY MATTERS AND RESPONSE

On the basis of the scientific evidences so far and the political response, global warming came to the United Nations agenda and during 1990s there were a succession of resolution aimed at marshalling efforts to counter it.

In 1992, Framework Convention on Climate Change (FCCC) was the first formal international statement of concern and agreement to take action to stabilize carbon dioxide concentrations. In this context the 1997 Kyoto Protocol was negotiated. It specified commitments by individual developed countries to reduce emission by on an average 5.2% below 1990 levels in the period 2008 to 2012. Decision beyond this period has generated a great deal of controversy.

The Kyoto Protocol involved several decisions:

- By 2012 developed countries would reduce their collective emission by 5.2% from 1990 levels, with each country being committed to a particular figure.
- The emission covered not only carbon dioxide but also methane, nitrous oxide, hydroflourocarbons (HFC), perfluorocarbon (PFC) and sulfur hexaflouride.
- These commitments would be reckoned on a net basis of, considering sinks as well as sources and each country must credibly measure its contribution and meet its commitment.
- Countries may fulfill their commitments jointly and they may improve the efficiency of compliance through "flexibility mechanism". While the countries that are party to the Protocol are expected to rely on reducing their own emission domestically, three "flexibility mechanism" were identified *viz.* Emission Trading, Joint Implementation and Clean Development Mechanism.

However, there is a dispute whether the industrialized world should be responsible for cleaning the mess. Many less industrialized nations are unable to afford action to prevent an increase in CO_2 and the fact they do not have incentives to reduce the emission. Several less industrialized nations argue that the developed world was allowed to use the nature creating welfare, and that it is now morally right for them to do it.

18.6.1 Suggestions / Remedies

There are several things can be done to decrease global warming. It is time now to look around and make our factories/working places and home more energy efficient.

First, scrutinize in and around and find several ways to decrease use of electricity. Plant several trees, which can give shade during hot summer days. Also install, an energy efficient thermostat with a day and night timer.

Just remembering to switch off electric gadgets, computer, air-conditions, lights, etc. when not in use can save a lot of energy. It costs nothing to follow this economical and simple rule.

Second, decrease the use of personal vehicle. Besides saving oil, tires, parts and repairs it could help reducing green house gases. Operate the vehicles efficiently, car pool, ride the bus, a bicycle or can walk for short distances.

Third, try following the strategy of "Reduce......Reuse......Recycle". Reuse of anything is easiest and best way to recycle. Also, use cotton cloth; towels and napkins instead of paper ones. Non-polluting, renewable solar, wind and hydrogen energy systems are the only potential solutions.

18.7 CONCLUSIONS

The debate and conflict continues and the earth and its system reacting to the warming also continues. Is Global warming a greatest challenge of 21st century or jeopardy to the living planet Earth? Goes on. The fact remains that the risk of global warming warns us in many spheres and those who do not see or a cynical about the concept itself should be perhaps living on some other planet not on Earth. Finally, let us make up our minds to whether the success, as an individual should be based on quantity of consumption or on quality of our environment. It is era to examine our moral values and scan our attitudes seeing that they relate to natural world. Let each one of us ask ourselves: What makes us really happy? What makes us feel secure? Has our striving for more and more materialistic consumption caused us to forget that we are living human beings? We have to realize a fact harder that "The value of life has a meaning beyond any materialistic value: a rat race of reaching there before others".

REFERENCES

Balling. R. (1992): The heated debate: Green house predictions versus climate reality, Pacific Research Institute for Public Policy.

Bhat. H. (2005): Silenced Spring, *Down to Earth* February, 2005, 13 :51-53.

Brown. P. (1996): *Global Warming*, Universities Press (India) Limited. Hyderabad.

Cunningham P. and Cunningham M. (2002): *Principles of Environmental Science, Inquiry and Application*, Tata McGraw Hill, India.

Hasnain. S. (2002): Melting into Oblivion *Down to Earth* May 15,2002,.10 : 24

IPCC I (1996): *The Science of Climate Change*, Cambridge University Press

IPCC II (1996): Economic and Social Dimensions of Climate Change, Cambridge University Press.

IPCC III (1996): Impacts, Adaptations and Mitigation of Climate Change, Cambridge University Press.

IPCC (United Nations Intergovernmental Panel on Climate Change) Climate Change (1990): The IPCC Scientific Assessment, Cambridge University Press.

Santra. S. (2001): *Environmental Science*, New Central Books Agency (P) Ltd. Calcutta.

Singh S. (2000): *Physical Geography*, Prayag Pustak Bhawan, Allahabad.

CHAPTER - 19

Bio monitoring of Air Pollution

Arun Arya

Botany Department, Faculty of Science,
The Maharaja Sayajirao University of Baroda, Vadodara 390 002
e mail : aryaarunarya@rediffmail.com

ABSTRACT

Plants are an important agents to be used as biomonitoring of air pollution. The plants can easily be grown in polluted area and pollution load can be assessed by organisms like Pathogens, Lichens, Bryophytes, and Higher Plants. The plants like lichens are perennial and absorb the pollutents from their whole thalli serves as better organisms than higher plants like Bell-W3, Clover and Popular etc. In most bryophytes the mineral uptake is from atmosphere and not from substratum. Airborne metal particles are removed from atmosphere by rainwater, gravitaional settings and impaction on various surfaces specially vegetation and they may also be actively or pasively accumulated by plant. The concentration of air born elements in plants is determined by their concentration in the air, the inherent ability of plants to absorb this elements, the ratio of plant surface to total plant mass, the ion ex-change capacity of the plants is the length of exposer. The analysis from plants can give the values of different pollutents or symptoms on the plants can be used to monitor the effect of pollutents. The monitoring of air pollution needs to be standardised before it can be used for urban and rural areas.

Key Words: Air Pollution, Bio-monitors, Lichens, Bryophytes, Phytopathogens, Tobacco.

Air is an important constituent of our atmosphere. We can survive with out food and water for some time but not in absence of air. 4th mantra of 1st part of 1st chapter of RigVeda says

Om swastay vavy mup brawamahe samoum swasti bhuvaransya yaspati Brahaspati sarvaganam swastay, Swastaya adityarao bhavantu nah.

Which means, O! Almighty God you may be beneficial to us. A wide variety of gaseous and particulate pollutants are being continually added to the environment through industrial and anthropogenic activities. It raises-serious concerns to the well-being of people and plants, may be the possible cause for global issues like climatic change global warming, ozone depletion etc.

It is desirable that the emission of pollutants be controlled; at least to an extent that their concentration at ground level is not hazardous. When we are concerned not only for present but want to leave this planet healthy for future generations. We should have specific methods to know the pollutants present in air this is termed as biomonitoring or 'Phyto monitoring' of Air Pollution.

In a computer "Monitor" is a vital component, without which the operations may not be possible. Similarly for the ecosystems, we need to have monitoring methods that can indicate the ecological status (normal, degradation or improvements) due to natural and biotic including anthropogenic activities for developing a better natural resource management system and in order to implement practicable conservation strategies. Monitoring devices for ecosystems can be classified as Physical (Remote sensing), Chemical (chemical, biochemical and molecular level) and Biological.

Earlier physical and chemical monitors were extensively used to detect the pollution levels in a location. However such monitoring instruments are very expensive, they can be used at only a limited number of sites and no distribution patterns or comparative data can be achieved etc. led to the search for biological indicators as natural monitors. 'Biological monitoring' is a term explaining the usage of organisms to indicate as well as to assess an ecological degradation or an improvement, due to a specific cause (pollution or degradation both natural and biotic including anthropogenic) in a specific or in similar locations, with minimal equipment at the field level by the non-specialists. After initial standardizations and establishment, the Biomonitoring techniques must be cost effective and become a people's movement in locating environment degradation. They serve as an effective early warning system.

Hence in Biomonitoring techniques, the choice of organisms and the method for data collection and analysis must be oriented to identify the specific problem (cause) responsible for the ecological degradation, less sophisticated in terms of equipment usage and lab level studies, statistical and other analysis suitable for non specialists.

Biomonitoring require regular data collection and it provide scientific basis for developing a conservation strategy. It is a prerequisite to have a data on keystone species or ecological changes in the natural composition of populations, organism's stress expressed as morphological symptoms due to the reactions of pollution mixtures for Biological monitoring. The basic aim of Biomonitoring is to indicate the emerging catastrophes at an early stage.

Currently a variety of subjects including ecology to highly specific chemical analysis or even molecular level damages in organisms are grouped under biological monitoring making it more complex to understand and confusive. Various clear-cut definitions are available on the terminologies used in this area of research (Hawksworth, 1992).

19.1 PLANTS AS FILTERS OF GASEOUS POLLUTANTS

Certain plant species are capable of filtering and absorbing gaseous pollutants. Studies have revealed that a green area of 500 m width surrounding a factory is capable of reducing SO_2 concentration by 70%. Sulphur dioxide and nitrogen dioxide are soluble and get absorbed readily by plants while nitric oxide and carbon monoxide are very insoluble and absorbed slowly and not by all vegetation. Moreover, damp and wet surface of the plants enhance pollutant removal rate by 10% because under such conditions stems, branches, twigs and leaves are engaged in absorption process. Light has also pronounced effect in foliar removal of pollutants by influencing physiological activities and stomatal opening (Smith, 1981). In "Updesh Tarangini it has been said about trees-

मंजरिभिः पिकं – निकरै, स्जेभिरंलिन फलैश्चपान्थगणम्।
मार्ग सहकार सतत, उपकुर्वन् नन्दति चिकरकालम्।।

Planted on both side of roads, O Tree! You live a long life and live happily, free from diseases, because you provide branches for stay of birds, nector to insects and shade to the travelers.

19.1.1 Pollution Tolerance Index

The tolerant species of plants function as pollution 'Sink' and therefore a number of environmental benefits can be derived by planting tolerant species of plants in polluted areas. For this purpose, evaluation of plants with respect to their tolerance level to air pollution is essential. To evaluate the tolerance level, the following formula has been used by Subramaniyum *et al.* (1985) for deriving the empirical number indicating the air pollution tolerance index (APTI).

$$APTI = \frac{[A (T + P) + R]}{10}$$

Where A is Ascorbic acid content, T is total Chlorophyll, P is leaf extract and R is relative water content of leaf. The whole sum is divided by 10 to get the APTI value.

19.2 SYMPTOMS PRODUCED BY BIOINDICATORS

Various bioindicators produce characteristic symptoms, which are listed below :

Table 1. Environmental degradation and symptoms produced on indicator plants/organisms

Bioindicators	Environmental degradation by	Symptoms
Lichens	Acid rain Ammonium Eutrophication Chlorinated hydrocarbons Heavy metal contamination Oil spills Sulphur dioxide air pollution Radionucleus	Lethal symptoms of thallus, death, reduction in occurrence, frequency and vitality
Lichens	Climate change Ozone layer depletion UV radiation Water levels	Bleaching of thallus and production of antisubmergence/ antidessication and anti-UV substances
Lichen diversity and its ecology	Ecological continuity Deciduous woodland Tropical Rain Forests Coniferous forest Fire, Stability of surfaces.	Modification in the occurrence and distribution pattern.
Bryophytes	Bioaccumulative indicators of aerial metal contamination, indicators of gaseous air pollution.	Lethal symptoms of thallus death, reduction in occurrence, frequency and vitality.
Fungi	Gaseous air pollution.	Lethal symptoms of thallus death, reduction in occurrence, frequency and vitality
Algae	Pollution in water bodies and in terrestrial environments.	Eutrophication/Reduction i occurrence.
Pine needles	Heavy metals, gaseous pollutants and Ozone layer depletion.	Loss of needles and damages

Rye grass	Lead, Fluorine, Sulphur, Cadmium, Copper and Zinc	Leaf blade damages
Tobacco/Beans/other experimental plants	Acid rain	Leaf blade damages
Benthic fishes	Organic load of the water body	Reduction or increase in occurrence, frequency and vitality
Earthworms	Heavy metals and other toxic materials in the soil	Reduction in occurrence

19.3 FUNGAL PATHOGENS AS BIO-MONITORS

Plant fungal pathogens would appear to have considerable potential as biomonitors, as different species have known to have both negative and positive responses to common phytotoxic pollutants at ambient levels (Bell *et al.,* 1993). Probably the best known example of such a biomonitor is *Diplocarpon roseae* – the cause of black spot of roses, which is as such a problem for rose growers. Anecdotal evidence suggested that this pathogen was absent from areas of the UK with high levels of SO_2. This was confirmed by Saunders (1966) who surveyed rose nurseries in different parts of the country and found a negative correlation between severity of disease and prevailing SO_2 concentration. The causality of this relationship was demonstrated subsequently by the same worker by means of fumigation experiments. Subsequently, following the massive fall in ambient SO_2 concentrations throughout the UK, anecedotal evidence has suggested.

The black spots have reinvaded the formerly polluted areas. This was quantified by Jarraud (2000) who resurveyed roses in the vicinity of the locations utilized by Saunders (1966). He discovered that black spot have recolonised all the areas from which it was absent in the mid -1960 s and there was no longer any relationship between levels of infection and ambient SO_2 concentrations. Thus black spots of roses has considerable potential for biomonitoring of both rising and falling levels of SO_2. A rather controversial fungal pathogen biomonitor is the *Rhytisma acernium*, cause of tarspot of sycamore (*Acer pseudoplantanus*). Bevan and Greenhalg (1976) observed that the intensity of infection of sycamore by this fungus in cities of northern England appeared to be lower in more polluted areas. They developed a tar spot index (TSI) based on the number of infections per unit area of leaf, which was found to be correlated negatively with prevailing SO_2 concentrations. On the basis of presence of tarspot was incorporated into a number of lichen surveys in northern England in 1970s. However unlike Saunders work with black spot, no attempt was made to demonstrate causality and Leith and Fowler (1987) concluded that the correlation was an artifact and was based on the amount of leaf litter available for over wintering spores to reinfect trees in the spring in the more polluted central areas of cities, most of the litter would be removed by street cleaners.

However, observations made by A.W. Davison (Pers. comm.) in the late 1980s suggested that the hypothesis of Bevan and Greenhalgh (1976) was correct: in the Colliery village of north east England smoke control orders were introduced at that time and as SO_2 concentrations fell sharply, not only did black spot reinvade the area, in and around Sheffield, where the presence / absence of tarspot has been recorded in the mid-1970s when SO_2 concentrations were still very high in some been located (South Yorkshire County Council 1976). This showed that tarspot was present at all sites where it had previously been absent, with the exception of the city centre, where no innoculum was present. The TSI showed no correlation with ambient SO_2 and all the evidence suggested that tarspot is indeed a good biomonitor for this bio pollutant. An interesting further study was carried out by Jarraud (2000) in which he measures TSI on sycamores along a transect from the countryside into central London. He found that TSI declined effectively to zero by the innermost site. It could be postulated that reinvasion by the fungus as SO_2 levels fell was too slow to reflect the impact of the latter. However, when he transplanted saplings and accompanying infected leaf litter the innermost sites. The only plausible explanation appears to be the effects of NO_x which is present at much higher levels in London than elsewhere in UK and which was negatively correlated with TSI along with transect into central London might render conditions less favorable for infection seems as tarspot is ubiquitous organism which occurs across the UK over a much wider range of climates.

19.4 LICHENS AS BIO MONITORS

It has been recognized since the mid-19th century that Lichens are particularly susceptible to SO_2. However, there are large inter-specific differences in degrees of susceptibility which has been exploited for many years as a tool for bio monitoring this pollutant. Various scales has been developed for predicting SO_2 concentrations based on Lichen community composition on different substrates probably the best known is that of Hawksworth and Rose (1970) who produced a lichen map of England and Wales. This system used a 10 point scale, ranging from a rich flora containing fruticose species (Fig. 1) in vary low SO_2 sites to a complete absence of epiphytes in the most polluted areas. It illustrated the destruction of lichen communities in the cities due to pollution. The lichen indicator scale evolved from observation by lichenologist of decreasing diversity from a point sources or a city center. These observation, when investigated in more detail, reveled that distribution correlated with concentrations of Sulphur dioxide and this facilitated the development of a semi quantitative scale linking key species with winter annual mean ambient concentration of Sulphur dioxide. Where sulphur dioxide and smoke from coal burning are still the main pollutants and appropriate conditions for lichen growth prevail the scale could be used for monitoring. However, where other pollution sources are common, different community, distribution patterns and abundance may be evident. James *et al.* (2002) have recorded the highest lichen diversity in central London for over one hundred years and members off Dr. Bell's team be currently studying these species in relation to present pollution climate. Upreti *et al.* (1998) have studied the lichen found in Antarctic region. A detailed study of 13 species of crustose lichens of the Maitri region of Schir Macher Oasis, in East Antarctica is undertaken. The heavy metal concentration of Antarctic lichen was not available, so far, therefore, to gain a base line information on recent levels of heavy metals for conducting environmental studies the heavy metal estimation for Cr, Cu, Pb, Fe, and Zn were conducted in 13 species of lichens inhabiting different ecological niches.

19.5 BRYOPHYTES AS BIO INDICATORS

Bryophytes are considered to be a very good biological indicator due to their sensitivity to environmental conditions. Bryophytes are the simplest and most economical tool for assessing air pollution features that make bryophyte suitable for monitoring air pollution are:

(a) Lack of well developed cuticle, especially in ectohydric species

(b) Absence of root system, which frees the bryophyte from reliance on the soil

(c) Flat structure of thalloid plant which provide large surface area for metal absorption

(d) And a vast range of sensitivity to air pollutants (Brown 1984).

Noticeable about these Kinney plants is that most mineral uptakes it from air environment not from soil. Metals continue pose serious threat to biota due to their acute toxicity. Metals are non degradable and once released in environment they become integral part of habitat. The bryophytes absorb pollutants by ion exchange method and it depends on the length of time of exposure a wide spread pollution can occur from urban transport which releases the metals like Cu, Zn, Cr, Pb, Fe, Ni, Cd and have their impact on all living organisms. There is an ample evidence that bryophytes can be used as reliable pollution monitors. Taoda (1973) suggested the use of bryophyte as bryometres, instrument for measuring phytotoxic air pollution.

Bryophytes are the simplest and most economical tool for assessing air pollution. Features that make bryophytes suitable for monitoring air pollution are: lack of well develop cuticle especially in ectohydric species, absence of root systems, which frees the bryophytes from a reliance on soil, flat structure of thalloid plants which provide a large surface area for metal absorption, and a vast range of specific sensitivity to air pollutants, greatly exceeding that of higher plants (Brown, 1984). The plants like *Plagiochasma appendiculatum* and *Marchantia paleacea* showed accumulation of heavy metals. High amount of Chromium was found in *Brachythecium kumaonense* followed by *Cratoneum commutatum* and *Funaria hygrometrica*. All the 3 plants are mosses. Zinc which is created due to tyre burning and combustion of fuel was present in the highest amounts in the plant like *Anobobryum filiforme*. *Asteralla mussooriensis* can accumulate zinc while *Brachythecium kumaonense* is a good accumulator of Ni and Cu (Nath *et al.* 2006).

Singh *et al.* (2005) studied the occurrence of bryophyte on the river Sai, in Raebareli. The presence of water helps to grow *Barbula juvanica*. With increase in pollution the density and biomass increase similar observation have been made in the studies of *Marchantcia polymorpha*, *Plagiochasma appendiculatum* L. et. L. (Romanova, 1965). *Hymanostelium recurvirostrium* (Hedw) Dix. grow well in clean water. The lowest population density was observed in highly polluted site of Jagdishpur. Similar observations were made for *Riccia frostii* growing on banks of river Ganges between Shuklaganj and Kalakankar (Kumar *et al.* 1994).

Nath *et al.* (2006) studied the use of three thalloid liverworts and 6 mosses growing in Massuri in pollution monitoring studies revealed that *Anoectangium strachyanum* is a good accumulator of Ni and Cu, *Brachythecium kumaonense* is a good accumulator of Zn and Ni, *Entodon luteus* is a good accumulator of Cu and *Asterella mussooriensis* accumulated Zn. Some times the emitted SO_2 lowers the pH and makes the heavy metals highly soluble. *Plagiochasma appendiculatum, Marchantia paleacea* exhibited the fair amount of metal accumulation. Metals are also found to exert strong toxic effect on bryophytes (Shaw 1990).

19.6 TOBACCO AS BIOMONITOR FOR O_3

The best known higher plant bio-monitor is the super sensitive Bell –W3 cultivar of tobacco (*Nicotiana tabacum*), which has been used for monitoring O_3 since the 1969's, initially in north America and later on elsewhere in the world. The threshold for injury is around 40ppb O_3 and thus Bell-W3 is particularly suitable for bio-monitor for risk assessment of O_3 impact on vegetation. A number of negative relationship between yield and NO_2 and SO_2 were detected with some indication of a similar pattern with Bell-W3 injury. *Trifolium repens* (white clover) is another O_3 sensitive plant.

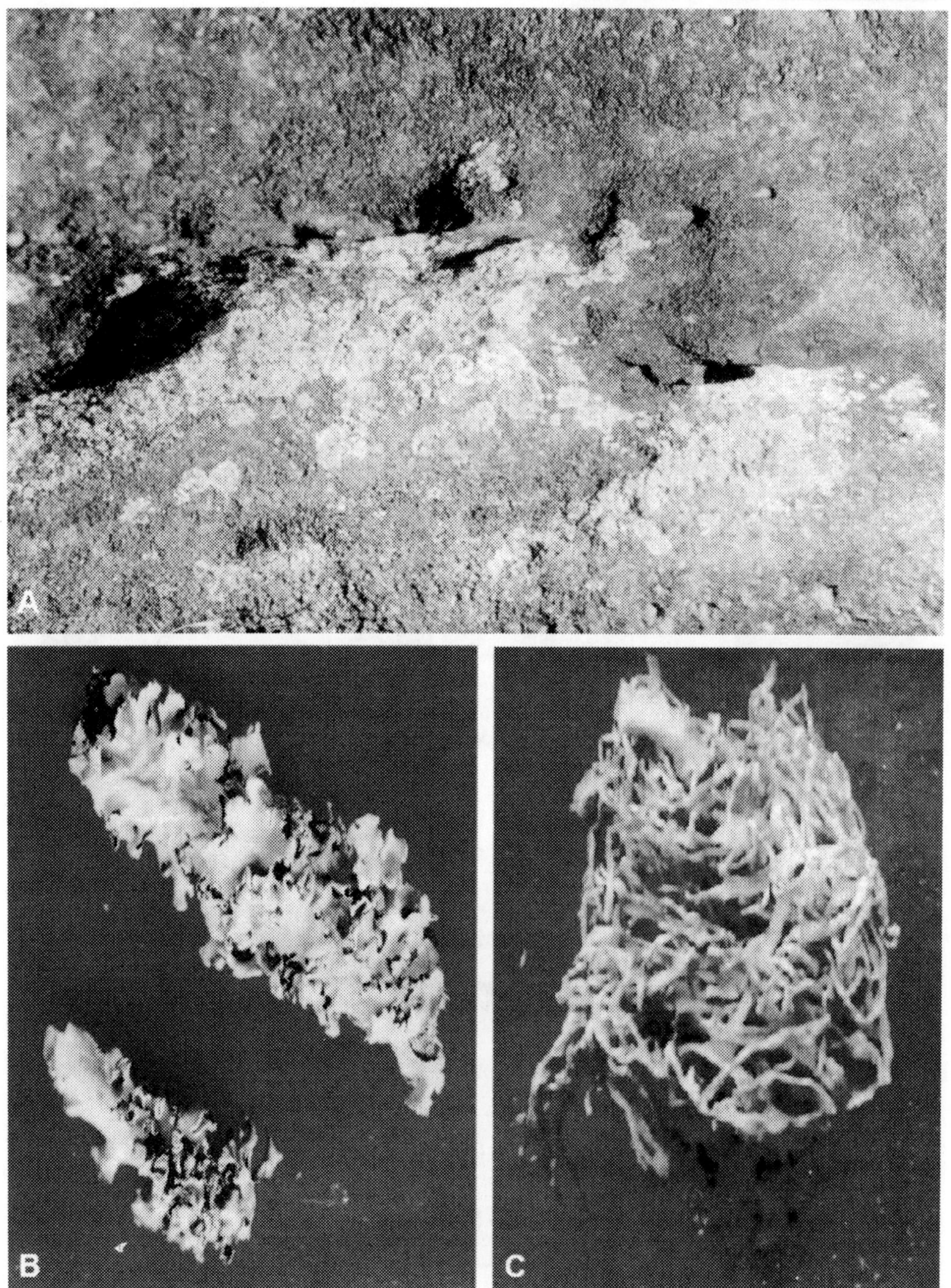

Plate I. A. Showing white coloured patches of crustose lichen; **B.** Foliose lichen - *Parmelia* is a good bio-indicator lichen; **C.** Branched thalli of *Rocella montegnii* – a fruticose lichen.

Plate II. A. Dichotomously branched thalli of *Plagiochasma;* **B.** Capsules of common moss *Funaria hygrometrica;* **C.** Gametophyte of *Polytrichum commune*; **D.** Green thalli of *Asterella* – a liverwort; **E.** Archegoniophore of *Asterella.* **F.** *Funaria* moss attached to branches of a tree

Plate III. A. A small tree of *Pinus longifolia;* **B.** A field of Tobacco (Tobacco plant is a good bio-indicator of ozone)

REFERENCES

Bell J.N.B., Mcneill S., Houlden G., Brown V.C. and Mansfield P.J. 1993. Atmospheric change: Effect on Plant pests and diseases. *Parasitology* 106:511-524

Bevan R.J and Greenhalgh G.N. 1976. *Rhytisma acerinum* as a biological indicator of pollution. *Environ. Pollut.* 10:271-285

Brown D.H 1984. Mineral nutrition. In: *Bryophyte ecology* A.J.E Smith (ed.) Chapman and Hall pp 383-444

Hawksworth D.L and Rose F. 1970. Qualitative scale for estimating SO_2 air pollution in England and Wales Using Epiphytic lichens. *Nature* 227: 145-148.

James PW Purvis O.W and Davies L. 2002 Epiphytic liches in London. Bull. British. Lichen Soc. 90:1-3

Jarraud N.S. 2000. The effect of Ambient Air pollution on leaf pathogens of Rose and Sycamore. Ph.D thesis, University of London.

Kumar A., Sinha A. K. and Pandey D.C. 1994. In Water...Management, Eds. Sinha A.K. *et al.* Gyanodaya Prakashan, Nainital, India p 309.

Leith I.D and Fowler D. 1987. Urban distribution of *Rhytisma acerinum* (Pers.) Fries. (tarspot) on Sycamore. *New Phytol.* 108:175-181.

Nath V., Asthana A.K. and Sharma S. 2006. Environmental studies on bryophytes of Mussoorie (Western Himalayas). In: Plant response to environmental stress. RD Tripathi *et al.* (eds.) International Book Distributing Company, Lucknow, India p 173-177.

Romanova E.A. 1965. In A.G.G Chikishew Plant indicator of soil rocks and surface water, New York. P 81.

Saunders P.J.W 1966. The toxicity of SO_2 to *Diplocarpon roseae* Wolf. Causing black spot of roses. *Ann. Appl. Biol.* 58: 103-114.

Shaw A.J. 1990. Metal tolarence in bryopytes. In: heavy metal tolerance in plants: Evolutionary aspect, A.J Shaw (ed.) CRC Press, Boca Raton, F.L pp 133-152

Singh M., Nath V. and Kumar A. 2005. the ecological studies on Bryophytes growing on the bank of pollute a river Sai (Rae Bareli) India. *Proc. Nat. Acad. Sci.* India. 75(B) : 41-50

Smith W.H. 1981. *Air pollution and Forests: Interaction between air contaminants and forests ecosystems.* Springer Verlag, New York.

South Yorkshire County Council 1976. Lichen servay of air pollution. South Yorkshire County Council Environment Department, Barnsley

Subramaniyum G.V., Rao D. N, Varshney C. K, and Biswas D. K 1985. Air pollution and plants: A state of art report. Published by Ministry of Environment and Forests, Department of Environment Govt. India, New Delhi.

Taoda, H. 1973. Bryometer, an instrument for measuring the phytotoxic air pollution. *Hikobia* 6: 224-228.

Uprti, D.K., Chandra S. and Nath V. 1998. Taxonomy and Morphological studies of lichens, bryophyte and pteridophytes. NBRI Annual report 57-60pp.

CHAPTER - 20

Microbial and plant growth posing threat to Buildings and Monuments

Arun Arya[1] and N.R. Shah[2]

[1]Botany Department, Faculty of science,
The Maharaja Sayajirao University of Baroda, Vadodara 390 002
e mail: aryaarunarya@rediffmail.com
[2] Department of Museology, Faculty of Fine Arts,
The Maharaja Sayajirao University of Baroda, Vadodara 390 002

ABSTRACT

Monuments and old buildings in India with natural stone have been exposed to decay for centuries. Growth of the microorganisms and plants on monument is not only dangerous but even fetal to them; the biochemical damage is due to extraction of organic acids which can erode the surface. If plants grow inside fissures or cracks present on the surface, the gap may increase further. The presence of the silica in the sand used for plastering the buildings may be helpful for the growth of members of Poaceae and Cyperaceae. Liverworts and mosses are non vascular plants which along with lichens are biodeteriogens. These prefer to grow on rocks and old buildings. They love a humid environment. Plants like *Calotropis, Lindenbergia, Eclipta, Azadirachta* and *Ficus religiosa* were seen on the walls and dome of Hazira- the mausoleum of Qutubuddin Mohammed Khan built in 1506AD. In a survey conducted at Vadodara, plants belonging to 17 different families were found on terrace of Arts faculty of the M.S. University of Baroda. Plants are described as green agents of destruction. Preventive and curative measures are suggested. The growing plants can be removed by use of Arsentites and arsenates of potassium and sodium in solution. Before the application of weedicides/ chemicals vegetation may be cleared by cutting.

Keywords: Microbes, Plants, Buildings, Monuments, Biodeterioration

The ancient monuments of India constitute her cultural heritage. Thy are situated between latitudes 8 south and 35 north & longitudes 68 west and 97 east. India has a vast area including dry regions, high Himalayan altitudes and humid sea coasts.

The eastern half of India and her west coast have thick vegetation due to tropical rainfall.

In the annals of Indian history and archeology stones have been generally used as a material for sculptures and monuments and living rocks have been used a material for rock cut caves and monuments. Monuments erected in India with natural stone have been exposed to decay for centuries. Micro-organisms such as fungi develop rapidly in an environment with relative humidity exceeding 65%. On surfaces that are persistently damp, even mosses, lichens and algae can grow. In addition to the physical disintegration that these growths can cause to the paint surfaces, the acidic secretions from some of them can have serious chemical consequences. When the plaster contains organic matter such as vegetable fibres and paddy husk, as at Ajanta, attack by insects such as silver fish, beetles etc. becomes a serious problem (Subbaraman, 1993). Vegetation growth on monuments is not only dangerous but even fatal to them. In a country like India where diverse kinds of climatic and ecological conditions are there, the plants found on the monuments is varied and includes all the visible classes of the plant kingdom. The problems of preservation of stone monuments all over the country are extremely complicated and their preservation is a very difficult task. Majority of the stone monuments are protected by archaeological survey of India and proper care is taken by the Department, but still many monuments are left uncared and they are damaged by the growth of plants.

20.1 TEMPLES AND MONUMENTS : SYMBOLS OF RICH CULTURAL HERITAGE

"And the whole earth was of one language, and one speech... and they said to one another, Go to, let us make brick, and burn them thoroughly. And they had brick for stone, and lime had they for mortar. And they said, Go to, let us build a city and a tower, whose top may reach unto heaven."

Genesis 11: 1-9

The monuments made up of stones and mortar not only show the architectural beauty but are good examples of cultural heritage prevailing at that time. We can know various things by peculiar designs of a monument. It is said that vast wealth of ancient Indian culture is still preserved in temples. The numerous temples have their own museums and collection of artifacts including paintings, utensils and armaments. Building materials are decayed by adverse environmental conditions and the extent of damage depends on both the materials and conditions. Traditional buildings which survive were soundly built in the light of tried and tested practices. The accepted practices were and are to minimize water penetration, ensure adequate air movement around all vulnerable structures, and to allow permeable materials to'breath'.

20.2 TYPES OF MONUMENTS

The monuments comprise rock cut caves, rock carvings, early stupas, temples, mosques, forts, and sculptures from Buddhist period to Muslim dynasties.

Buddhist period monuments are stupas at Sanchi, Sarnath, Barhut, Nagarjunkonda, Amaravati and Devni mori. Chaity Halls and monasteries at Guntapalli, Nagarjunakonda and Devni mori.

Rock cut monuments of Buddha, Jain and Hindu, period of Bhaja, Pitalkhora, Ajanta, Karla, Kanheri, Kondana, Udagiri, Badami, Aihole, Ellora, Elephanta, Jogeshari. Cut-out Monolithic temples are Mahabalipuram, Ellora, Masur and Dhamnar.

Temples of Gupta period, Chalukyan period, Rastrakuta period, Hoysalas period, Pallava period, Chola and Nayakas period. Muslim monuments include tombs, forts, mosques and pillars. They are Qutb-minar, Sultan ghari's tomb, Siri fort, Hazrat Nizamuddin Tomb, Lodi Mosque, Sher Shah Suri tomb, Purana Quila, Humayun Tomb, Agra Fort, Fatehpur Sikri, Taj Mahal, Red Fort and Jama Masjid.

20.2.1 Stones used in monuments

The more common rocks used for monuments and sculptures in India are granite, sand stone schist, marble, lime stones, quartzite, basalt, laterite and gnisses. Granite and banded gneisses found in Rajasthan, Madhya Pradesh, Bihar, Andhra Pradesh, Karnataka and Tamil Nadu, have been used as excellent building material for temples, Forts, mosques etc. Knondalite and sand stone were used at Konark and Puri. Marble of Rajasthan has largely been used as building materials in many monuments of Agra. Delhi and Calcutta. Quartzite has been used in Purana Quila and the Qutub Minar at Delhi. Vindhyan sand stone was used at the Buddhist stupas of Sarnath, Barhut and Sanchi. Temples at Khajuraho, Mosques at Agra, Delhi and Fatehpur Sikri, were also built from sand stone. Basalt was used at Ambarnath, Karle, Bhaja, Elephantas, Ajanta and Ellora caves. The Indian artists have made use of the stones and rocks available to them within the easy reach of the area where they were working.

20.2.2 Composition of stones

1. Granite - Granite is an igneous rock with granite texture. It is strong, durable and hardest of all the rocks.
2. Gneisses - It is a metamorphic coarse grained roughly foliated rock. It is generally light coloured.
3. Schist - Schist is a medium to coarse grained metamorphic rock.
4. Marble - Marble is a metamorphic rock composed essentially of Calcite and or dolomite. It is soft and produce effervescence with acid. Generally white in colour.
5. Sand stone - sand stone is a consolidated sedimentary rock composed mostly of mineral or rock fragments cemented together.
6. Lime stone - Lime stone is a sedimentary rock consisting chiefly of calcium carbonate and is the consolidation of limy mud, calcarious sand or shell fragments.
7. Basalt - Basalt is a fine grained volcanic rock. It is hard, tough, even grained and dense.
8. Laterite - Laterite is a mixture of Iron and aluminium in porous and concretionary form.

20.2.3 Bacteria and different group of Plants growing on Monuments

1. **Schizomycetes:** Commonly known as bacteria. Bacteria are microscopic, simple, unicellular plants. They are found on stone monuments. They are Nitrifying bacteria, sulphur bacteria, thio-bacteria, silico-bacteria, *Sarcina lutea*, *Cytophaga cellvibrio, Desulfo vibrio* and *Styphylococcus citreus.*
2. **Algae:** Algae are chlorophyll bearing thallophytes. Several coloured algae have been observed growing on historical buildings and monuments. Two types of Cyanophyceae for Blue green algae are found growing on old brick walls or stones of monuments during the rainy season. They are *Oscillotaria irrigua* and *Phormidium autumnale. Nostoc* and *Anabaena* are also found growing on brick floor of the passage Chlorophyceae or green algae are also growing on old walls and monuments during rainy season. They are *Protococcus vulgaris* and *Ulothrix subtillissima.*
3. **Fungi:** Fungi are the non-chlorophyllous plants living on dead or live organic matters. Few species of Fungi are also growing on monuments. They are *Candida albicans, Lipomyces neoformans* and *Blastmyces* sp.
4. **Lichens:** Lichen is a unique combination of two different organisms living in close contact with each other. It is a combination of Fungi and

algae. Lichens are pioneers in the growth of vegetation on rocks and on old walls. Then it is followed by other plants. The lichens growing on stones or rocks are saxicolous. Two types of Lichens are found on the monuments. Crustose and Folliose types. Crustose types are *Graphis duplicate*, *Graphis scripta*, *Lacanora chlarona* and *L. distans*. While Foliose type are *Gyrophora cylindrical*, *Parmelia cirrhata*, *P. furfuracea*, *Peltigera scutata* and *Physcia aegiliata*.

2. **Bryophytes**: The simplest and the most primitive member of the sub-kingdom embryophyta. They are commonly known as Liverworts and Mosses. They are found on stone walls and brick walls of monuments. They are *Plagiochasma*, *Fimbriaria*, *Riccia*, *Notothylas*, *Cyathodium*, *Fossombronia* and mosses like *Fissidens*, *Merceya*, *Hyphilla, Bryum* and *Tortula* sp.

1. **Pteridophytes**: They are highly developed cryptogams. Two plants of this group are found growing on walls of Fort and old buildings. They are *Cheilanthes* sp. and *Adiantum* sp.

3. **Angiosperms:** Many flowering plants are growing on stone monuments. They are *Woodfordia fruiticosa* of Lytheraceae family *Trachyspermum stictocarpum* of Apiaceae family. *Canscora diffusa* and C. *decurrens* of Gentianaceae family. *Trichodesma amplexicaule* of Boraginaceae family. *Kickxia incana* and *Lidenbergia indica* of Scrophulariaceae family. *Didymocarpus pygymea* of Gesneriaceae family. *Rungia pectinata* of Acanthaceae family. *Leucas biflora* and *Nepeta hindostana* of Lamiaceae family. *Fleurya interuupta* of Urticaceae family. *Ficus religiosa, Ficus benghalensis* and *Ficus arnottiana* of Moraceae family *Melanocenchrus-jacquemontii* and *Tripogon* sp. of Poaceae family.

Table 1. Angiospermic plants found on two different buildings of Vadodara

Sr. No.	Plant species	Family	Hazira monument	Arts Faculty
1.	*Acalypha indica*	Euphorbiaceae	+	+
2.	*Achyranthus aspera*	Amaranthaceae	-	+
3.	*Amaranthus tricolor*	Amaranthaceae	+	+
4.	*Azadirachta indica*	Meliaceae	+	+
5.	*Calotropis procera*	Asclepediaceae	-	+
6.	*Cassia tora*	Caesalpiniaceae	+	-
7.	*Cyperus rotundus*	Cyperaceae	-	+
8.	*Eclipta alba*	Asteraceae	+	+
9.	*Ficus benghalensis*	Moraceae	+	+

10.	*F. carica*	Moraceae	+	+
11.	*F. religiosa*	Moraceae	+	+
12.	*Lindenbergia urticifolia*	Scrophulariaceae	-	+
12.	*Trichodesma amplexicaule*	Boraginaceae	+	-

20.3 EFFECTS OF PLANTS ON MONUMENTS

Algae, Fungi, Lichens, Liverworts and mosses propagate with tremendous speed. Nitrogen fixing bacteria which convert nitrogenous material into ammonia to form nitrous and nitric acids, which may attack and damage rocks like granite, gnesisses, schists, limestone sandstone etc.

Algae attack the stone masonry of the building and also pollute the water of the tanks and ponds nearer to the monuments. Algae growing on brick floor make the floor slippery during the rainy season. The worst action of algae takes place when they grow on monuments, stones and brick walls and absorb moisture and thus making them brittle. Action of algae also attacks the matrix of stones and thus making them weak and friable.

Fungus found on monuments disfigures the monuments and make the stone and masonry work weak and friable.

Lichens growing on monuments are most dangerous. It will etch the surface of stones particularly marble and lime stone. The outer surface of stone becomes rough due to etching. Liverworts and mosses growing on monuments do much damage due to its ability of forming soil and conserving and making it fertile and suitable for the growth of other vegetation. The expanding and branching rhizome of pteridophytes penetrates into cracks and crevices of stones and loosens up the joints due to pressure.

Roots of flowering plants by penetrating and enlarging disorganizes the structure of monuments (Plate II) The breaking of old wall can be seen in this drawing. Indian Peepal and Banyan tree in particular are the most destructive. Their root penetrates into crevices, loosens and dislodges stones. Various plants growing on stone monuments produces complex organic acids. They are gluconic acid, gallic acid, itaconic acid, kojic acid, aspergillic acid, gyrophoric acid, salacinic acid and usnic acid. These organic acids produced by the monuments, thus weakening the structure and pitting the smooth and polished surface.

Plate I. Fig. A. Photograph of Hazira- the mausoleum of Qutubuddin Mohammed Khan built in 1506AD in Vadodara; **Fig. B.** Diagram to show exchange of different ions by roots of a plant; **Fig. C.** *Ficus religiosa* growing over the tomb of Hazira, breaking the top layer

Plate II. Fig. Growth of a tree showing breaking of wall; roots are exposed.

Plate III. Fig. A Dome of faculty of Arts, The maharaja Sayajirao University of Baroda; growth of Cyanobacteria and plants was observed on this historical building; **Fig. B** Front side of the building showing luxuriant growth of plants.

Plate IV. Fig. A. Front side of Performing Arts building showing blackening of wall due to growth of Cyanobacteria

Plate V. Fig. A. Building of Nayamandir in Vadodara showing growth of small plants; **Fig. B.** Wall of Chapaner Fort showing damage due to growth of plants.

Plate VI. Fig. A-C. Removal of plants from Arts faculty dome and restoration work initiated by Mayor of Baroda, Mrs. Bharti Vyas, Dr. N.R. Shah and students.

20.4 PREVENTIVE MEASURES

Vegetative growths are most dangerous for stone monuments. Hence preventive measures are more important. The surface is to be brushed or swept clean for accumulation of dust and dirt. They form a background for accumulation of spores and seeds of plants and help to germinate. Excess of water on the surface of stone causes the spreading of growth of Algae, Fungi and mosses. Monuments are kept as dry as possible by proper draining system. Study of direction of wind will help in checking migration of spores and seeds. Joints of stone must be frequently cleaned and young plants should be removed as soon as they appear.

20.5 CURATIVE MEASURES

Old buildings and monuments are found covered with plants which during the rainy season look green or greenish blue and becomes dark and black during summer. This looks extremely ugly and their removal is necessary. Several methods for the eradication of plants are in practice. The dry plant from rocks or monuments is usually removed with coir or nylon scrubbing brushes. Flat brushes are used for plain walls while round brushes are used for carvings, joints and crevices. Cutting, drying, firing and steaming are other methods for removal of plant growth.

20.5.1 Treatments

Cryptogamic plants are removed by using 3-5% Ammonium Hydroxide solution and gently brushing with brush. The ammonia solution softens the plants and comes out easily by brushes. Then cleaned surface is treated with 1-2% aqueous solution of Zinc silicofluoride. This will also act as a fungicide and does not allow the spores or seeds to germinate. Fungicides like pernox and diethane may be used. Formaldehyde, copper sulphate, mercuric chloride, sodium arsenite are also used but most effective will be Zinc silicofluoride. Fungicides used can be effective for 5-7 years.

Flowering plants are removed by using Arsentites and arsenates of potassium and sodium in solution. Solution is injected in the roots. Tree-killer or weedicides available in the market can be used in case of large area of vegetation. Before the application of chemicals vegetation may be cleared by cutting and racking out. Stumps and other remains must be carefully removed, bit by bit and crevices, cracks and joints sealed with suitable materials. Finally 1-2% solution of Poly-methyl-methacrylate or 1-3% solution pf vinyl acetate in toluene is applied. 2-3 layers of the preservative will give sufficient protection to the surface layer.

REFERENCES

Agrawal, O.P. 1977. Care and preservation of Museum objects, N.R.L.C. New Delhi.

Agrawal, O.P. 1979. Cleaning of stone problems and Further Research. N.R. L.C., Lucknow.

Archaeological Survey of India 1964. Archaeological Remains Monuments & Museums. Part I & II.

Blatter S.J. and d'Almeida 1922. The ferns of Bombay. D.B. Taraporwala, Bombay.

Chavan A.R. and Oza G.M. 1966. *The flora of Pavagadh, Gujarat*, M.S. University of Baroda.

Chavan A.R. and Mahabale T.S. 1954. Distribution of liverworts in Gujarat, *J. of M.S. Univ.* III (2): 13-16

Subbaraman S. 1993. Conservation of Mural paintings. *Curr. Sci.* 64 (10) : 736-756.

CHAPTER - 21

Pollution control in India – Approaches and achievements

R.C. Trivedi

CPCB, Parivesh Bhavan, East Arjun Nagar, Shahadara, New – Delhi 110 032

ABSTRACT

To monitor different types of environmental pollution central and state pollution control boards are constituted after 1974 under the water (prevention and control of pollution) act, 1974. The pollution control board are responsible for setting up national standard and promote cleanliness of soil, water and air. The paper describes basic approach to pollution control. CPCB is a national apex body responsible for preparation of pollution control action plan for the country. Bio-monitoring of water quality is being undertaken at 173 station a cross the country. Air quality monitoring programme (NAAQM) started in 1984 with 28 stations spread over 7 cities which has now increased to 290 stations covering 99 cities. Air quality studies are being conducted in collaboration with All India Institute of Medical Sciences, the pollution hazards and environmental strategies are described in the paper in great detail.

Key Words : Pollution control, Approach, CPCB, Ganga Action Plan, NAAQS

The environmental concerns in India are as old as the Indian civilization. Although the focus was mainly on the conservation of forests, wild life and natural resources, the period after independence in 1947 and up to 1972 witnessed continuous deterioration of the environment due to rapid industrialization, accelerated urbanization, population growth and increasing poverty. During this period of degradation of environment in terms of deforestation, loss of wild life, growth of slums, occurrence of droughts and floods and increasing air, water and soil pollution continued.

21.1 AN OVERVIEW OF ENVIRONMENTAL CONCERN

Although, initially the Government made provisions for supply and sanitation but control of pollution and conservation of environment as such were not emphasized. The concern for preserving the quality if life and protecting environment was stressed for the first time in the Fourth Five Year Plan (1969-74) (Planning Commission, 1970, IV Five Year Plan), wherein it was felt necessary to regulate pollution in a stricter manner. The year 1972 marks a landmark in the history of environment management in India. Prior to 1972 environment programs were dealt with by different ministries of the Government of India, and each pursued its own objectives without proper – coordination. When the 24th United Nations General Assembly decided to convene a Conference on Human Environment in 1972, a National Committee on Environmental Planning and Coordination (NCEPC) was established in the Department of Science and Technology. The NCEPC was an apex advisory body in all matters relating to environmental protection and improvement.

The water (Prevention and Control of Pollution) Act, 1974 (The water act 1974) was passed for restoration and maintenance of wholesomeness and cleanliness in our national aquatic resources. To implement the provisions of the Act, Central and State Water Pollution Control Boards were formed. The Air (Prevention and Control of Pollution) (The air prevention and control act, 1981) Act was passed in 1981 for the control of air quality in the country conferring water pollution control boards such powers and functions as necessary to air pollution prevention and control. The continued decline in the quality of the environment together with the tragedy at Bhopal in 1984 in which a leak of Methyl-isocyanate (MIC) from a pesticide factory killed more than 2,500 people injured several thousands has prompted the Central Government and few state governments to adopt stronger environmental policies with enactment of new laws. The Department of environment was expanded as the Ministry of Environment and Forests. The most significant Central legislation in recent times is the Environment (Protection) Act, 1986. (EPA) which empowered the Central Government to take all necessary measures to protect and improve the environment. The amendments to the Air Act in 1987 and water Act in 1988 empowered the enforcing agencies to close polluting industries and to stop their electricity or water supply. The penal provisions in both these laws were also strengthened.

The most challenging problem in protection of environment is the need for pursuing economic growth. This challenge is manifested in policy pronouncements of the Government of India. The major policies are focused on integrating the environment with developmental activities. These concerns are reflected in the national planning process, constitutional provisions, and the administrative machinery set up to accommodate to development policies that focus on environmental conservation. In February 1992, the Policy statement includes strategies for Abatement of Pollution was tabled in the Parliament. The statement includes strategies, such as adoption of clean process technology, change of concentration based standards to mass based standards, incentives to industries for pollution control, public participation, environmental auditing and distinguishing environment friendly products through eco-marks labeling etc. Control of pollution at source is the immediate short-term objective adopted by all State Pollution Control Boards. For this, the industries must know the extent up to which their effluent or emission must be treated/controlled, so that they can discharge the treated effluent to receiving environment without significant effect. The Central Pollution Control Board (CPCB) has initiated evolving industry a specific Minimal National Standards (MINAS) as early as in 1977-78. Effluent standards for 23 categories of industries and emissions standards for 21 categories have been notified under the Environment (Protection) Act, 1986. Government of India. Several rules have been notified under, the EPA. The State Boards use these guidelines for necessary follow up. While no State Board is permitted to relax the MINAS, but they may make them more stringent depending upon the location.

21.2 CENTRAL AND STATE POLLUTION CONTROL BOARDS

Central and State pollution Control Boards have been established under the Water Act 1974. The functions of the central Board include advising the Central Government on water pollution issues, coordinating the activities of State Pollution Control Boards, sponsoring/organizing research and training programs related to water pollution, and developing a comprehensive plan for the prevention and control of water pollution. State boards perform these functions in the respective states. The Central Board was also functioning as states boards for the Union Territories till 1991. During 1991 these functions were transferred to local administration of respective Union Territories.

The Central and State pollution Control Boards was constituted in September 1974 under the Water (Preservation and Control of Pollution) Act, 1974. All the twenty five states of the Union of India have adopted the Act and the respective State Pollution Control Boards have been constituted. Since May 1981, the CPCB and the State Pollution Control Boards have been entrusted with the added responsibility of air pollution control under the provisions of the Air (Prevention and Control Pollution) Act, 1981. Enactment of the environment (Protection) Act 1986, which is an umbrella legislation for enforcement of measures for protection of environment, has further widened the scope of activities of the Boards.

21.3 FUNCTIONS OF THE CENTRAL BOARD

Main function of the Central and State Pollution Control Boards as spelt out in Water Act 1974 and Air Act 1981 are : (i) to promote cleanliness of wells and streams in different areas of the states by prevention and control of water pollution and (ii) to improve the quality of air and to prevent, control or abate air pollution in the country.

21.3.1 Functions of the Central Board at the national level are

- Advise the Central Government on any matter concerning the prevention and control of water and air pollution and the improvement of the quality of water and air.
- Plan and to execute a nation wide program for the prevention, control or abatement of water and air pollution.
- Coordinate the activities of the State Board & resolve disputes among them provide technical assistance and guidance to the State boards, carry out and sponsor investigation and research relating to problems of water & air pollution and for their prevention, control or abatement.
- Plan and organize training of persons engaged in the prevention, control or abatement of water and air pollution.
- Organize through mass media, a comprehensive mass awareness program on the prevention, control or abatement of water and air pollution.
- Collect, compile and publish technical and statistical data related to water and air pollution and the measures devised for their effective prevention, control or abatement, prepare manuals, codes and guides relating to treatment and disposal of sewage and trade effluents as well as for stack gas cleaning devices and stacks and ducts.
- Disseminate information in respect of matters relating to water and air pollution and their prevention and control.
- Lay down, modify or annul, in consultation with the State Governments concerned, the standards for streams or wells and lay down standards for the quality or air and water.
- Perform such other functions as may be prescribed by the Government of India.

21.3.2 Functions of the Central Board as State board for the Union Territories are

- Advise the Government of Union Territories with respect to the suitability of any premises or location for carrying on any industry which is likely to pollute a stream or well or cause air pollution.

- Lay down standards for treatment of sewage and trade effluent and for emissions from automobiles, industrial plants and any other polluting source.
- Evolve efficient methods for disposal of sewage and trade effluents on land.
- Develop reliable and economically viable methods of treatment of sewage, trade effluent and air pollution control equipment.
- Identify any area or areas within Union Territories as air pollution control areas or areas to be notified under the Air Act, 1981; and
- Assess the quality of ambient water & air & inspect wastewater treatment installation, air pollution control equipment, industrial plants or manufacturing processes to evaluate their performance and to take steps for the prevention, control or abatement of air & water pollution.

21.4 BASIC APPROACH TO POLLUTION CONTROL

The basic objective of water Act is to maintain and restore the wholesome ness of water by prevention and control of pollution. The Act does not define the level of wholesomeness to be maintained or restored in different water bodies of the country. The task of defining these levels for different water bodies is given to the CPCB. Which has initially taken use of water as a base for identification of water quality objectives for different water bodies. Hence it has classified the national aquatic resources according to their uses. The CPCB has also identified primary water quality criteria for different uses to act as a yardstick for the preparation of different pollution control programs. The use based classification system is given in table 1.

Table 1. Use based classification of surface water in India.

Designated best use	Quality Class	Primary Quality Criteria
Drinking water source without conventional treatment, but with chlorination.	A	6.5 to 8.5(1); 6 or more (2); 2 or less (3); 50, 5% - 200 and 20%-50 (4); NIL (5-8)
Outdoor bathing (organized)	B	6.5 to 8.5(1); 5 or more (2); 3 or less(3); 500, 5% - 2000 and 20%-500 (4); NIL (5-8)
Drinking water source with conventional treatment	C	6.5 to 8.5(1); 4 or more (2); 3 or less(3); 5000, 5% - 20000 and 20%-5000 (4); NIL (5-8)
Propagation of wildlife and fisheries	D	6.5 to 8.5(1); 4 or more (2); NIL (3-4); 1.2(5); NIL (6-8)
Irrigation, industrial cooling and controlled waste disposal	E	60 to 8.5 (1); nil (2-5); 2250 (6); 26(7); 2 (8).

(1) pH, (2) dissolved oxygen, mg/l (3) BOD, (20c) mg/l (4) total Coli form (MPN/100ml) (5) Free ammonia mg/l, (6) Electrical conductivity in μmho/cm, (7) sodium adsorption ratio, (8) boron mg/l

21.5 USE OF WATER QUALITY CLASSIFICATION SYSTEM

For the preparation of a pollution control plan for any water body one has to define its water quality objectives to identify the extent of pollution control needed. The present status can be used for prioritization.

The CPCB is a national apex body responsible for preparation of pollution control action plans for the country. CPCB, therefore needs a scientifically sound yardstick. The "best designated use" yardstick is based only on primary criteria, and needs further scientific support. The criteria should be such that they can support not only the designated uses but also the ecological sustainability of water body (wholesomeness), which is the prime objective of Water Act 1974. Hence the CPCB is evolving a methodology for water quality evaluation which can integrate all the water quality parameters. This methodology is known as biomonitoring methodology.

21.6 WATER QUALITY MONITORING

Water quality monitoring is one of the important activities of the Central pollution Control board. It helps in the identification of water bodies which are in need of quality improvement, by taking pollution control measures. It also helps in performing the formulation of national pollution control programs in rational and scientific manner. Giving proper advises to Central and State Government in various pollution control policies is also supported by monitoring activities. Inland Water quality monitoring is currently being done under three major programs namely:

Global Environmental monitoring System (GEMS)

Monitoring of Indian National Aquatic Resources (MINARS)

Ganga Action Plan (GAP)

At present there are 480 WQM stations (locations), of which 51 are under GEMS 402 under MINARS and 27 under GAP

Out of the 480 water quality monitoring stations distributed throughout the country, 422 are on rivers 32 on lakes and creeks 2 6 are groundwater.

21.6.1 Automatic Water Quality Monitoring Station on River Ganga and River Yamuna

On the recommendation of the Central Pollution Control Board, the Ganga project Directorate (GPD) is monitoring the quality of River Ganga at nine Automatic Water Quality Monitoring Stations (AWQMS) with the following objectives:

To collect baseline data on the river water quality at sensitive and non sensitive points on river Ganga, which will also be complimentary to manual data;

To collect data to be utilized in strategic planning models and in assessing the impact of sewage treatment plants on river water quality by monitoring up-streams and down streams of major cities; To detect the compliance with water quality criteria established for various designated best uses; To provide early warning to downstream users about the adverse water quality conditions upstream.

21.6.2 Biomonitoring of water quality

The present WQM program takes into account only a limited number of physico chemical parameters (about 19), which is often not sufficient to assess actual water quality. Hence the CPCB has recently developed a methodology to use biomonitoring techniques for water quality evaluation, under Dutch collaboration. The biomonitoring methodology is more informative and can integrate long term effects of all the adverse environmental factors. The methodology which was developed in a three year pilot study on Yamuna river is now being tested in other river systems. Once the methodology is found suitable, it is proposed to be introduced on a large scale.

21.6.3 Coastal Water Quality Monitoring

A monitoring network of 173 stations was established during the seventh five year plan over the entire coast line of the country at varying distances from the coast to assess the quality of the coastal and estuarine waters. This program was jointly taken up by the Department of Ocean Development (DOD) and Department of Environment (DOEn). Out of these 173 stations, the Central Pollution Control Board co-ordinates monitoring at 107 stations through the State pollution Control Board of Gujarat, Maharashtra, Kerala and Tamil Nadu and East Zonal Office of the Central Board, Calcutta. The remaining 66 stations were monitored by the department of Ocean Development through various agencies.

21.7 AIR QUALITY MONITORING

21.7.1 National Ambient Air Quality Monitoring Program

The National Ambient Air Quality Monitoring Program (NAAQM) started in 1984 with 28 stations over 7 cities. Over the years, a number of major cities and industrial towns have been covered. The total number of sanctioned stations at present is 290 covering 99 cities. Air quality monitoring at these locations is conducted by the respective state boards with financial assistance from CPCB. There basic parameters, namely, sulphur dioxide (SO_2), nitrogen dioxide (NO_2) and suspended particulate matter (SPM), are monitored for 24 hours on alternate days. Data collected are analyzed statistically by the CPCB and published.

21.7.2 Automatic air quality monitoring

The ambient air quality is being continuously monitored 24 hourly at the integrated monitoring station at Bahadur Shah Zafar Marg in Delhi throughout the year. Nitrogen dioxide and carbon monoxide are measured using automatic analyzers. A new SO_2 analyzer has been acquired and is installed recently at the station. The manual monitoring of SO_2, NO_2, and SPM are being carried out simultaneously using High volume sampler (HVS).

21.8 ENVIRONMENTAL PROBLEMS AND CONTROL MEASURES

Water component of the environment is degraded by municipal wastewater at many locations and by industrial wastewater at few locations. Counter measures to deal with municipal wastewater have taken shape in the Ganga Action Plan and the proposed National River Action Plan. In respect of industrial wastewater, categories of industries, significant from water pollution angle, have been identified and compliance with standards in a time bound manners is being ensured.

Air component of the environment is also degraded in several locations. In metropolitan cities, exhausts from automobiles cause pollution in high traffic density areas. To counter the problem, it is considered that strict standards for exhausts at manufacturing stage is the most important measure. Accordingly, such standards for 1995 and 2000AD have been evolved and forwarded to the Government for its consideration. Emission from industries are sought to be countered in a manner similar to wastewater discharge as stated above.

In an attempt to counter the environmental problems in an integrated way at areas seen to be problematic, twenty two such areas were identified. A time targeted action plan to control the pollution and improve the environment is prepared and being implemented.

21.8.1 Industrial Estate Survey

In India over 2 million small scale industries are existing and about 2000 industrial estates have come up. Collecting full details of each individual small scale unit in the country is quite difficult at any stage; and hence it is decided to collect data of small scale units existing in an organized industrial estate or on cluster basis. Though small scale industry as a unit may not be severely polluting, but a cluster of these industries may severely pollute the environment.

There are certain industrial estates which are critically polluting the environment. Such critical industrial estates need a detailed survey to establish the feasibility for setting up of common effluent treatment plant (CETP). With these objectives the CPCB has taken up the study of critically polluted industrial estate with the assistance of state boards. Dry inventorization of various industrial estates in the country has been taken up. After the completion of dry inventorization, critically polluted industrial estates will be identified following which detailed survey will be conducted.

21.8.2 SODAR Inversion Studies

Sonic detection and ranging (SODAR) or acoustic echo sounder is a unique remote sensor for mapping the atmospheric stability, convective plumes, low level disturbances, turbulence and depth of planetary boundary layer using naturally occurring fluctuations in atmospheric temperature and wind velocity and is useful for air pollution modeling and dispersion studies.

21.8.3 Studies on heavy metals in Particulate in Ambient Air

The determination of heavy metals in airborne suspended particulate is important in order to assess the anthropogenic air pollutants which may lead to emission source identification, to assess existing levels to which human population is exposed, and application of strategies in air pollution control program.

Measurement of lead (Pb), Cadmium (Cd) and Zinc (Zn) in air borne particulate matter in Delhi is being carried out. The lead concentration in air borne particulate matter has been determined as between 42ng/m^3 to 3617 ng/m^3 with a distribution trend as residential area < residential cum commercial area < industrial area.

The concentration of cadmium in air borne particulate matter ranged between 1ng/m^3 to 46 ng/m^3, which is mainly attributable to fly ash emissions from thermal power plant and industrial process emission while zinc was recorded between 346ng/m^3 to 3313 ng/m^3 with same distribution trends as for lead.

21.9 AMBIENT AIR QUALITY MONITORING IN RELATION TO HUMAN HEALTH

The study on effect of ambient air quality on human health had been initiated in collaboration with the All India Institute of Medical Sciences (AIIMS) from October 1, 1989. The aim of the project was to evaluate the health status of the population residing in the areas in relation to ambient air quality which is being continuously monitored. Data on population about age, sex, income, smoking habits and health status have been collated and later on clinical examination carried out. The study involved survey of 607 families comprising 2511 members.

21.9.1 Analytical Quality Control

The quality of analysis is the backbone of the pollution control boards because all the decisions are based on the data provided by the laboratory analysis and then interpretation. Hence, the analytical quality control has been introduced among the pollution control boards. Both intra and inter laboratory analytical quality control have been initiated. Several practical training programs for quality control procedures have been organized where the scientists of pollution control boards participated.

21.9.2 Pollution control in major polluting industries

Compliance by the industries with the environmental standards has not been adequate in spite of the best efforts including financial incentives made over the years by the pollution control boards/state govt./Central Govt.

The central pollution control boards selected 17 categories of major polluting industries for priority action and in February 1991 directed the state boards to ensure compliance of the standards in these industries by December 31, 1991. Accordingly most of the state boards amended the consent order of all the industries in these categories so that no defaulting unit shall have consent of the board with effect from January 1, 1992. Extensive discussions were held with the state boards in groups in a scheduled manner, to asses the problem and identify the defaulting units and initiation of suitable actions.

Pollution control status of industries in 17 categories, in 23 states and union territories (including all major industrial states) have been collected, and compiled and defaulting units have been identified. Time target of December 31, 1991 for compliance continues to be in focus but if adequate steps have been taken to display proof of commencement of physical work, then the latest dates by which units could be permitted to comply beyond the date, have been notified by the central Government as follows:

It was suggested that an industry, operation or process which has commenced production on or before the day of 16th May, 1981 and has shown adequate proof of at least commencement of physical work for establishment of facilities to meet the specified standards within a time bound program to the satisfaction of the concerned state pollution control board, shall comply with such standards latest by the 31st day of December, 1993.

An industry, operation or process which has commenced production after the 16th May 1981 but before the 31st day of December 1991 and has shown adequate proof of at least commencement of physical work for establishment of facilities to meet the specified standards within a time bound program, to the satisfaction of the concerned state pollution control board, shall comply with such standards latest by the 31st day of December 1992.

21.10 ENVIRONMENTAL AUDITING

Environmental auditing is defined as a basic management tool comprising a systematic documented, periodic and objective evaluation of two well environmental organization, management systems and equipment are performing. The aim of the audit is to facilitate compliance with company's policies including meeting regulatory requirements. Environmental auditing is seen as an internal process that should become a necessary and routine part of industrial management. Environmental audit has been made mandatory by the central government and covers all industries which fall under the water act 1974, air act 1981 and hazardous waste rule of 1989. The CPCB has conducted such audit for several industries and developed a methodology. The objectives of this indepth study is to evaluate the performance of the pollution control systems to identify good pollution prevention and control systems for demonstration besides imparting on the job training to industry personnel in environmental monitoring including sampling and analysis of effluent/emissions.

21.11 DEVELOPMENT OF LOW AND NO WASTE TECHNOLOGY FOR SMALL SCALE INDUSTRIES

The CPCB in collaboration with various research institutes and state pollution control boards in the country is developing demonstration project on low waste/no waste technology for small scale industries.

21.11.1 Scheme of labeling environment friendly products

A scheme of labeling of 'environment friendly' products was instituted with a view to providing accreditation and labeling of house hold and other consumer products which meet standard for products. The label shall be known as the ECOMARK". Any product which is made, use or disposed off in a way that significantly reduces the harm it would otherwise cause to the environment, could be considered as 'environment friendly' product.

A steering committee has been set up in the ministry of environment and forest (MOEF) to determine the product categories for coverage under the scheme and also formulate strategies for promotion, implementation, future development and improvements in the working of the scheme. The government of India has constituted a technical; committee in the central pollution control board to identify the specific product to be selected and prepared criteria/guidelines for the individual product category. The criteria notified by the MOEF for individual product category will be adopted in Indian standards as optional environmental criteria by the Bureau of Indian standards and the scheme will be finally implemented by them. The ECOMARK label shall be awarded for a minimum period of one year and shall be renewed annually. The time period of the award may be reviewed for time to time

21.11.2 Incentives for control of pollution

From the initial year of pollution control the expenditure on providing pollution control equipment were termed as non productive expenditures and the regulatory authorities had to face a tough task in convincing the industry for providing and running of pollution control device. Although a few industrialists understood the need of pollution control as a social obligation and were to discharge their responsibilities but the main constraint was non availability of cheaper technologies and finance. It was felt if the expenditure on the Prevention of pollution could either be made productive or compensated most of the industry may come forward to provide the same to comply with pollution control regulation. Pollution Control technologies which are self paying and sustainable were developed and publicized. The concept of recovery of energy and by-products received appreciation. At the same time government showed its willingness to extend fiscal and financial incentives to encourage adoption of pollution control measures by the willing industries. A prime example of this is the exemption on fly ash bricks from excise duty.

The concept of common treatment plant and providing financial assistance upto 50% of the cost of common effluent treatment plant will greatly benefit the small scale industries located in clusters and having space, technological and financial constraint in controlling the pollution individually. Provisions of rebate in water cess for these industries who comply with the prescribed effluent standards have also helped indirectly the industries who run the effluent treatment plant regularly and properly (Policy Statement, 1992).

Following are the fiscal incentives currently available to the industry for control of pollution.

(a) *Depreciation Allowance:* A depreciation of 50 per cent is provided on devices and systems installed by manufacturing units to control pollution. Specific equipment has been notified by government of India for this purpose.

(b) *Water Cess:* There is a provision for a rebate of 70 per cent in the cess levied on water use if the industry concern has installed equipments for treatment of sewage or effluent.

(c) *Concession Custom Duty:* Concession Custom Duty is levied on notified equipments and spares for pollution control. Currently the Government is examining the possibilities of including more items under this act.

(d) *Excise Duty*: Concessional excise duty is provided for manufacturing goods used for pollution control. In 1989 manufacture of fly ash bricks containing more than 50 per cent of fly ash was completely exempted from excise duty.

(e) *Financial Institutions*: Financial Institutions have the provision for extending soft loan facilities for installation of pollution control equipment. Industrial Finance Control of India (IFCI) has a specific scheme for subsidizing cost related to pollution control in the medium and small scale sectors which come under the category, and if a capital outlay upto Rs. 3 crores. The scheme is mainly to subsidize the cost of assessing the individual requirements for equipment, analyzing existing effluent advising appropriate methodology among other things.

The technical consultancy organization whose fees to prepare the pollution control equipment are to be subsidized should have one or more financial institutions or development agencies as either share holders or members of the board.

The industry may not remain unaffected if the environment around the industry is affected. Large industries located in urban area may adopt small area in the cities for improving its environment and thereby the quality of life. In the process many industries may join hands to adopt bigger areas. The industry must take all steps necessary to keep the environment clean and to contain pollution. Only few years back the attitude of industries was antagonistic. The industries were not willing on the pretext of finance, space and technical constraint to provide Effluent Treatment Plants (ETPs). With the generation of mass-awareness, development of technology, offering of fiscal and financial incentives, rigorous persuasion and stricter supportive legislation, many of the industries have not only understood the necessity of pollution control but have also taken necessary steps to control pollution. Now a trend is setting where the industry is interacting with surrounding settlements. Many of industries have not only made sincere efforts to keep the surrounding environment clean but also did service to the society by providing schools, parks, hospitals and other amenties as a compensation to their inconvenience.

21.12 ENVIRONMENTAL TRAINING

The water (Prevention and Control of Pollution) Act, 1974 and Air (Prevention and Control of Pollution) Act, 1981 emphasize that training is one of the functions of the Central Pollution Control Board. Trailing in various aspects of prevention, amendment and control of pollution to the identified target groups is important. The target groups are important. The target groups includes officials dealing with planning, funding and implementing programs for prevention and control of pollution in the central Government, the state Governments, the Central and state pollution Control Boards, Universities, individuals and the local bodies. Courses are also organized for operators of industrial and wastewater treatment plants for proper operation and maintenance of treatment plants.

21.12.1 Major challenges

Legal Considerations

The basic objective of the water (Prevention and Control Pollution) Act, 1974 is to maintain and restore the wholesomeness of the national aquatic resources.

Due to large variation in type, size, shape, quality of water available of our aquatic resources, each one has got a very specific waste load receiving capacity. This implies the need for prescribing different effluent standards based on assimilative capacity of recent systems. Notwithstanding its merits, it is difficult to administer compliance of relating effluent quality to ambient water quality the concept of Minimum National Standards (MINAS) was evolved by the Central Pollution Control Board whereby minimum effluent limits are prescribed for each category of discharge, regardless of receiving of water requirements. Where water quality standards cannot be reached by imposition of a standard level of treatment alone, then and only then will the conditions of the receiving waters dictate more stringent controls (State Boards can make the MINAS more stringent). The conventional methods of treatment cannot cope with these specific situations and the polluters are required to go for introduction of new methods for specific requirements. The number of such situations in our country is gradually rising due to water scarcity in many water bodies and some times due to the nature of effluents.

Water Quality V/S Quantity Considerations

Although, the MINAS, is mainly technology based, but some considera-tion is given to the receiving water body based on the assumption that in the receiving water body at least 10 times dilution is available to achieve the ambient water objective in immediate vicinity of the downstream. In the recent past due to over use of surface waters to fulfill the ever increasing demand of agriculture, industrial and domestic uses, the water level in our aquatic resources is gradually decreasing. As a result, the assumption of 10 times dilution is no more valid in many parts of our country. This poses great difficulties to achieve the ambient water quality objective (Wholesomeness) even after implementing MINAS. This is a challenge before the pollution Control Boards and the polluters. Thus, it has become necessary to achieve progressive reduction or elimination of effluent discharges. It implies adoption of no waste technologies through new treatment methods which can fulfill the above objectives.

Experience from Industries

Many industries inspire of adopting conventional treatment, are not able to achieve the MINAS either due to high BOD or presence of xenobiotics like lignin, phenols, cyanides, pesticides, PAH, PCB etc. or toxic substances which are not easily degradable. In such situations specific techniques could achieve the targeted quality of final effluent. Specific genetically engineered microbes are now a days becoming popular to cope up with such specific wastes. Many techniques are being employed wherein microbes are fixed on a media and are subjected to the effluent treatment in a defined specific condition. Many such microbial systems are being imported in our country by the industries and consultants on trial basis for treatments of effluents from pharmaceutical and organic pesticide industries. The effluents are thus treated to the targeted level. Several modifications are developed in the conventional treatment methods to tailor the specific requirement of the industrial effluents.

Small scale industries in India are facing great difficulty in achieving the MINAS due to variety of problems. The most important one is the financial hardship. It is difficult for small scale industry to afford the cost of the conventional treatment. Secondly, even if they establish a treatment system they cannot run it properly due to lack of skill which is very important for running such plants. Moreover, many of the small-scale industries operate on job basis. Hence, the wastewater is generated intermittently. But conventional treatment plants are generally effective when operated on continuous basis. Interrupted operation results in conditions where microbes to die-off which may take long time to regenerate. Till such time, the final effluent is unable to achieve the targeted quality. The problem is more acute in handling of specific wastewaters like the effluents from H- acid plants, dye intermediate plants, electroplating, textile hand processing units etc.

21.12.2 Experience from the Ganga Action Plan

Under the Ganga Action Plan, several sewage treatment plans have been constructed, besides renovating the existing ones. During operational phase, several problems are being faced. Apart from the administrative problem of operation and maintenance, there are technical problems. Amongst others, inadequacy in the design of treatment plants leads to a host of problems that are being faced in construction, operation and maintenance. This is due to the fact that the characteristics of sewage actually reaching the plant are much different than the design criteria. The plant designs are based on population based estimated BOD load, which is much higher than the BOD actually to the plants. Thus the plants were over designed. The treatment plants need skill operation and maintenance, besides being energy and cost intensive. All these factors may result in manufacturing of the ETP and jeopardize the very objective of the GAP. Therefore, now in the implementation of second phase of GAP and National River Action Plan more focus is placed on such systems which are less expensive, simple to operate and capable of functioning without high operational skill.

REFERENCES

Policy Statement for Abatement of Pollution, 1992. Ministry of Environment and Forests, Government of India.

Planning Commission, Government of India 1970.: IV th Five Year Plan.

Rosencranz, A., Shyam Divann and Martha L. Noble 1991. Environmental Law and Policy in India. Cases, Materials and Statuis. Tripathi, Bombay.

The Water (Preservation and Control of Pollution) Act, 1974; No.6 of 1974 Gazette (Extraordinary) of India.

The Air (Preservation and Control of Pollution) Act, 1981; No. 14 of 1981. Gazette (Extraordinary) of India, Part II, Sec. 1 30th March 1981.

The Environment (Protection) Act, 1986; Gazette (Extraordinary) of India, part II, Sec. 1 29th may 1986.

The Environment Protection Rules, 1980. Notification No. S.O. 844 (E) 19th.

CHAPTER - 22

Understanding about an air environment among school children

S.J. Bedi

Botany Department, Faculty of Science,
The Maharaja Sayajirao University of Baroda, Vadodara 390 002

ABSTRACT

Education may be defined as the process of bringing desirable change into the behavior of human beings by acquiring knowledge through institution or study in order to achieve definite goal. Educational programmes are the means of communication, which enable people to share knowledge ideas about a communicated message. Environmental education helps top to provide every person with opportunities to acquire knowledge, values, attitudes, commitment and skills needed to protect and improve the environment. UNEP, WWF and United Nations conference on Environment and Development, (UNCED), 'The Earth Summit held in Brazil 1992 have emphasized the need for Clean environment. Indian Association for Air Pollution Control (IAAPC) was formed at B.H.U Varanasi (U.P) on 18th September 1976 to bring together all those interested in the problems of Air Pollution. The association has organized seminars and exhibitions on issue related to Air Pollution and its abatement. The paper describes the views of certain students about environment. They have actively participated in debates, elocution, slogan writing competition, drawing and painting competition. Selected paintings will give you a glimpse of their understandings about some complex environmental issues.

Keywords: Air pollution, Environmental education, Awareness, School children

Air quality is steadily deteriorating and in industrial pockets of the country air pollution levels have exceeded the WHO and CPCB standards. In the absence of integrated abatement strategies, increased air pollution will pose even greater threat to the health and well being of India's urban populations, eventually under-cutting the prospects for further economic growth. Dynamic partnerships among government, industry and the public will be crucial in making air quality management a part of urban development plans across India (Dubey, 2005). Jahangir is credited with the laying out of the Shalimar garden in Srinagar. Alexandar Rogers (1909) writes that Jahangir does not like the city Ahmedabad due to dust pollution. He mentions "since all this country has sandy soil, when any movement takes place, so much dust rises that the faces of people are seen with difficulty, so that one should call Ahmedabad 'Gardabad' (i.e. abode of dust) (Nene, 2000). Mughal emperor Jahangir during his reign of 22 years visited several places. About Agra he describes that Mass of people (in Agra) is so great that moving about in the lanes and bazaar is difficult.

Indian Association for Air Pollution control has organized poster and drawing competition for school children to create awareness about pollution hazards (Fig. 9 & 10). The topics for on the spot poster event were 'Clean and green Vadodara' 'Air pollution effects and mitigation', 'Pollution and Health, Eco degradation due to Air Pollution, the topic "protect the whale shark" was given to conserve the wildlife in the event held in 2005. The event was jointly organized by IAAPC along with CPCB west zone, WWF-India, NEEDS, Vigyan Parishad and Zoo Outreach organization, Coimbatore. Students of large number of schools have showed their interest and actively participated in these events. Some of the posters are depicted in this chapter.

A slogan writing competition was organized on 5th June 2005. The event was for both school children as well as common public. The Environment can be more lively if we conserve it with these messages.

Clean Environment – For a better future ahead – Sarita Das.

Environment is in our hands - Devashree A. Vyas

Keep green our lands

Save forests – Save life - Devashree A. Vyas

Save future generations

Save Environment – Save Life – Anshul Arya

"Pollution"

If you won't kill it

It will kill you. - Chirag K. Shukla

Do not forget, earth is warming

Grow more trees and make it charming - Abhinav Pratap

Man says 'I am Supreme'
Yes, the supreme parasite
He rides the steed of progress
With his demon pollution
Pining for blood of life.
As he tramples the fragile
Fabric of life
In Evil Smog the earth suffocates as man irradiates
The wounded skies bleed
The colourful tapestry of life fades as grey pervades
We gasp for breath
The earth is parched for pure water
Now a cocktail of poisons
Time is running out Do something ! -Ankush Sabharwal, 1997

A student from Baroda High School, Alkapuri, Ms. Prajakta Modak (6 years, 1st standard) has used two colours blue and green to colour Earth, the land area which is rich with vegetations is painted green. The planet is surrounded by two more colours to show troposphere and stratosphere. The girl has clear cut concept of atmospheric layers and all around the planet black clouds of pollution are depicted. She has appealed in the poster to "Save the Earth from pollution" (Fig. 1). Mr. Abhinav Varma (9 years, 6th standard) of City Montessori School, Lucknow has depicted various sources of pollution present on this earth like industries, pesticides, motor or vehicular exhaust, aeroplanes, noise pollution and pollution due to wars. To overcome the pollution due to varied sources he has suggested massive tree plantation (Fig. 2).

Amongst diversified group of plants the trees/ herbs/ shrubs with higher toxic scavenging potential are better suited for plantation. For example the *Bougainvellia* is preferred on road side because of its better SO_2 absorbing potential. The *Nerium* can survive the high pollution load because it has many layered epidermis.

Ms. Anuradha Shrimali from Gorwa, Vadodara has found industries as major source of pollution. This may cause acid rains and loss of fertility of soil. A child is shown with a drum to warn that stop this pollution (Fig. 3). She also advocates the use of green belts around industries. A green belt of excellent and best categories of trees having a width of 50 to 100m between residential and industrial zone can decrease the dust level up to 52%, noise up to 40 to 60%, approximately, A 500m belt of trees around the industrial area can mitigate about 76 per cent of SO_2 and 62 per cent NO_x from the atmosphere and one ha. of trees absorb about 3.7 tonnes of CO_2 from the air and generating 2.5 tonnes of life sustaining oxygen (Srivastava *et al.* 2006)

Mr. Hardik Patel (9th standard) of Bright School, Vadodara has shown a city with industry, beautifully coloured houses, and greenery. He has depicted following four messages in the poster

1. Plant trees and save the country
2. Save petrol stop air pollution
3. Small family is a happy family
4. No smoking

Plants prevent air pollution is well known. They provide clean air and shade to the pedestrians, Oxygen, besides many valuable things to us. Plants do have life was shown by Sir Jagdish Chandra Bose a well known Physiologist of the country. Mr. Sarjan A. Dalal (8 years, 3rd standard) of Bright school, Vadodara has painted the earth with orange colour which shows the prosperity, and jubilant mood of planet. It has industries with smoke and car moving on highways on both sides of industries. The sky is painted blue with blazing sun smiling on earth. Two trees are visible in background (Fig. 5). But the centre place of the poster is occupied by a personified tree. An axe has cut the main trunk and it is bleeding. The tree has five branches of these two are depicted with eyes. These eyes are full of tears the red coloured tears of the tree shows cry of the plant.

Plants are multicellular organisms with green pigment chlorophyll. Chlorophyll traps the energy from sunlight which, together with a water and gas called carbon-dioxide converts into simple sugars which allow it to grow. They grow at one place and require minerals for growth. Scientists found that tiny plants were growing 3,000 million years ago. At present there are more than 2,50000 species of flowering plants which are beneficial to us or the ecosystem. Hamburger buns and all the bread we eat, come from the ground up seeds of plants such as wheat or corn. Milk comes from cows, which again feed on grass. Even the fish feed on phytoplankton (smaller plants) in river or sea.

Not only food or feed, the plant fibers are used to make cloth. Now we are using not only cotton but plants like *Agave, Musa* and *Bambusa* for fibers. Trees provide timber for building and furniture and paper for newspaper and books, many medicines, botanical pesticides and perfumes are extracted from plants.

The world can be divided into different belts of vegetation, from the poles to the equator. Each belt consists of plants that require similar conditions of rainfall and temperature. If you start at the Artic the first vegetation you find is short, turfs tundra with no trees. Then comes a wide forest belt of coniferous (cone bearing) followed to the south by forest of broad – leaves trees, then grassland, scrub, desert and possibly tropical forest near the equator. Climbing a high mountain provides a rather similar effect, as you pass through the different vegetation zones. Near the summit or close to the snow-line if it is a really high mountain, you may find only some tiny, short stemmed flowers the high alpines that have adapted to life in extremes of wind, sun and snow. These are the equivalent of the tundra plants of world vegetation belts (Black and Huxley, 1985).

Ms. Gargi Khiste (15 yr. 10th standard) of Shreyas Vidyalaya, Vadodara has named her poster as "Let me breath". She has drawn a picture of green tree with human face and has shown that how polluting smoke from industries can enter our lungs (Fig. 4). The picture depicts very clearly harmful effect of polluting gases and plant as an alternative source to reduce this damage.

Ms. Tarun Patel (12th standard) has shown an imaginative scene in which industry is emitting smoke it is labeled as Devil and due to pollution tree is bleeding, the blood droplets can be seen coming from it. The birds and cattles are in bad conditions A child is crying and requesting to save the nature from the devil pollution (Fig. 6).

Ms. Shraddha Arya (10th standard) of Baroda High School, ONGC, Vadodara has depicted two situations on right hand side the polluting industry is shown and on the left hand side the natural green vegetation is depicted (Fig. 7). There is no pollution, large number of trees is surrounding the residential huts. On the other hand industry is causing soil and water and air pollution, smoke from stacks is seen. Mr. Kunal Joshi (8th standard) from MGM School, Vadodara has tried to show the agony of present human being, He has tried to show the industrial pollution, pollution from vehicles like scooter, auto car, truck, aero plane, and pollution from waste water and pesticides. All these may cause serious problems to him and he may suffer with various complications and even mental tension.

Mr. Zuber Pirzada (9th standard) of MES High School, Nagarwada, Vadodara has suggested some methods that how Baroda can be made clean and green in 5 years? He advocated tree plantations and use of better fuel like CNG and electric driven vehicles on roads (Fig. 8). Ms. Nikita Desai (9th standard) of Navrachna School, Vadodara has painted Gujarat as Pollutiondham (house of pollution), the future of our towns. The small girls has thought various situations like city has traffic jam, suffocating shopping mall and industries producing CFC, Industries are polluting atmosphere with sulphur rich fumes and acid rains has deserted the forest all the trees are naked. City is facing smog problem.

The damage to the ozone layer and the enhanced green house effect are two global environment problems which have attracted a great deal of attention in recent years from scientists, policy makers and general public. Both phenomenon are commonly regarded in isolation, which is correct as a first approach. The ozone layer is primarily damaged by active chlorine or bromine compounds in stratosphere, originating in the anthropogenic emissions of such compounds as CFC's and halons. The enhanced green house effect is to great degree caused by emissions of CO_2, CH_4, N_2O and CFCs. Green house emission can affect the ozone layer in variety of ways: via chemical reactions with ozone, via reactions with chlorine compounds, via UV radiation, and via reaction in the troposphere.

In scientific terms, the enhanced green house effect is studied in a separate context than the damage to the ozone layer. Research into the enhanced green house effect, together with its social and economic consequences, is assessed in international frameworks as the IPCC (Intergovernmental Panel on Climate Change). Damage to the ozone layer is investigated in the UNEP/WMO framework. Both groups regularly publish reports on the current status of research (WMO, 1994; IPCC, 1995). In policy terms, too both environmental themes are treated separately: the enhanced greenhouse effect comes under the damage to the ozone layer falls under the Vienna Convention and he Montreal Protocol.

The concentrations of compounds that damage the ozone layer have increased sharply since 1930s. Fig. 1 shows the atmospheric concentration of CFC-11 one of the most important compounds that damage the ozone layer. The use of such compounds started to decline at the end of 1980s due to measures adopted under the Montreal Protocol. This led to a maximum in the tropospheric concentration of CFC -11 in 1992-93, followed by a slight decline thereafter. Damaging compounds will reach their maximum concentration in the stratosphere in roughly 1988-1999, and the ozone layer can start to recover after that.

Ms. Monali Patel (9th standard) has shown two faces of earth one with polluting industrial units and vehicles and other with green trees free from atmospheric pollution.

Gaseous emissions are one of the major urban pollutants. Auto mobiles vitiate the environment in the process by emanating obnoxious toxic pollutants in the surrounding atmosphere and thereby pose serious health hazards to biotic community both urban population and plants life. According to Srivastava *et al.* (2006) the pollution load of Kanpur city due to vehicular pollution is 2550 kg h^{-1}. This pollution can be reduced by use of better quality fuel and use of filters, catalytic converters lead free petrol in vehicles.

Plate - I

Plate - II

Plate - III

Plate - IV

Eco-Education UGC Directive

With environment degradation generating grave concern worldwide during the last few decades, many organizations have started exploring effective and meaningful ways to impart eco-education to the masses, especially to younger generation. A fall out of this is Supreme Court intervention in December, 2003, directing all state governments to include environmental studies as compulsory subjects in the University curriculum.

The University Grants Commission (UGC) – the key regulatory body for higher education – is also involved in the process. The organization has evolved a six month long mandatory core module course for all the universities and colleges under its jurisdiction. Provision has been made to combine theoretical knowledge and field work in a balanced manner. The course will have eight units, covering a wide array of issues (like linkages between social issues and the environment). The priority areas are as follows: create awareness and sensitivity about, environmental and allied issues; enhance knowledge and understanding about environmental issues and challenges; imbibe an attitude for environment; and motivate students to maintain or improve the quality of environment.

REFERENCES

Ahmed K.J. 2004. High rises – A solution for urban pollution. *Environews* 10(3):1

Black D. and Huxley A. (1985) *Colour library of Science – Plants* Pub. By Orbis Pub. Ltd. London 64 pp

Dubey P.S. 2005. Urban Air pollution and its management. In : *Urban Pollution issues and solutions* (eds) A. Arya, S.J. Bedi, Y.T. Jasrai and V.S. Patel, Nidhi Book Centre, Tri Nagar, New Delhi:1-9

Nene Y.L. 2000. Jahangir : A Naturalist : In Ancient and Medieval History of Indian Agriculture (S.L. Chaudhary, G.S Sharma and Y.L. Nene, eds) Rajashan College of Agriculture, Udaipur 160 - 195

Rogers, A. and Beveridge, H. 1909. The Tuzuk –i – Jahangiri (Memoirs of Jahangir) Vol – I. Royal Asiatic Society, London, U.K 478 pp

Srivastava R.J. Dubey P., Kumar A. and Prasad K. 2006. Bioremediation of urban pollution through green belt

Velders G. 1987. The effect of green house emissions on the ozone layer. *Change* 39:4-6.

CHAPTER - 23

Fungal biopollutants prevailing in fruit markets of Baroda and associated diseases of certain fruits

Chitra Arya and Alpa Parmar

Department of Botany, Faculty of Science,
The Maharaja Sayajirao University of Baroda, Vadodara – 390 002
e mail: caarya@yahoo.co.in

ABSTRACT

Numerous aerobiological surveys have revealed that air may be contaminated by bio-particles like pollen grains and fungal spores. These bio-particles are potential cause of allergy and skin diseases in human beings. Many of the fungal organisms cause diseases in plants and deterioration of organic objects. Scientists have found that not the occurrence but quantum of inoculum or number of spores per unit area has direct effect on development of diseases in field as well as in markets. Usually the number of fungal colonies increases in the months of July-August and March- April due to favourable temperature & RH, which results in increased disease incidence.

During regular survey of Kadak Bazaar, ISKON Temple market and Khanderao markets of Baroda infected cherry fruits were noticed in first week of June 2004. Different samples were collected from these two markets in clean sample bags and isolations were made in PDA slants. A total of 24 fungi were isolated from air and surface washing technique. The isolation from infected fruit resulted in axenic cultures of 3 fungi. Pathogenicity of *Alternaria alternata* (Fr.) Keissler and *Fusarium moniliforme* J. Sheld. was confirmed. Third organism *Rhizopus stolonifer* Went, belonging to Mucorales group failed to cause infection in healthy cherries thus it was a saprophyte.

A comprehensive survey of fruit markets of Vadodara was undertaken to determine various fungal pathogens associated with sweet red cherry. Frequent isolations from surface washing technique yielded twelve micro fungi. Artificial inoculations with the isolated propagules were also performed in mature cherry fruits and were able to cause rot.

Key words: Surface mycoflora, Aeromycoflora, *Alternaria alternata*, *Fusarium moniliforme*, Fruit rot, Cherry

Bio-pollutants like fungi, yeast and bacteria may cause diseases in agriculture fields, orchards or in vegetable markets. They may attack during harvesting, packaging or during transport to the fruit markets. Fungi constitute major share of bio-pollutants, as well as contaminant present in our atmosphere. Usually they reproduce by sexual and asexual means and their spores as well as conidia form component of aerospora. Their presence in environment may pose threat to plants animals and human beings. Fungas may cause biodeterioration of organic materials. Gregory (1973) recorded spread of fungi and bacteria by the action of wind and rain splash. In the past Phytopathologists could identify spores stages and predict infection time and probable damage in certain areas (Stakman and Christensen 1946). The success in case of cereal rust infection patterns attracted wide attention. Gregory (1973) observed that plants can act as source of biopollution and microorganism may reach in the markets with fruits and vegetables. Unhygienic condition of markets in most parts of the country is responsible for huge economic losses.

In tropical and sub tropical climate diseased plants can act as sources of air borne biopollution. These propagules may cause plant diseases and even allergy to human beings. The ability of a pathogen to infect the host and not the known host continues to be enigmatic. A fungus may fail to cause infection or show its reduced activity in spite of formation of infection structure. The influence of cell walls on synthesis of polysaccharides of their activity may play an important role in determining the resistance to invasion.

23.1 AEROBIOLOGICAL STUDIES IN INDIA

India has unique distinction of being one of the earliest countries where Aerobiological studies were initiated. As early as 1873 Cunnigham attempted to relate air borne organisms to the incidence of cholera and other fevers, using an aeroconiscope in Kolkata prison (Cunnigham 1873). Nearly about half a century after Cunnigham's once again studies were initiated, but this time by Plant Pathologists (Mehta, 1952; Padmanabhan *et al.* 1953). Prof. K.C. Mehta is known for his classical work on rust fungi. Dr. S. Nagarajan as well as Dr. L.M. Joshi and his associates discovered the 'Puccinia Path'.

Air borne microorganisms have been recognized as major bio-pollutants capable of causing deterioration and complete degradation of food materials in markets. The fruit and vegetable markets serve as the important source of fungal growth and spore dissemination into the atmosphere. Study of post harvest diseases deteriorate is known as market pathology. Prevailing aeromycoflora can drastically deterioatrte the quality of fruits. Scientists including Narayan *et al.* (1982), Verma (1989), Verma and Khjare (1988), Bellad and Reddy (1989), Verma & Sheorey (1994) Ghani (1996), Kakade *et al.* (2001) Kulshreshtha and Chauhan (2002) Sarma and Bhattacharjee (2002) Sahney and Punwar (2002) have studied the aeromycoflora of vegetable and fruit markets of Bangalore, Jabalpur, Gulbarga, Bhagalpur, Nagpur, Guwahati, Agra and Allahabad, in India respectively.

23.2 FUNGI AND POST HARVEST LOSSES

Of all the losses caused by plant diseases those which occur after harvest, are most costly whether measured in monetary terms or in man hours. World food shortage demands for reducing post-harvest losses and developing improved systems for transporting and distribution of produce to international organizations responsible for monitoring food resources have recognized that the most economically feasible and expedient means to increase the world food supply is to reduce losses in food crops that occur after they have been harvested (Kelman 1984). Post harvest losses are much higher for highly perishable fresh fruits and vegetables than cereals and other field crops. Because of their high moisture content, fresh fruits are very susceptible to attack by pathogenic fungi and bacteria during the period between harvest and consumption.

It is amazing that post harvest losses of perishable crops in developing countries range from 5-50% or even more (Salunkhe and Desai 1984). A major reason for this has been the lack of awareness at all stages right from fruit production to fruit consumption. Harvested fruits and vegetables become vulnerable to attack by microorganisms because of (1) the high moisture content (2) high sugar and nutrients (3) harvesting, packing and transportation facilitate entry of certain pathogens (4) during ordinary storage a single infected fruit can successfully spread the infection in the whole lot.

Fungi have been identified as a major cause of biodeterioration of all kinds of stored products throughout the world leading to both qualitative and quantitative losses (Meronuck 1987). Eckert, 1978 has enumerated the following less evident consequences of post deterioration of perishables.

(1) Partial or total loss of consumer packages with only one or several diseased units.

(2) Reduced post-harvest life of the product due to accelerated ripening or senescence triggered by ethylene released from a few diseased fruits in a package or a storage room.

(3) Possible contamination of the edible product with mycotoxins elaborated by the disease inducing microorganisms e.g. Aflatoxins (produced by *Aspergillus* spp.) Patuline (*Penicillium expansum*) in diseased apple and stones fruits, furanoterpenoid metabolites as in sweet potatoes infected with *Ceratocystis fimbriata* (Boyd and Wilson, 1972).

(4) Softening of processed fruits by heat tolerant pectolytic enzymes of post-harvest pathogens *e.g. Rhizopus* sp. and *Sclerotinia* (Strand *et al.*, 1981).

Cherries (*Prunus avium* L.) are temperate climacteric fruits that are usually picked at peak maturity for optimal taste and appearance. The fleshy red cherry which are rich in vitamin C, sugar, amino acids, phosphorus, iron and other minerals. However, in India, and other countries that produce sweet red cherry for fresh consumption, the cherries are often stored for up to 2 weeks at cold temperatures to increase the seasonal supply. At the time of storage these nutritional cherries get infected by majority of the microorganisms that cause post harvest deterioration. The pathogens invading the tissues directly or more usually through wounds attained during harvesting and poor handling practices (Dennis 1983).

Propagules of the pathogenic fungi causing post harvest fruit rot are found abundantly in the atmosphere (Meredith 1961) and even on the surface of fruits as they mature in the orchards (Dennis 1975). Thus post harvest diseases may be influenced by the location of orchards and the husbandry of the fruit trees. Some literature is available on the surface microflora of freshly harvested fruits and their possible role in spoilage and marketing (Singh and Kainsa 1983, Gupta *et al.*, 1983a, 1983b, Kalafatolgu and Karapinar 1989). Singh and Sumbali (1997, 1998) reported post harvest fungal diseases of Indian Jujube. Arya (2004) reported different fruit rots in markets of Vadodara. Madan (2000) reported diseases of crop plants and human allergies due to air born fungal spores like – *Cladosporium, Alternaria, Curvularia and Aspergillus.* A direct correlation was observed between percentage rot of fruits and occurrence of fungal spores in different months in fruit market in Baroda (Arya and Arya, 2007).

23.3 QUANTIFICATION OF AIR BORNE BIOLOGICAL PARTICLES

Quantification of air borne biological particles forms the basis of aerobiological research related to agriculture, horticulture, forestry, air borne pathogens of plants, animals and humans. The fruits and vegetables are important constituents of our diet. The post harvest spoilage may further reduce their availability to common man at economical rates. According to Mayee (2005).

"Aeroflora diagnostics shall has to go molecular way to understand not only taxa, genus, species but also virulence, avirulence and hence frequency of type of pathogen if at all, this knowledge is to be used for management. Air has been a very major dispersal mechanism of large number of pathogens like fungi, vector borne viruses, bacteria, mycoplasma, protozoan etc. aerobiologists have used air sample traps, balloons, helicopters, Vaseline coated slides as easy means of catching spores. Use of Global information systems (GIS) and Pest Distribution Map (PDM) is gaining momentum all over the world. Efforts have been made to develop calendar for pathogens which will help to correlate the incidence of plant diseases in particular season.

23.4. STUDIES UNDERTAKEN

Baroda is situated on western part of the country on Delhi-Mumbai train route. It is an important city of Gujarat, India. The place is situated 72° – 45' - 74°30' E and 21°-45 – 23°0' N and altitude 35 mt. Three markets selected for study were Khanderao market, situated in heart of city is a whole sale vegetables market. Kadak bazaar is another market situated near railway station. Iskon temple is the third market selected. It is situated on northern side of the city and is comparatively a small market.

23.4.1 Materials and Methods

Regular visits were paid to these 3 markets to collect the diseased fruits. Aeromycoflora was studied by petriplate exposure (gravity fall) method.

23.4.2 Isolation of aeromycoflora from different fruit markets

The study was done three times a day, morning 8:30am on 1st & 15th of every month. Five petriplates with PDA medium were taken to the fruit market and exposed at a height of 1mt. The petriplates were exposed at different places in market for 10 min. After exposure they were brought to the laboratory and kept in BOD at 25 ± 2°C for 7 days till the colony start developing. Number of the individual colonies, and type of the colonies were counted and the fungi were identified (table 1) Percentage abundance of each fungal isolate was calculated as follows

$$\text{Abundance (\%)} = \frac{\text{Total No. of colonies of a particular fungal species in all the replicates}}{\text{Total number colonies of all the replicates}} \times 100$$

To observe the effect of different time of a day the Petri dishes were exposed at 3 different times in a day and data were recorded for 1 market (Table 3).

Table 1. Percentage abundance of fungal organisms present in aeromycoflora of fruit markets of Baroda during May-August 2006

Fungi isolated	Percentage abundance of Fungi					
	15 May	1 June	15 June	1 July	15 July	1 August
Alternaria alternata	2.3	2.0	2.1	1.5	1.5	2.5
Aspergillus niger	7.8	7.0	8.9	12.0	15.5	10
A. flavus	5.7	7.5	9.2	10.0	8.5	8.5
A. ochraceus	1.6	1.6	-	-	-	-
A. tamarii	1.5	-	1.2	-	1.0	2.5
Chaetomium globosum	0.16	0.5	0.5	-	-	-
Cladosporium oxysporum	6.8	6.0	7.2	2.0	5.0	-
Colletotrichum gloeosporioides	1.4	2.0	1.8	2.0	2.5	5.0
C. capsici	5.4	6.6	6.6	6.0	5.5	2.5
Curvularia lunata	5.9	6.0	6.4	5.9	5.0	2.5
Drechslera sp.	8.7	-	7.3			
Emericella nidulans	3.5	2.0	2.6	-	-	-
Fusarium moniliforme	6.9	5.0	7.2	2.2	2.2	-
F. oxysporum	5.3	5.0	2.6	2.0	5.0	-
F. solani	6.5	2.5	5.4	6.5	6.0	2.5
Lasiodiplodia theobromae	1.6	2.5	1.3	1.5	1.5	2.5
Monascus ruber	-	0.5	0.5	-	-	-
Mucor sp.	8.6	8.5	5.2	2.0	2.0	2.5
Myrothecium roridum	0.4	1,5	0.5	1.5	-	-
Penicillium citrinum	1.3	1.5	1.0	2.5	1.0	2.5
P. expansum	-	-	1.3	1.5	-	-
P. oxalicum	-	-	1.2	1.5	2.0	-
Phoma herbarum	0.2	-	-	2.0	2.0	-
Rhizopus stolonifer	9.7	8.0	8.4	10.0	10.0	10.0
Syncephalastrum racemosum	-	-	-	1.5	2.0	1.5
Thielaviopsis paradoxa	-	-	1.5	0.5	-	-
Trichoderma oryzae	-	-	2.0	2.0	1.0	-
T. viride	-	1.5	1.4	1.4	-	-
Unidentified	2.3	5.0	5.0	2.5	5.0	2.5

Data based on 5 petridishes (100 mm dia) exposed for 10 min.

Table 2. Study of aeromycoflora in Kadak bazaar at different times during the day

	Sampling day	2.11.06	20.12.06	
Time	Fungi isolated	No. of colonies	Fungi isolated	No. of colonies
Morning	**8:30 a.m.**	**37**	**8:30 a.m.**	**31**
	Aspergillus awamori	4	*Penicillium* sp.	2
	A.niger	10	*A.niger*	12
	A.flavus	4	*A.flavus*	6
	A.fumigatus	6	*A.fumigatus*	-
	Cladosporium sp.	6	*Cladosporium* sp.	3
	Mucor haemalis	5	*Mucor haemalis*	7
	Fusarium oxysporum	2	*Fusarium oxysporum*	-
Noon	**1:00 p.m.**	**26**	**1:00 p.m.**	**22**
	Aspergillus flavus	2	*A. flavus*	
	A. fumigatus	5	-	-
	A. niger	11	*A. niger*	9
	Cladosporium sp.	4	*Cladosporium sp.*	3
	Mucor haemalis	4	*Mucor haemalis*	3
Evening	**6:00 p.m.**	**28**	**6:00 p.m.**	**27**
	Aspergillus flavus	2	*A. flavus*	3
	A. niger	14	*A. niger*	11
	Cladosporium sp.	2	*Cladosporium sp.*	2
	Mucor haemalis	8	*Mucor haemalis*	4
	Penicillium citrinum	2	*Fusarium sp.*	2
			Penicillium sp	2
			Rhizopus sp.	3

Table 3. Fungal bio pollutants isolated from fruit washing of cherry (*Prunus avium*)

S.N	Fungi associated with fruit surface	No[a]
1	*Alternaria alternata*	4
2	*Aspergillus niger*	8
3	*A.flavus*	2
4	*A.tamarii*	2

5	*Cladosporium oxysporum*	5
6	*Fusarium moliniliforme*	1
7	*F.oxysporum*	2
8	*Myrothecium roridum*	1
9	*Nigrospora oryzae*	2
10	*Penicillium expansum*	2
11	*Pestalotiopsis palmarum*	1
12	*Rhizopus stolonifer*	5

[a] - colonies record by washing 25 g fruit in sterile distill water. **Total – 35**

Table 4. Percentage fruit rot recorded in different months

Fruit	15 May	1st June	15thJune	1st July	15 July	1st August
Mangifera indica L.	2.6	10	15	15	7	2
Carica papaya L.	2	2	2.6	10	15	10
Prunus avium L.	7	7	2.6	5	5	5
Achras sapota L.	5	5	2	-	-	-
Prunus domestica L.	-	5	5	2	5	-
Musa paradisiacal L.	5	7	7	10	15	10

(Data based on results obtained by Arya & Arya 2007)

23.4.3 Isolation of surface Mycoflora

For isolation of the surface Mycoflora associated with red cherry fruits, samples were taken at regular interval of 7 days during the period of available in fruit market, subjected to surface washing technique. In this method, 25g of whole fruit was taken in Erlenmeyer flask (500 ml) containing sterilized distilled water and given horizontal shaking for 30 min. The liquid was then centrifuged at 3000 rpm for 20 min and the residue thus obtained was mixed with 10 ml sterilized distilled water and shaken vigorously to obtain a homogenous suspension that was poured in sterilized petriplates at the rate of 1 ml/plate. PDA modified by adding Streptomycin sulphate (0.06 g/l) was used and replicates were maintained for each sample. The inoculated petriplates were incubated at 25 ± 2°C for 7 days and the resulting fungal colonies were sub cultured on potato dextrose agar (PDA) slants and identified. Data is recorded in Table 3.

23.4.4 Isolation of fungi from diseased fruits

Fifty fruits of each variety were collected from 3 fruit markets in clean polyethylene bags and brought to the laboratory. Percentage rot was calculated on the basis of visual symptoms of whole fruit and cut fruits (Table 4). The organisms responsible for different fruit rot were isolated, purified by subculturing culture and maintained on PDA slants. Fungal isolates were identified on the basis of their macro and micro-morphological characters, using the relevant literature and keys.

23.4.5 Pathogenic potential of surface Mycoflora

Tests were conducted to check the pathogenic ability of the isolated surface mycoflora assess their relative importance in the spoilage of cherry fruit. For pathogenicity test, mature healthy fruits of similar size were surface sterilized with 70% ethanol and then inoculated with the test fungus by the pin – prick method (Tomkins and Trout 1931). Inoculum used in each case was 7 days old culture grown on PDA at 25 ± 2°C. After inoculation, the fruits were placed in sterilized polythene bags maintained at 25°C and 100% relative humidity. Extent of spoilage in each case was evaluated after 8 days of inoculation.

23.5 RESULTS AND DISCUSSION

23.5.1 Fungal spores in Fruit markets

A large proportion of aerobiological studies are devoted to study air borne spores in the markets of different cities and towns by employing viable plate or gravity slides, and a few have used air samplers. Irrespective of place, duration and technique of sampling the predominance of deuteromycetes fungi in air spore have been confirmed by all the investigators with *Cladosporium* as more predominant at all places. The total number of spore types identified varied from 35-65 (Vittal, 2005).

Percentage abundance of species comprising the mycoflora is depicted in Table 1. *Aspergillus flavus* Link was the most prevalent component of the surface Mycoflora appearing in all the samples investigated. Other frequently isolated species were *Alternaria alternata* (Fr.) Keissler, *Aspergillus niger* Van Tieghem, *Cladosporium oxysporum* Berk., *Curvularia lunata* (Wakker) Boedijn, *Fusarium solani*, *Penicillium expansum* Thom. Of these fungi *Alternaria alternata* was found responsible of spoilage of cherry during post harvest period, indicating thereby that propagules of these fungi possess the potential to cause rot of mature fruits. Fungi which were absent from environment during 1st July to 1st August included *Aspergillus ochraceous, Chaetomium globosum, Drechslera* sp. *Emericella nidulans, Monascus ruber*. Spores of *Myrothecium roridium, P. expansum, Thielaviopsis paradoxa, Trichoderma viride* were absent on 15th July and 1st August. May and June were hot summer months with maximum temperature reaching 42^{0}C and July and August were rainy months.

It is likely that during harvesting, packaging and transportation injuries of various kind are caused which facilitate the entry of surface mycoflora resulting in disease development, even after harvesting, as the fruit continue to respire, the resultant heat accelerate respiration and ageing which in turn make the fruit susceptible to attack by the surface mycoflora (Das Gupta and Mandal 1989). Certain fungi *A. alternata, A. flavus, A. niger C. oxysporum, C. lunata* and *F. solani* over others towards the harvesting time, similar observations have been recorded on chillies (Gupta *et al.,* 1983). In contrast some of the fungal species (*Aspergillus, Fusarium* and *Mucor*) that were isolated from the fruit surface throughout the growth period were not observed to be associated with post harvest fruit rot during survey period. Similar behaviour has been cited earlier also in reference to Strawberry spoilage by same zygomycetes member (Harris and Danis, 1980).

Kulshreshta and Chauhan (2002) conducted aeromycologial survey of Chhipitola vegetable and fruit market of Agra, the second biggest market of Uttar Pradesh in 2000. Total numbers of 27,755 fungal spores per meter cube of air were trapped by Rotorod sir sampler of Perkins modified by Harrington. A total of 38 fungal spores were recorded. The proportion of the 4 taxonomic groups of fungi was Phycomycetes (2.63%), Ascomycetes (7.78%) and Deuteromycetes (68.42%) the group of Phycomycetes showed only one fungal type *i.e Phytophthora*. Among the Ascomycetes 8 fungal spores types were trapped with *Chaetomium* (7.00%) and *Pleospora* (2.45%) as dominant types. A total of 26 types of Fungi imperfecti members were seen. *Aspergillus* was the most dominant type followed by *Cladosporium* and *Alternaria.* Ghani (1996) and Kakade *et al.* (2001) also found that spores distribute amongst 40 types, *Alternaria* (675.8) constitutes the pre-dominant spore followed by *Cladosporium* (367.1) *Helminthosporium* (246) *Epicoccum* (127.9) *Drechslera* (117) *Curvularia* (99.4) and *Fusarium* (88.7).

Post harvest fruit rot causing potential of associated aeromycoflora which is also associated with the surface of cherry. As many as 19 fungal species inhabiting the surface of cherry fruits were able to develop rot out of 24 fungi found in aeromycoflora shown in Table 1. Healthy cherry fruits when inoculated individually these fruits showed high spoilage by pathogenic fungi which were present in aeromycoflora (Table 2).

All these fruits allow a conclusion that a definite relationship exists between the aeromicrobial population and post harvest fruit rot of cherry fruits the extent of deterioration may vary depending upon the harvesting procedures practiced physiological capabilities of the surface mycoflora and nutritional qualities of the fruit.

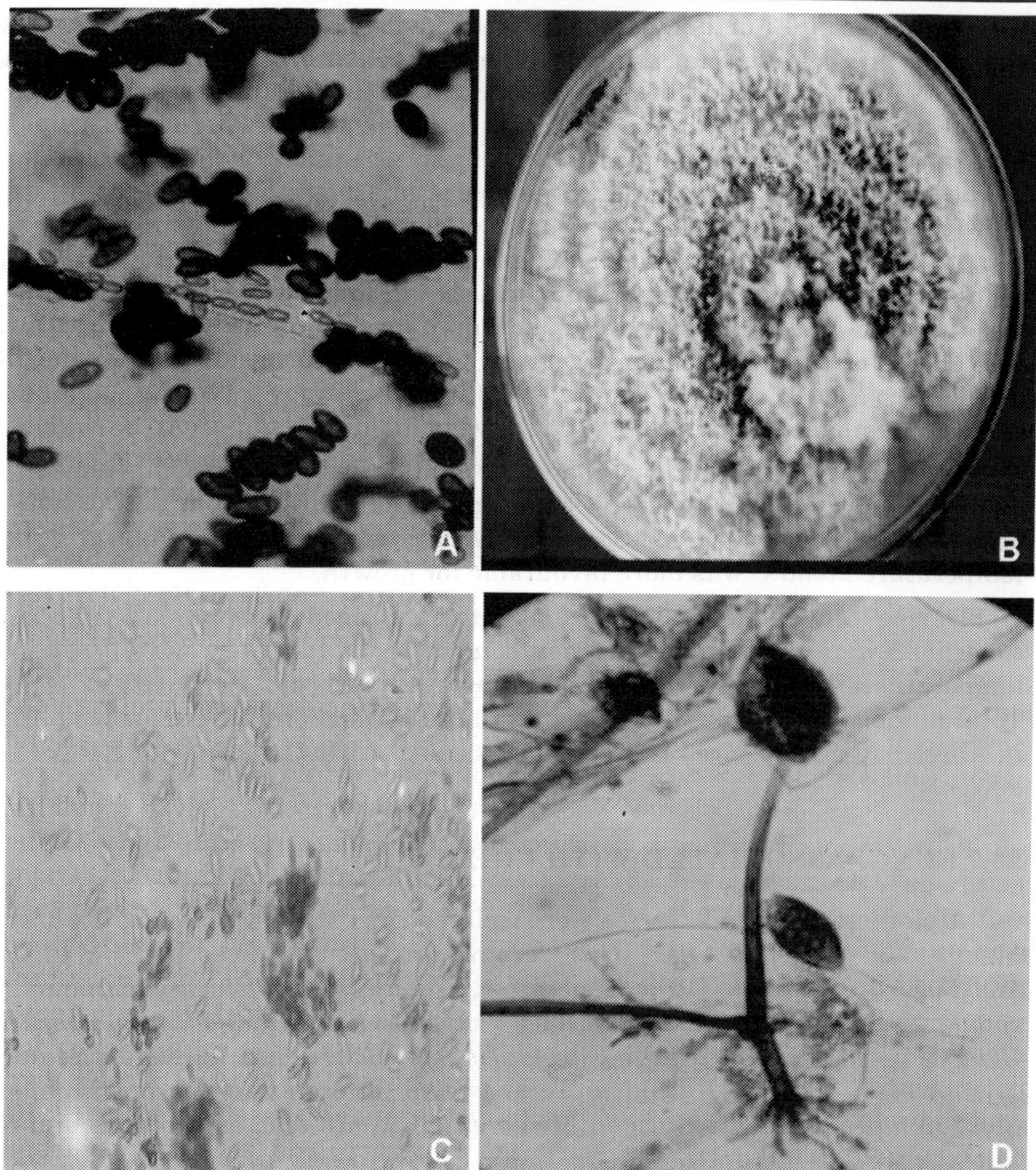

Plate - I

23.6 EFFECT OF ENVIRONMENTAL CONDITIONS ON OCCURRENCE OF FUNGAL SPORES

Abundance of fungal spores in the markets may be due to unhygienic conditions which prevail in the market area. Rotten fruits, vegetables, and eatables, decaying paper bags and packing materials in the market are good substrates for fungal growth and become the chief source of dissemination of spores in the air. Further semi enclosed environment of the crowded and nearness of the sampling sites to the source of contamination may be accounted for the higher spore concentrations. In most of cases maximum colonies were encountered during rainy season (July – August). This was probably due to congenial climate like high rainfall with corresponding increased humidity and moderate temperature might have added in the growth and sporulation of these microorganisms. In real sense a micro climate existing between fruit packing boxes with no air circulation, hot and humid, encourages pathogen to flourish and disperse species when fruits are taken out for sale. In the present study number of fungal species was 21 on 15th May22 on July and 25 on 15th June, 2006. The percentage abundance of fungal species was more on 1st and 15th July, which may be explained on the basis of the increased humidity in the environment.

Meteorological conditions play an important role on incidence of aerospora. More fungal spores were recorded in months of February to April and October to November (Sahney and Purwar 2002). In February the temperature 20-30°C was more favourable for growth.

During May and June the temperature ranged between 31-42°C and conditions were dry which were non conducive for fungal growth. Very low temperatures during December & January was recorded in Northern India. Total numbers of colonies were 99 and 165 in December and January months in Allahabad as compared to 277 colonies in April 1997 (Sahney and Purwar, 2002).

23.7 FUNGI ASSOCIATED WITH CHERRY FRUIT SURFACE

A plethora of fungal species were isolated from the fruit surface of cherry Members of the genus *Aspergillus* were predominant followed by *Penicillium* and *Fusarium* species. Other genera isolated by fruit washing technique as components of the fruit surface mycoflora included *Chaetomium, Cladosporium, Colletotrichum, Curvularia, Drechslera, Lasiodiplodia, Mucor, Myrothecium, Phoma, Rhizopus.* Total twelve different fungi were isolated. More number of colonies encountered were of *A. niger, C. oxysporum* and *R. stolonifer* (Plate I; Fig. D).

Fungal spores may settle on fruit surface by impaction or sedimentation under the influence of gravity, and during rain by splash droplets. The presence of saprophytic mycoflora may influence the pathogenicity of an organism by an antagonistic effect. The nutritional status of particular fruit and available moisture has direct correlation on occurrence of disease. Post harvest losses can be reduced by monitoring the aeromycoflora in different markets, and by providing the conditions so that pathogenic fungi can not come in direct contact of fruits. Wattal *et al.* (2003) reported consumer losses due to post harvest diseases of cherry. Isolation of *Cercospora cercumcsissa, Wilsionomyces carpophylum, Penicillium expansum* and *Tricothecium roseum* was done from cherry fruits for the first time. The studies further revealed that except *C. cercumcsissa* all species were pathogenic. Earlier *Alternaria alternata, A. denritica, F. semitectum, Geotrichum candida* have been found associated with cherry fruits.

23.8 POST HARVEST DISEASES OF CERTAIN FRUITS

Percentage fruit rot of six different fruits were calculated every fortnight and data are presented in Table 4. It is evident from the table 4 that percentage rot was more in *Musa paradisiacal* L. and *Carica papaya* followed by *Mangifera indica* and *Prunus avium*. The losses were less in chiku and plum. Banana showed infection due to *Gloeosporium musarum, Aspergillus niger, Lasidiplodia theobromae* and *Thielaviopsis paradoxa* (Plate I Fig. A) *Fusarium, Colletotrichum* and *Aspergillus* rots were common on papaya, while mango suffered due to *Aspergillus, Pestalotiopsis* and *Phomosis* sp. (Plate I; Fig. B) *Prunus domestica* had infection due to *Aspergillus, Penicillium* and *Colletotrichum* sp. *P. avium* showed infection due to species of *Aspergillus, Penicillium* and *Myrothecium roridum* (Fig. C) Presence of most of these fungi was recorded during sampling from three markets of Baroda.

Acknowledgements

Sincere appreciation is expressed to the Head, Division of Plant Pathology, IARI, New-Delhi for identification of fungi, to Head Department of Botany, The Maharaja Sayajirao University of Baroda for providing various infrastructural facilities.

REFERENCES

Bellad, S. and Reddy, C.N. 1989. Aeromycoflora of the vegetable market and its relation to the occurrence of market diseases. *Indian J. Aerobiol.* 2: 22-30.

Boyd M.R. and Wilson B.J. 1972 In : *Post-harvest pathology of perishables*, M.K. Dasgupta and N.C. Mandal, Oxford and IBH Publishing Co. Pvt. Ltd. 3

Dasgupta M.K. and Mandal N.C. 1989 *Post-harvest pathology of perishables*. New Delhi: Oxford and IBH Publishing Co. Pvt. Ltd.

Dennis C. 1983 Post-harvest pathology of fruits and vegetables. London: Academic Press.

Eckert J.W. 1978 Pathological diseases of fresh fruits and vegetables. In *Post-harvest biology and biotechnology*, (H.O. Hultin and M. Milner, eds.). 161-209, Food and Nutrition press , USA.

Ghani, T. 1996. Mycoflora of fruit and vegetable markets at Bhagalpur, Bihar. *Ind. J. Allergy Appl. Immunol.* 10(1): 35-41.

Gregory P.H. 1973 *The microbiology of the atmosphere* (2nd ed) Leonard Hill Books Aylesbury, Bucks U.K. 377p.

Gupta J.S., Agarwal S.P. and Dxit R.B. 1983a. A comperative study of air spora and surface microflora of chilli (*Capsicum annuum*) during summer. *Ind. Phytopath* 36: 138-139.

Gupta J.S., Agarwal S.P. and Dixit R.B. 1983b. Incidence of pathogenic conidia of *Colletorichum capsici* on the surface of chilli. *Ind. Phytopath.* 36: 357-359.

Gupta R.C. 1983 Studies of the effect of the endosulphan on *Vigna radiate* (L.) Wilczek. Ph.D. Thesis, Lucknow University, Lucknow India.

Harris J.E. and Danis C. 1980 Distribution of *Mucor piriformis, Rhizopus sexualis* and *Rhizopus stolonifer* in relation to their spoilage of strawberries.*Trans Br Mycol Soc.* 75:445-450.

Kalafatolgu H. and Karapinar M. 1989 Investigation of spoilage microflora in selected apple cultivars during storage. Universitiesi Ziraat Dargisi 26: 347 -356.

Kelman, A. 1984 In Post harvest Pathology of Fruits and Vegetables: Post-Harvest losses in Perishable crops. *Univ Calif Agaric Exp. Stn. Bull.*, ed. Moline H.E. 1914pp.

Madan, P.K. 2000 Fungal aerospora – A review. In : Glimpses in plant sciences (eds K.R. Aneja M.U. Charaya, A. Aggarwal, D.K. Hans and S.A. Khan). Pragati Prakashan, Merrut. (PP 333-340).

Meredith, D.S. 1961 *Botryodiplodia theobromae* and *Nigrospora* species in the air of Jamaican banana plantation. *Nature* 196:555-557.

Meronuck, R.A. 1987 PL. Dis.71: 287-291.

Salunkhe, D.K. and Desai B.B. 1984 Post-harvest Biotechnology of fruits, 1 Boca Rata, FL: CRC., 168p

Singh, J.P. and Kainsa R.L. 1983 Microbial flora of grapes in relation to storage and spoilage. *Ind. Phytopath.* 36:72-76.

Singh, Y. and Sumbali G. 1997 Some new additions to the postharvest fungal diseases of *Zizyphus maritiana* from India and their influence on ascorbic acid content. *J. Ind. Bot. Soc.* 76, 127-128.

Singh, Y. and Sumbali, G. 1998 Preharvest mycobial population of Indian jujube fruits (*Ziziphus mauritiana* Lamk.) and their implications in postharvest pathogenesis. *Mycopathologia* 142, 77-80.

Strand, L.L., Ogawa, J.M., Bose, E. and Rumsey, J.W. 1981 Bimodal heat stability curves of fungal pectolytic enzymes and their implication for softening of canned apicots. *J. Food Sci.*, 46: 498-500, 505.

CHAPTER - 24

Pollution Prevention Pays (PPP) – A profitable investment for industry

V. Mane

Botany department, Faculty of science,
The Maharaja Sayajirao University of Baroda, Vadodara 390 001

ABSTRACT

Rapid industrialization and exploitation of natural resources to achieve the quick economic development has enhanced the pollution load alarmingly in the country. Pollution prevention is a profitable investment. It deals with elimination of waste or reduction in pollution, reduction in energy consumption by utilizing raw materials and other resources efficiently. There is a need to add innovative features to manufacturing technology or product design. PPP can be profitable if we reduce wastage by improving efficiency, converting waste into raw materials or products of value. Left over wastes can be sold too. Environmental management offers great potential for a better and clean environment for the future generations.

Key words: Pollution Prevention, Environment, Industry.

Pollution prevention from all industry is desirable because sound environment planning and management in industry is concomitant with the development of the community as whole.

Industries operate within societies and their economic structures, rather than distract from them. This relationship produces benefits, such as the creation and expansion of markets; it also produces liabilities, such as environment impacts. As a consequence industries are constrained by governmental policies and regulations and more broadly, by social morals and economic and technological conditions. Therefore, it is necessary for industry to arrive at the realization that business activities and environmental concerns are not antithetical, but are essential components of sound management.

(a) The PPP Concept demonstrates that :

Profitable opportunities exist for a industry to invest in pollution prevention and resource conservation.

Environmental safeguards cost are necessary in the long – term environmental management of the enterprise.

A positive effort by the top management and operating management is imperative to seek and develop such investment opportunities.

(b) Pollution as an industry problem can be tackled by an input approach, an output approach or both, thus:

The input approach focuses on the amount of matter and energy flowing into the process and attempts to reduce this flow by eliminating waste materials and conserving energy *i.e.* a "preventive" approach (cleaner technology options).

The output approach attempts to reduce pollution as a product of the system mainly after the pollution has been generated *i.e.* the "end – of – pipe or "corrective" approach.

(c) The input approach is far more efficient because it not only reduces pollution at the initial stages but also ensures a lower cost of inputs; hence it is normally more profitable. The out put approach results in inflated costs due to internalization of the costs of environmental damage and results in reduced productivity and productive investment.

24.1 Industry and Environmental damage

(a) Industries continuously face conflicting short term and long term goals regarding:

Economic survival (Cash flow, Profit, Growth etc.).

Human and social responsibilities to employees, community, region, state and environment.

(b) Environmental damage by industry can become a significant social cost for the total community. It can eventually result in the closure of the industry besides incurring additional costs such as compensation for the affected community as well as a bad public image (*e.g.* Bhopal Gas disaster).

(c) Immediate economic pressures on industries make it difficult for them to feel responsible for social costs.

(d) Industry should recognize that poor environmental management (social conflict, jobs lost, materials wasted, health problems, pollution etc.) imply profit lost through a failure of non investment in PPP.

24.2 CRITERIA FOR PPP PROGRAM

(a) Eliminate waste or reduce pollution.

(b) Reduce energy consumption by utilizing raw materials and other sources efficiently.

(c) Add innovative features to manufacturing technology or product design.

(d) Increase profit.

24.3 STRATEGIES FOR PPP

(a) PPP usually entails

It includes Product change or re-design, process modification, equipment redesign, waste material recovery for re-use etc.

Table 1: Showing the industries and use of waste products

S.N.	Industry	Process Modification	Raw Material Recovery Recycle	Equipment Redesign	Product Redesign	Wastewater Recycle/ reuse
1	Thermal Power plant		Fly ash utilization for building materials	Need to increase efficiency		Filtration & softening of cooling tower blow down for reuse
2	Chloralkali	Ion – Exchange & Electrolysis for production of Cl & NaOH				

(b) The following checklist should be considered in formulating the PPP strategy to reduce liquid emissions in industry.

If trace metals, nutrients, solvents or oils, organic species of concern are used in any manufacturing process in connection with this assessment, is their use minimized, and have substitutes been investigated? Are acids in any manufacturing process in connection with this assessment? Have alternatives to liquid species of concern been thoroughly investigated and incorporated where possible?

Where liquid species of concern are used, do they have the lowest possible water pollution potential? Have markets and uses been identified for liquid species that leave the facility? Have the processes been designed to utilize the maximum amount of recycled liquid species from outside suppliers rather than virgin/solid residues minimized to the extent possible? Are pumps, valves and pipes inspected regularly to minimize leaks?

24.4 PPP AND PROFIT MAKING

(a) Creative profit making is demonstrated in

(b) Profits result from

cutting costs through improved material efficiency

using your own staff and know how to make improvement

selling by –product or residues (waste exchanges)

selling acquired know – how

24.5 IMPLEMENTATION

(a) A three step approach is given in Fig. 5.

Analysis of the current situation

Search for alternatives

Decision for change

(b) Responsibilities of top management

Adopt and set constructive policies which relate the industry to the community in which it operates.

Consider environment regulations and legislation and strict observance.

Evaluate staff performance by environment as well as economic criteria, establish an information system to support this evaluation.

Plan for emergencies and accidents (Disaster plan management).

Encourage environment awareness and involve staff at all levels in the PPP programme.

(c) Responsibilities of production/projects managers

Trained personnel in the skills needed to:

Comply with regulation for safe operations

Protect against health hazards

Control pollution and use preventive measures

Assign responsibilities and train personal to implement hazard management and emergency procedures.

Re-examine – existing processes with a view to introducing low-waste technologies which will minimize consumption of energy (insulation, heat recovery, fuel substitution).

Establish maintenance and good house keeping procedures and check on their implementation.

Review analytically all proposed project stages, or changes in products, production packaging or transportation for their impacts on health, safety and the total environment.

Involve personal

In a consultative process (joint committees)

Installing a complaints/redressal of grievances system

Establishing a regular information system

Rewarding outstanding contribution.

24.6 CONDUCT LIFE CYCLE ASSESSMENT (LCA) AND PREPARE AN ENVIRONMENT STATEMENT (ES)

LCA is an objective process to evaluate the environmental burdens associated with a product, process, or activity by identifying and quantifying energy and material usage and environmental improvements. The assessment includes the entire life cycle of the extracting and processing raw materials, manufacturing transportation and distribution, use/re-use/ maintenance, recycling and final disposes.

An Environmental Statement (ES) is essentially a management tool comprising a systematic documented, periodic and objective of how well environmental Organization management systems and equipment are performing with the aim of –

— Waste prevention and Reduction

— Assessing compliance with regulatory requirements

— Facilitating control of environmental practices by the company's management and

— Placing environmental information in the public domain.

An ES provides information about water and raw material consumption, pollution generated, composition and quality of hazardous and solid wastes along with their disposal practices, impact of pollution control measures taken on conservation of natural resources and consequently on the cost of production, additional measures/investments taken for environment protection including abatement and prevention of pollution. The information generated thus would aid in sectoral performance assessment and infer relative performance of units within a particular structure.

To sum up, organizations should treat the environment as an asset rather than a freely available commodity. This approach would usher in major changes in the corporate management. The need for change must be recognized as the highest corporate priorities. The first condition for successful implementation of the PPP strategy is a clearly stated top management commitment. By incorporating the goals and principles outlined earlier, organizations can gradually bring their operations in tune with environmental conservation.

Table 1: PPP Investment Opportunity in Waste Reduction in a Pulp and Paper Mill

Mill Production Data	Before PPP	After PPP
Waste solids (ash) to dispose	150 tons per day	3 tons per day
Consistency of waste	20% solids	30% solids
Unit cost of disposal	$14 /ton	$ 14/ton
Annual disposal cost	$ 765,000	$ 15,000
Fuel value of solids (dry)	9200 BTu/ton	
Consistency of ash		90% solids

PPP INESTMENT OPPORTUNITY

Capital cost of improvements		$2,900,000
Savings:		
Fuel	$ 365,000/yr	
Disposal costs	$ 750,000/yr	
Total	$ 1,115, 000/yr	
System operating costs		$ 305,000
Net annual savings		$ 810,000
Estimated return on investment		30%
Payback		4 years

Table 2 PPP Ecological and Economic Pay-Offs: Example for the Agro – Industry

TECHNOLOGIES	RESIDUES USED WITH TECHNOLOGIES
1. ENERGY	
Methane generation	Animal manures, food processes, liquids and solids, crops residues, palm oil sludge, fish wastes, meat slaughtering and processing residues.
Pyrolysis	Animal manures, wood processing residue
Burning	Dried animal manures, wood forestry residues, bagasse, nut husks, rubber wood, rice husk, cotton stalks, coconut leaf
Ethanol	Molasses, sugar cane, cassava
Charcoal	Coconut shell, coconut stem and husk, rice husk
2. HUMAN FEED	
Mushrooms	Rice straw compost, manure compost
3. ANIMAL FEED	
Direct refeeding, single cell Protein production microbial or chemical processing	Maize stalk, cobs and husks, coconut meal, cassava residues, molasses, treated straw, rubber seed meal, fish silage, edible oil refining residues, separated fruit and vegetable processing solids, ensiled manure, blood, feather meal, animal manures
4. FERTILIZERS	
Nutrients and soil Condiments	Sugar cane filter mud, distillery stillage, manure, rice, straw ash, coconut coir
5. CONSTRUCTION MATERIAL AND PAPER	Bagasse, rice straw, coconut stem
6. HANDICRAFTS	Coconut tree parts
7. CHEMICALS	
Vinegar, glycerine, Furfural, Silica, Organic acids, activated carbon, glue	Agro –industrial wastewaters

Source: F.Walker, Industry and Environment, Jan/Feb/March 1980.

24.7 CREATIVE PROFIT MAKING WITH PPP

Reducc waste by improving efficiency. Convert waste into raw materials or products of valuc. Sell what is left over.

Analyse the self cleaning and dispersing level of the environment. Work out the maximum amount of effluent that can be safely discharged in compliance with the emission standards. Involve other organizations for constructing joint facilities.

Make the local community aware of the environmental management practices being carried out within the organization.

24. 8 IMPLEMENTATION OF PPP - A THREE STEP APPROACH

Step – 1. Analysis of current situation

- — Site analysis
- — Flow sheets, operating conditions and flexibility
- — Material balance
- — Water requirements
- — Legislative trends
- — Waste treatment costs (actual and potential)

Step – 2. Search for alternatives

(a) For reduced water consumption including measures to:

- — Reduce wastage
- — Re-use waste waters
- — Treat and recycle water in the plant
- — Use separate drains
- — Substitute other solvents, etc.

(b) For reduced material use including measures to :

- — Increase conversion
- — Increase yields
- — Recycle materials
- — Avoid spillages
- — Re-use wastes from other processes
- — Reduce wastes, etc.

(c) For reduced energy use including measures to:

- Re-use waste heat
- Couple endothermic and exothermic processes
- Integrarate heat and power production
- Reduce temperatures
- Reduce pump and compressor loads
- Reduce heat losses, insulate, etc.

Step – 3. Decision for change

- Examine the impact of the proposed alternatives on the capital and operating costs of the site (Process plants and waste treatment plants) taking into account practicability of the suggestions and their interactions.

CHAPTER - 25

Phytoremediation and Bioremediation: Emerging Technologies for Pollution Control

Sayantani Ghosh

Botany Department, Faculty of Science,
The Maharaja Sayajirao University of Baroda, Vadodara 390 002,
email : sayantani04@yahoo.co.in

ABSTRACT

Phytoremediation is the use of plants for the removal of contaminants and metals from soil and water or to render them harmless. It is used to remove, metabolize or degrade toxic environmental materials. This technique has applications in both the outdoor and indoor environments. Phytoremediation is really description of what floral plants do naturally as part of their metabolisms, the uptake of moisture and dissolved elements, minerals and compounds from the soil and the removal of air borne pollutants under the normal conditions. It is an aesthetically pleasing mechanism that can reduce remedial costs, restore habitat and clean up contamination in place rather than transporting the problem to another site.

Major forms of pollution and their effects are highlighted in the chapter. The paper discusses in detail Phytoremediation of Mercury, Arsenic and Zinc etc. Bioremediation is one of the technologies that involves various microbial processes. Bioremediation and process of bisorption are described.

Keywords: Phytoremediation, Bioremediation technology, Pollution control, Biobattery

Pollution is the introduction of pollutants (whether chemical substances or energy such as noise, heat or light) into the environment to such a point that its effects become harmful to human health and other living organisms.

The EPA defines pollution as the presence of a substance in the environment that, because of its chemical composition or quality, prevents the functioning of natural processes and produces undesirable environmental and health effects. Any material that causes the pollution is called a pollutant. Pollution is not usually the result of deliberate mistreatment of the environment. The additions that cause pollution are almost always the by – products of otherwise worthy and essential activities – producing crops, creating comfortable homes, providing energy and transportation and manufacturing products and of our basic biological function (excreting wastes). Pollution problems have become more précising over year. Many materials now widely used like heavy metals can be extremely toxic as they damage nerves, liver, kidney and bones and also block functional groups of vital enzymes (Carson *et al.* 1986).

Stringent environmental laws demand for new and innovative technologies to remove metal contaminants from waste water. For more than a decade researches have been looking for cheaper and more effective methods to remediate heavy metal contaminated water and reduce the growing public health risk.

Phytoremediation

Studies on metal tolerant in plants are gaining importance owing to the possible utility of metal tolerant plants in Phytoremediation. Phytoremediation is a pollution abatement technology using algae and higher plants. Scientist have used plants like *Lemna, Eichhornia, Cynodon, Chloris* etc. Efforts have been made to grow some plants in waste water and other in soil waste. The high concentration of metal contents like Cu, Cd, Pb. Phytoremediation is sustainable and cost effective technology.

Biosorption

Biosorption is proven to be quite effective at removing metal ions from solution in a low cost and environment friendly manner (Volslky, 1990). The major advantages of biosorption over conventional treatment methods into low cost high efficiency of metal removal from dilute solution, minimization of chemicals and/or biological sludge, no medicinal nutrient requirement, regeneration of biosorption and the possibility of metal recovery.

25.1 Major forms of pollution and major polluted areas

The major forms of pollution are listed below along with the particular pollutants relevant to each of them:

- **Air pollution**, the release of chemicals and particulates into the atmosphere. Common examples include carbon monoxide, chlorofluoro carbons (CFCS) and nitrogen oxides produced by industry and motor vehicles. Photochemical ozone and smog are created as nitrogen oxides and hydrocarbons react to sunlight.
- **Water pollution** is caused via surface run off, reaching to ground water, liquid spills, waste water discharges, eutrophication and littering.
- **Soil contamination** occurs when chemicals are released by spill on underground storage tank leakage. Among the most significant soil contaminants are hydrocarbons, heavy metals, MTBE herbicide, pesticides and chlorinated hydrocarbons.
- **Radioactive contamination**, added in the wake of 20th century discoveries in atomic physics.
- **Noise pollution**, which encompasses roadway noise, aircraft noise, industrial noise on well, as high intensity sonar (is a technique that uses sound propagation under water to navigate, communicate or to detect other vessels).
- **Light pollution** includes light trespassing over illumination and astronomical interference.
- **Visual pollution** can refer to the presence of overhead power lines, motorway billboards, scarred landforms, open storage of trash or municipal solid waste.
- **Thermal pollution** is the temperature change in natural water bodies caused by human influence such as use of water as coolant in a power plant.

There are many substances in the air, soil which may impair the health of plants and animals including humans. These arise both from natural processes and human activities. Substances not naturally found in the air, soil, water or at greater concentrations or in different locations from usual are referred to as pollutants. Pollutants can be classified as primary and secondary pollutants. Primary pollutants are substances directly emitted from natural processes. Secondary pollutants are not emitted directly, rather they form in the air when the primary pollutants react or interact. An important example of secondary pollutant is ground level ozone- one of the many secondary pollutants that make up photo chemical smog.

The term heavy metals refer to any metallic chemical element that has a relatively high density and is toxic or poisonous at a low concentration. Examples of heavy metals include mercury, cadmium, arsenic, chromium, thallium and lead. Heavy metals can enter a water supply by industrial and consumer wastes or even from acidic rain, breaking down soils and releasing heavy metals into streams, lakes, rivers and ground water. Mercury is a global pollutant with complex and unusable chemical and physical properties. The major natural source of mercury is the degassing of the earth's crust, emissions from volcanoes and evaporation from natural bodies of water.

25.2 Effects of Pollution

Pollution produces physical and biological effects that vary from mildly irritating to lethal. The more serious of the two are the biological effects. Physical Effects The physical effects of pollution are those that we can see, but they include effects other than actual physical damage. One obvious physical effect of pollution is the result of oil spills that are caused by ship collisions or other accidents. Oil spilled into the sea coast everything it touches. It fouls boat hulls, pier pilings, and shore structures; spoils the beauty of nature by killing fish and birds; and makes beaches unusable. In addition to the physical effects, oil spills require costly clean up operations. Air pollutants damage a wide variety of materials. Burning oil and coal produce sulfur oxides, which cause steel to corrode two to four times faster than normal. When combined with other pollutants (soot, smoke, lead, asbestos and so on), sulfur oxide particulates cause corrosion to occur at an even faster rate. By themselves, particulates damage and soil materials, structures and equipments. Air pollutants speed the erosion of statues and buildings, which in some instances, destroys works of art.

25.2.1 Biological Effects

The most serious result of pollution is its harmful biological effects on human health and on the food chain of animals, birds, and marine life. Pollution can destroy vegetation that provides food and shelter. It can seriously disrupt the balance of nature, and in extreme cases, can cause the death of humans. Pesticides, which include herbicides and insecticides, can also damage crops; kill vegetations if sprayed in high concentration; and poison birds, animals and fish. Most pesticides are nonselective; they kill or damage life forms other than those intended. For example, pesticides used in an effort to control or destroy undesirable vegetation and insects often destroy birds and small animals.

Some life forms develop immunity to pesticides to destroy them. When that happens, we develop more potent chemicals and the cycle repeats itself. The wide spread use of pollutants, such as oil, chemicals, and fertilizers, pollutes our water ways. The biological effect of water pollution is dangerous to our water supplies; we require water to survive. Water pollutants are also dangerous to all forms of marine life.

All naval activities must meet both federal and state standards for preventing air pollution. The Navy has begun taking steps to help meet the nation's goal of reducing air pollution. Oil is especially a harmful pollutant. It kills surface swimming animals and sea birds and once it settles on the bottom, harms shellfish and other types of marine life. The primary pollution concern of Navy personnel involves pollution produced by shipboard wastes. In addition to oil, shipboard wastes include sanitary wastes, gallery and laundry waters, chemicals, solid wastes, and combustion by products of oil and gasoline-driven engines. Pollutants produced by ships are similar to those generated by municipal and industrial operations. Based on an Executive Order, all government agencies must provide leadership in the protection and enhancement of the quality of our air and water resources. They also must comply with all environmental laws and regulations. Accordingly, the Secretary of the Navy, Naval operations, and other authorities have issued several pollution control instructions. Those instructions cover the abatement of air, water, and noise pollutions.

25.3 Pollution Control Measures

- Regulations, treatment techniques and disposal method for hazardous wastes.

- Direct steel making processes that reduce pollution.
- Modified glass making, furnaces that decrease pollution.
- Non- chlorine, pulp bleaching sequences that curtail production of toxic substances such as dioxin.
- Secondary fiber utilization and reduction of PCB emissions.
- Resource recovery from sludge and ashes
- Chemical spills containment and clean up
- Uses of degradation and recycling to reduce plastic wastes.
- Plantation is also important for pollution control

27.4 Plants to reduce Pollution

Despite all of today's present technology, it seems that foliage plants and trees may be the best means of improving outdoor air quality. They are capable of removing gaseous and particulate pollutants from the air, reduce energy expenditure, lower air temperatures and make our communities more attractive to live.

25.5 Phytoremediation – A clean up Technology

Phytoremediation is an advantageous process, when plant is able to produce abundant biomass in a short period of time.

- It circumvents the removal of soil, low cost and is less disruptive and enhances ecosystem restoration.
- Contaminants/ pollutants will be transformed into less toxic form.
- It reduces the amount of wastes to be filled inland.
- It has Low costs compared to other treatment methods. The cost of cleaning up one acre of sandy loam soil at a depth of 50 cm with plants is estimated at US$ 60,000 to US$ 100,000 compared to US$ 400,000 for the conventional excavation and disposal method. One reason for this low cost is that Phytoremediation may not require expensive equipments or highly specialized personnel and can be relatively easy to implement.

25.5.1 Phytoremediation Process

There are four different types of Phytoremediation mechanisms:-

- **Phytoextraction:** The removal of contaminants and metals from the soil and transport them to leaves and stems for harvesting and disposal of metal recovery through smelting process.

- **Phytodegradation:** The uptake and degradation of organic compounds. Organic contaminants are absorbed inside the plants and metabolized to non toxic molecules by natural chemical processes within the plants.
- **Phytovolatilization:** The volatilization of pollutants in the atmosphere. Plants take up water and organic contaminants through the roots, transport them to the leaves and release the contaminants as reduced form of detoxified vapor into the atmosphere.
- **Phytostabilization:** The transformation of one species of molecule into less toxic species on the reduction of mobility. Plants prevent contaminants from migrating by reducing surface erosion and ground water flow rates. "Hydraulic pumping" can occur, when the roots reach to the ground water and take up large amount of water, control the hydraulic gradient and prevent lateral migration of contaminants within a ground water zone.

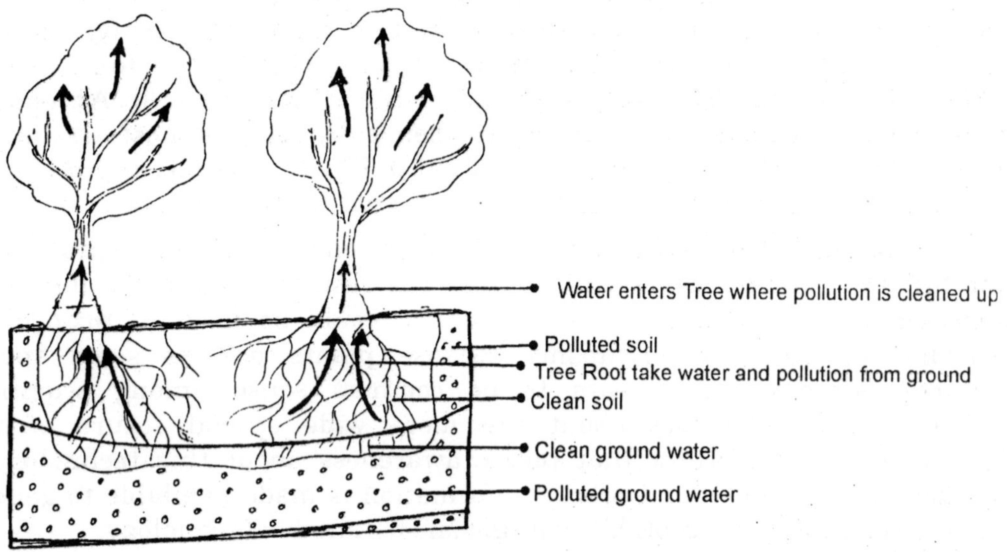

Fig. 1. Process of Phytoremediation

25.5.2 Phytoremediation of Mercury and Arsenic

Mercury (Hg) pollution is a global environmental problem. Numerous Hg contaminated sites exists in the world and new techniques for remediation are urgently needed. Plants used for Phytoremediation of Hg must tolerate high levels of mercury. A large variation (upto 30 fold difference) was detected among the six investigated clones of Willow in their sensitivity to Hg as reflected in their empirical toxicity threshold. These give us a possibility to select Hg tolerant Willow clones to be successfully grown in Hg contaminated soil for Phytoremediation. Release of Hg into air by plants is a concern when using Phytoremediation in practice. No evidence was found in this study that Hg was released to the air via shoots of Willow, *Pisum sativum, Triticum aestivum, Beta vulgaris, Trifolium*. Thus it was concluded that the Hg burden to the atmosphere via Phytoremediation has not increased. Scientists successfully engineered *Arabidopsis thaliana* to use the bacterial MerB gene (responsible for methyl mercury lyase) to convert methyl mercury to less toxic ionic mercury. The MerB transgenic plants can store to level Hg (II) produce (<0.5 ppm) in their tissues and continue to grow. Secondly, a modified bacterial mercury ion reductase gene MerA, was examined in transgenic tobacco and *Arabidopsis*. MerA enzyme detoxifies ionic mercury reducing to metallic Hg(O). Plant expressing MerA gene can tolerate ionic mercury 25-2500mM (5-500 ppm) in media or soil than normal one. This MerA and MerB gene have been transformed into and partially tested for Phytoremediation activity in *Brassica, Liriodendron, Oryza* etc.

Mercury and arsenic pollution are serious world wide problems affecting the health of millions of people. The toxic form of mercury is methyl mercury (CH_3 Hg^+). It is more toxic than Hg(O) or ionic Hg(II). The toxic form of arsenic is arsenate (AsO_4-3), more toxic than arsenite (AsO_3-3). An endogenous arsenate reductase activity electro chemically reduces most arsenate to arsenite. Some wetland species are involved in Phytoremediation process like *Scirpus validus, Carex lacustris, Phalaris, Typha, Eichhornia* etc.

Phytoremediation technology is still in its earlier developmental stages and full scale application is still limited. The results already obtained have indicated that the plants are effective and could be used in toxic metal remediation than microorganisms because plants possess some special features. Although it appears to be common sense among scientists, engineers and regulators about the more wide spread future of this technique; it is important that public awareness about this technology is considered and cleared and precise information is made available to general public to enhance its acceptability as global sustainable technology.

25.5.3 Phytoremediation of Zinc

Zn is potentially toxic metal when transferred from plants via food chain to human zinc shows the toxicity in medicinal and aromatic plants cultivation (Misra and Ramani 1991). The Zn efficient genotype *Mentha spicata* L. cv MSS-5 Arka and Neera genotype were grown in controlled glass house condition in CIMAP, Lucknow. Fe- efficient genotype showed more toxic level of Zn in plant than Arka and Neera genotype. Phyto-chelating and Zn accumulation were more pronounced in spearmint-Zn, by a variety of mechanism, such as reduced uptake of Fe, active efflux and intracellular or extracellular sequestration of Zn. Predominant molecules in intracellular detoxification are Glu-Cys peptide or the phytochelatins. These phytochelatins are activated by sublethal metal tolerance.

27.5.4 Uptake of Zn by AM inoculated plants

Darshini (2007) studied the uptake of Zn from potted plants of cowpea and maize inoculated with AM consortium. Arbuscular Mycorrhizal Fungi is known to enhance uptake of P, Al and Zn etc. When cowpea was planted in pots containing 500 and 1000 ppm the concentration of Zn was reduced to 133ppm and 366 ppm respectively. The reduction was148ppm and 500ppm in case of maize at these two concentrations. Thus cowpea a leguminous plant was more effective as a phytoremidient.

25.5.5 Metal tolerance of *Chloris barbata*

Studies on metal tolerance in plants are gaining importance owing to the possible utility of metal tolerant plants in phytoremediation and abatement of environmental metal pollution (Ernst 2000). Metal tolerance in plants very often has been shown to be a heritable trait (Mac Nair 1993, Tilstone *et al.* 1997). Cd tolerance in *Holcus lanatus* L. has, however, been shown to be due to physiological plasticity rather than to genetics (Barker *et al.* 1986). Antioxidant responses constitute one of the major biological responses to metal stress. Metal stress is also associated with triggering of synthesis of phytochelatins (PCs) or metallo thioneins (MTs) in plant cells. Among several metals, Cd has been shown to deplete glutathione (GSH) pool with concomitant accumulation PCs in plants. Metal tolerance of *Chloris barbata*. a grass growing in Ganjam district of Orissa was studied by Baisakhi *et al.* (2006). The population of the grass from the contaminated site, which showed increased tolerance to Cd and Hg was designated as tolerant population. The plants failed to show tolerance to Cu or Pb. This tolerant population of *C. barbata* is interpreted as a metal tolerant transient ecotype.

25.6 Biosorption

Biosorption is a low cause and environment friendly pollution abatement technology. The process involves removal of toxic metal ions and other toxic substances from dilute solutions by applications of biosorbent. It may be living or dead.

Biomass produced by bacteria, fungi or marine macro algae is used for biosorption. Certain other plants waste as Coconut husk and plant bark have been also used as biosorbent. The process also makes use of same biosorption again and again after slight treatment or gentle washing.

Table 1. Recommended tree species for industrial pollutants

Type of pollutants/ refractive sites	Recommended species	Planting techniques as developed by F.R.I., Kanpur, U.P.
Sewage effluents	*Eucalyptus tereticomis, Leucaene leucocephala, Casuarina equisitefolia, Acacia auriculiformis, Dalbergia sissoo*	Ridge and Furrow method; where plants are raised on ridges and sewage effluents is disposed in furrow for irrigation.
Tannery effluents	*Acacia nilotica, Terminalia arjuna, Cordia dichotoma, Cassia auriculata*	1-2 yer old nursery seedlings planted at 1 m x 1m spacing, after establishment stage irrigated with effluents@ 4 lit/ plant/ week.
Dairy effluents	*Acacia nilotica, Terminalia arjuna Casuarina equisitefolia, Syzygium cumini*	Two year old nursery seedling planted at 2m x 2m spacing. Seedlings irrigated first three months with ordinary water followed by treated effluents.
Distillery effluents	*Cassia siamea, Eucalyptus tereticornis* (Syn. *E. hybrida), Casuarina equisitefolia, Acacia nilotica, Dalbergia sissoo, Azadirachta indica*	1-2 yrs old nursery seedling planted at 2m x 2m spacing. Irrigation with ordinary water @ 20 lit. up to establishment stage followed by treated effluents @ 20 lit/ plant fortnightly during the rest period.
Textile effluents	*Eucalyptus camaldulensis, Acacia nilotica, Hardwickia pinnata, Prosopis juliflora.*	1-2 yrs old nursery seedling planted at 2m x 2m spacing. Irrigated with ordinary water up to establishment stage followed by food irrigation of textile effluents (fortnightly) during the rest of period.
Fly ash dumps/ ash ponds	*Acacia nilotica, Leucaene leucocephala, Pithecelobium dulce, Albizzia lebbek, Dalbergia sissoo*	Planting in pits (Size 45 x 45 x 45cms) at a spacing of 2.5 x 2.0 m. and treatment of 50% fertile soil F.Y.M. (2.0 kg) and DPA (150 g) per pit before planting. Frequent irrigation (weekly during summer months and fortnightly during winter months) up to first two years of establishment stage.

Data as presented by Srivastava *et al*. (2006)

25.6.1 Biosorption using *Ganoderma* and other macrofungi

Microorganisms alive or dead or their derivatives are used for sequestration of metals. A detailed study on the uptake of uranium and thorium on the *Rhizopus arrhizus* as a waste fungal biomass was carried out by Tsezos and Volesky (1982 a, b). They reported binding capacity of *R. arrihzus* was 20 times higher than that of commercially available ion exchange resins.

Ganoderma lucidum, a naturally occurring wood rotting fungi can effectively bind Cu (II) (Muralidharan *et al.* 1995). The mushrooms possess physical characteristic required for the use in continues flow reactors, metal binding potentials of these biomass have been investigated. Nine different non edible macrofungal species were screened for Cu (II) uptake. Batch equilibrium studied were conducted with each of these biomass, which followed a typical saturation profile. Langmuir relationship, which has been extensively used for the comparison of the metal uptake capacities of biosorbents was used in this study

Table 2 : Q max of different Fungi

Speices	Qmax (m mol g-1)
Coriolopsis strumosa	0.115
Daedalea tenuis	0.109
Lentinus strigosus	0.167
Lenzites malaccenis	0.143
Phelinus xeranticus	0.173
Rigidoporus lineatus	0.173
Rigidoporus microporus	0.104
Tremetes lactenia	0.048
Ganoderma lucidum	0.375
Fitrasorn - 400	0.030

$$Ce/qe = Ce/Qmax + 1/b\ Qmax$$

Where : qe and Qmax are Cu (II) sorbed at equilibrium concentration of Ce and maximum uptake capacity in m moles per unit weight of sorbent respectively, b is a constant

Results presented in table 2 clearly indicate that Qmax varied from 0.048 to 0.38 m mol g^{-1} sorbent. *G. lucidum* exhibited the maximum uptake potential which was higher than the capacity of few biosorbents reported in literature (Muzzarell *et al.* 1980; Tobin *et al.* 1984). All further studied were carried out with *G. lucidum*. Biosorbent can bind a wide range of metals. *Ganoderma* also showed its affinity with metals like Zn, Cd, Ni Cu, Cr, lanthanum and uranium, although there was variation in the uptake capacity. EPR, EDAX as well as proton/calcium release after Cu (II) binding indicated that mechanism of biosorption to be predominantly ion exchange/ complexation.

Presence of strong chelating agents significantly decreases Cu (II) binding to *G. lucidum*. Apart from Cu (II), *G. lucidum* could effectively bind many transition, lanthanide and actinide metal ions (Pandey *et al.* 2006).

25.6.2 Use of bark and other plant products in removal of heavy metals and dyes

Low cost sorbents of plant origin used are bark / tannin rich materials, lignin, peat moss, leaf mould, moss, modified wool and cotton (Sanghi and Bhattacharya 2006). Bark is effective because of its high tannin content. The polyhydroxy polyphenol groups of tannin are thought to be active treatment with acidified formaldehyde(Vazquez *et al.* 1994) that have been shown to eliminate the bleeching of colored compounds with appreciably affecting capacity. Morais *et al.* (1999) used bark of *Eucalyptus* as adsorbent. Randall *et al.* (1974) compared bark adsorption to that of peanut skis, walnut meal, and coconut husk. Orhan and Buyukgungov (1993) compared nut and walnut shell, waste tea and coffee to activated carbon and observed the capacities of the tannin containing products to be only slightly less than that of activated carbon. Another product of plant origin is sunflower stalks.

Another waste product from the timber/ agro industry is sawdust and rice husk. Bryant *et al.* (1992) showed adsorption of Cu and hexavalent chromium by red fir sawdust. Adsorption of Pb, Cd, Cu and Ni by sawdust is seen. Use of chemically treated withered tree leaves have been done for the removal of basic dyes malachite green, methylene blue, crystal violet and safranin from aqueous solution. Orange peel, root and shoot of water hyacinth plant were also found to be suitable adsorbents. Laszlo (1996) used bagasse as an adsorbent. The polyol structure of this material makes it strong chemical adsorption to cations such as metal ions and organic bases, as well as physical adsorption to other acidic and anionic compounds.

25.6.3 Effect of sludge in soil remediation

Remedial effect of industrial sewage sludge was seen in Lamarao do passé in Brazil. The sludge originated from petrochemical complex. In different plots the plants like *Brachiaria decumbens, B. humidicola* and *Canavalia ensiformis* and *Crotalaria* sp. were introduced. In different treatment plots 18 spp. of plants belonging to 7 botanical families were collected. Introduced species *Crotalaria* was collected in large amount (Lima *et al.* 2006). Uptake of Fe, Zn, Cu, and Mn from sludge was reported.

25.6.4 Bioremediation by microorganisms

Bioremediation is one of the technologies that involves almost exclusively microbial process. Bioremediation makes use of microorganisms like bacteria and fungi. The white rot fungi which utilize lignin as energy source possess the ability to degrade a wide spectrum of environmental pollutants using peroxidase enzymes. These microorganisms has catabolic characteristic under nutrient limiting condition.

Rafie and Helmy (2001) studied efficacy of 4 bacterial strains on degradation of crude oil obtained from Cairo refinery in Egypt. The bacteria (*E. coli*) NRR B3704 could only reduce the phenol content while fungi like *Phanerochaete crysosporium, Helminthosporium turcicum, Diplodia oryzae, Saccharomyces, Candida* were found effective to reduce the phenol content from crude oil. The use of *Candida* and *Diplodia oryzae* have been suggested by Amund and Igri (1990).

Yeast strains displayed the greatest reduction in concentration of normal alkalines in the order of C>E>B for crude oil samples. The n – alkalines degradation for oil in water emulsion is represented by strain D. It is interesting to note that the concentration of the more resistant and - alkalines (n – C $_{11}$- n - C$_{16}$) inoculated with C, E and B increased over the control concentration. This phenomenon may be attributed to the lose of longer chain alkalines to shorter alkalines by degradation and production of lower fragments and/ or low molecular weight oxidation products (Rafie and Helmy 2001).

25.6.5 Biosorption of metals using seaweed

Marine macro algae are harvested or cultivated in many parts of the world and are therefore readily available in large quantities for the development of highly effective biosorbents materials. Kumar and Kaladharan (2006) investigated the biosorptive potential of *Sargassum wightii* (Class Pheophyta) on Cadmium and Lead. Based on the results an experimental bio battery with sea weed as biosorbent was designed to treat a solution containing Cd, Pb, Cu, Zn. The biobattery consisted of simple cylindrical PVC bottles (10x 5 cm) each containing a bolting cloth bag filled 5 g ground *S. wightii* inside (Fig. 2). There were provisions at the bottom of the bottles for fitting aeration tubes, and with the help of a screw mechanism the bottles were temporarily attached to the aeration tubes. The bottom and topsides of the bottles was well perforated to effect continuous circulation of water through them. The efficiency of the biobattery was tested in a multimetal ion solution (to stimulate industrial effluents containing more than one metal) containing four metals, Zn, Cd, Pb and Cu, each having 100 mg/l concentration, at pH 4.5, 11 metal solution was taken in 10 l capacity long cylindrical glass container. Five bottles were used to treat metal solution, one after another, each operated for 30 min. After each 30 min, the bottles were changed manually by lifting up the aeration mechanism. Next 15 ml of the water samples was collected every 30 min after the operation of each bottle, filtered and preserved for analysis through anode stripping voltammetry.

In the biobattery experiment, the maximum metal reduction was observed in the case of Cadmium and Lead. The reduction in meal uptake observed during the successive battery operations could be attributed to lower metal/ biosorbent ration in solution. According to Fohrest and Roux (1992), Reduction in biomass concentration in the suspension at a given metal concentration enhance the metal biosorbent ratio, and thus increase the metal uptake/ unit biosorbent as long as the latter was not saturated. Thus reducing the biomass concentration with reduction in meal concentration can be suggested to improve metal uptake in final stages of bio battery operations. Bio battery can be considered as an eco friendly technique for treating the affluent.

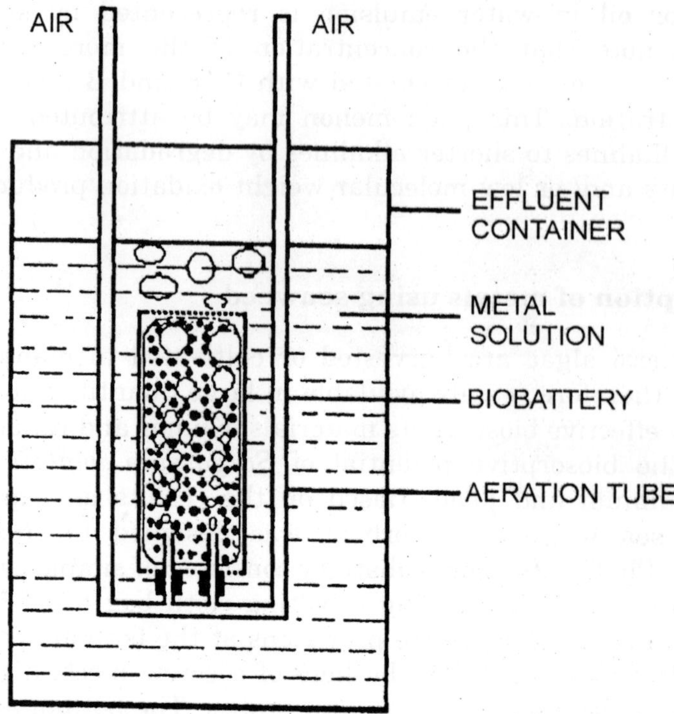

Fig. 2. Design of a biobattery

25.7 Bio-filtration vs. Bioremediation

Biofiltration implies physically separating particles based on their sizes. Bioremediation by contrast, implies a biochemical change as contaminants or pollutants are metabolized by micro- organisms and broken down into harmless, stable constituents, such as carbon dioxide, water, and salts disposed off. This adsorptive compost technology is more suitable for some types of VOC-containing products than is activated carbon, which is a poor adsorber of acetone. Vapor-phase biofilters using compost are gaining increasing attention as an alternative technology for treating aerosol cans. This growth is due, in part, to the high cost of conventional treatment and disposal methods, as well as to new regulations concerning VOC emissions from hazardous waste storage tanks and containers. Unlike conventional VOC control technologies, such as activated carbon, biofilters actually break down hazardous contaminants into harm-less products. They also offer low capital, life-cycle and operating costs—and require minimal maintenance and energy. The energy required to power a 100 cfm airflow unit, according to the manufacturer, is rated at 20 amps. At 8 cents per kilowatt hour, the cost of the requisite electricity is estimated at $1.80 per day. Additionally, according to one manufacturer, vapor-phase biofilters maintain a consistent VOC removal efficiency of 99.6 per cent, even when exposed to heavy or uneven surges of toxics.

25.7.1 Control of Composting Odors

Rockland County, New York, recently announced construction of a composting facility with America's first large, industrial-sized, odor- control bioremediation system. The enclosed 55,000-square-feet facility will be fitted with a compost filtration system that can process 82,000 cubic feet of air per minute. The air will be treated using ammonia scrubbers, then forced into an enclosure stacked with compost and other organic materials that function together as an air filtration system. The system binds odorous compounds, which the micro-organisms in the compost then degrade. This system has allowed the Rockland County Authority to obtain a contractual guarantee of no detectable odor at or beyond the site.

25.7.2 Biofilters in Municipal use

By converting its disposal operation from strictly land filling to one that utilizes a vapor-phase biofilter, the Metro Central Household Hazardous Waste collection facility in Portland, Oregon, saved nearly $47,000 in hazardous waste disposal costs over an18-month period. The facility used vapor-phase One application of biofiltration technology. It involves placing punctured cans, contaminated rags, or other items in the lower chamber of the biofilter. Next, the entire unit is heated to vaporize the contaminants. A small amount of air is then injected into the system to draw the now gaseous contaminants through two separate layers of a compost-rich biomatrix. The bulk of the contaminants are absorbed by the biomatrix at the first level, where most of the microbial activity takes place. The upper level serves as a surge control layer (to treat heavy or uneven surges of VOCs). Microorganisms living in the biomatrix metabolize the absorbed organics as food, converting the pollutants into carbon dioxide and water vapor.

References

Amund O.O. and Igiri C.O. 1990 Biodegradation of petroleum hydrocarbons under tropical estuarine conditions. *World J. of Microbiol. Biotechnol.* 6: 255-262.

Baisakhi B., Patra J., Vijayan V. and panda B.B. 2006 Metal tolerance of *Chloris barbata,* a grass obtained from chloralkali plant site in ganjam, orissa. In: *Plant response to environmental stress* (eds) Tripathi R.D, Kulshreshtha K., Agrawal M., Ahmad K.J., Varshney C,K, Krupa Sagar V. and Pushpanagadan P. Internation Book Distrubuting C., Lucknow. pp 189-202.

Baker, A., J.M., Grant, C.J., Martin M.H., Shaw S.C. and White Brook, J. 1986. induction and loss of cadmium tolerance in *Holcus lanatus* L. and other grasses. *New Phytol.* 102:575-587.

Bryant P.S., Petersen J.N., Lee, T.M and Brouns T.M. 1992. Sorption of heavy metals by untreated red fir sawdust. *Appl. Biochem. Biotechnol.* 32: 749.

Carson, B.L., Ellis, H.V. and Mc Cann, J.L. 1986. *Toxicology and biological monitoring of metals in humans.* Lewis Pub., Michigan pp 71-133

Darshini 2007 Uptake of Zn by cowpea and maize inoculated with AM fungi. M. Sc. Dissertation, The M.S. University of Baroda.

Ernst, W.H.O. 2000. Evolution of metal hyperaccumulation and phytoremediation hype. *New Phytol.* 146: 357-358.

Fourest E. and Roux J.C. 1992 Heavy metal biosorption by fungal mycelial by products: Mechanism and Influence of pH. *Appl. Microbiol. Biotechnol.* 37: 399-403

Kumar V.V and Kaladharan P. 2006 Biosorption of metals from contaminated water using sea weed. *Curr.Sci.* 90(9): 1263-1267.

Laszlo S.A. 1996.Preparing an ion exchange resin from sugarcane bagasse to remove reactive dye from waste water. *Textile Chemist and Colorist* 1: 13-17.

Lima J.S, Santo A.A., Gomes S.S., Aguiar A.C. and Salles P.A. 2006 Preliminary evaluation of the effects of sludge in the remediation of soils polluted by industrial atmospheric emissions. In: *Plant response to environmental stress* (eds) Tripathi R.D, Kulshreshtha K., Agrawal M., Ahmad K.J., Varshney C.K , Krupa Sagar V. and Pushpanagadan P. International Book Distributer C., Lucknow. pp 121-128.

Mac Nair, M. R. 1993. Transley Review No.49. The genetics of metal tolerance in vascular plants. *New Phytol.* 124: 541-559

Misra, A. and Ramani, S. 1991. Inhibition of iron absorption by Zn- induced Fe-deficiency in Japanese mint. *Act Physiol. Plant.* 13: 37-42.

Morais L.C.Freitas, O.M. Goncalves E.P. Vasconceles L.T. and Gonzalez Beca C.G. 1999. Reactive dye removal from waste water by adsorption Eucalyptus bark; variable that define the process. *Wat. Res*, 33: 979-988

Muralidharan T.R., Iyengar L. and Venkobachar C. 1995 Screening of tropical wood rooting mushrooms for copper biosorption. *Appl. Environ. Microbiol.* 61: 3507-3508.

Muzzarelli R.A. Tanfini F. And Scarpini G. 1980 Chelating film forming and coagulating ability of the Chitosan glucan complex from *Aspergillus niger. Industrial Wastes Biotechnol. Bioengg*. 22: 885-896

Orhan Y. and Buyukgungov H. 1993. the removal of heavy metals by using agricultural wastes. *Water Sci. Technol.* 28(2): 247-255.

Pandey A. Iyengar L. and Venkobachar C. 2006 Wood rotting fungi as biosorbent for toxic heavy metal removal. In: *Plant response to environmental stress* (eds) Tripathi R.D, Kulshreshtha K., Agrawal M., Ahmad K.J., Varshney C.K., Krupa Sagar V. and Pushpanagadan P. Internation Book Distrubuting C., Lucknow. pp 189-202.

Rafic S.E and Helmey S.M 2001 Bioremidiation of Egyptian crude oil using different microorganisms. *Res J. of Chem and Environ*. 5(2): 45-53

Randall J.M, Hautala E. and Waiss A.C. Jr. 1974. Removal and recycling of heavy metal ions from mining and industrial waste streams and with agricultural by-products. Proccedings of the fourth mineral waste utilization symposium, Chicago, TL, May 7-8, 1974, pp. 329-334.

Sanghi R and Bhattacharya B. 2006 A review of low cost adsorbents for the removal of heavy metals and dyes. In: *Plant response to environmental stress* (eds) Tripathi R.D, Kulshreshtha K., Agrawal M., Ahmad K.J., Varshney C.K., Krupa Sagar V. and Pushpanagadan P. Internation Book Distrubuting C., Lucknow. pp 137-143.

Srivastava R.J., Dubey P., Kumar A. and Prasad K. 2006 Bioremediation of urban pollution through green belt development: A case study In: *Plant response to environmental stress* (eds) Tripathi R.D, Kulshreshtha K., Agrawal M., Ahmad K.J., Varshney C.K., Krupa Sagar V. and Pushpanagadan P. Internation Book Distrubuting Co., Lucknow. pp 203-209.

Tilstone, G. H., Mac Nair, M. R. and Smith, S.E. 1997. Does copper tolerance give cadmium tolerance in *Mimulud gittatus*. *Heredity*. 79: 445-452.

Tobin J.M., Cooper D.G. and Neufeld R.J. 1984. uptake of metal ions by *Rhizopus arrihizus* Biomass. *Appl. Environ. Microbiol.* 47: 821-824.

Tsezos M., and Volesky B. 1982a. the mechanism of uranium biosorption by *Rhizopus arrhizus. Biotechnol. Bioengg* 26: 305-401.

Tsezos M., and Volesky B. 1982b. The mechanism of uranium biosorption by *Rhizopus arrhizus*. *Biotechnol. Bioengg* 25: 965-969.

Vazquez G., Antorrena G., Gonzalez J. and Doval M.D. 1994 Adsorption of heavy metal ions by chemically modified *Pinus piraster* bark. *Bioresource Technol*. 48: 251-255.

Volesky, B. 1990. Biosorption and biosorbents. In: *Biosorption of Heavy Metals*, CRC Press, Florida, pp 3-6.

CHAPTER - 26

Effect of Air Pollution on Breeding Performance of Birds

Geeta Padate

Division of Avian Biology, Department of Zoology, Faculty of Science, The M.S. University of Baroda, Vadodara -390 002

ABSTRACT

Birds are one of the most important indicators of health of an ecosystem which respond to secondary changes resulting from primary causes and can be monitored inexpensively. Their high mobility enables them to respond very rapidly to the changes in their surroundings. Phenomenon of bird migration has fascinated ornithologists for more than 150 years. Birds usually leave the area where they wintered to breed elsewhere to find safe nest site, and breeding season food. Birds are usually resident in order to maintain (through site dominance) a social advantage in contests for resources during either the breeding season or the non breeding season (but not necessarily both). Habitat of birds in polluted area may influence their breeding behaviour. The small insectivorous passerine birds are considered to be good indicators for heavy metal pollution because they are high in the food chain and are characterized by a high metabolic rate. Hence changes in the bird population in a habitat are often suspected to be related to environmental modifications that are many a time caused by man.

Direct and indirect influence of air pollution on egg formation and breeding success of adult birds include: decline in size of egg which may provide less nutrient to growing embryo and decline in thickness of shell which may lead to breakage of shell resulting in embryo death. Impaired growth can lead to deformed, weak, unhealthy young that may not survive to adulthood. Under the influence of pollution impaired calcium metabolism can result in fragile skeleton of mother as the Calcium is taken from the bone. Calcium is also required for various physiological processes in the body. The biochemical mechanisms leading to thin shell is described. Pollution and calcium metabolism in birds is discussed.

Keywords: Air pollution, Breeding, Birds, Calcium

Food and shelter are the two most important factors for the survival of an individual. These also influence the successful breeding of a population and continuation of a species. Majority of the birds depend on these two factors which show annual fluctuations in their availability, especially in the regions with profound seasonal variations. Hence most species breed at such time of the year when there is maximum possibility of the survival of the young and minimum stress on the parents (Immelmann, 1971). This generally coincides with maximum seasonal availability of the food as tremendous increase in the population of a species occurs, depending on the clutch size, that puts huge demand on food supply. In birds, the rapidly growing young as well as hyperactive parents which have to search and bring energy rich food to the fast growing young in the nest, need extra food supplies. When a young spends minimum time in the nest the possibility of exposure of helpless young to environmental hazards including predation decreases. Adequate the food supplies faster the growth of the young. In addition, being oviparous, the female parents have to accumulate large supplies of nutrients that can be incorporated in the egg yolk for developing embryo as well as deposition of egg shell for its protection. The altricial species, those in which young ones depend on parents for beak to beak feeding and all other needs, parents have to search and bring food to them by spending more energy. A faster growth needs good nutrition that in turn develops healthy body. A healthy body ensures faster escape to security from the environmental hazards including predation.

Birds are one of the most important indicators of health of an ecosystem which respond to secondary changes resulting from primary causes and can be monitored inexpensively (Koskimies, 1989). This is because of their high mobility that enables them to respond very rapidly to the changes in their surroundings (Hilden, 1965; Fuller *et al.*, 1995; Morrison, 1986; Louette *et al.*, 1995). The small insectivorous passerine birds are considered to be good indicators for heavy metal pollution because they are high in the food chain and are characterized by a high metabolic rate (Scheuhammer, 1988; Root, 1990). Hence, changes in the bird population in a habitat are often suspected to be related to environmental modifications that are many a time caused by man (Furness and Greenwood, 1993). Environmental modifications produce series of chained effects at various trophic levels that may lead to the local extinction of a species (Osenberg *et al.*, 1994).

26.1 POLLUTION AND BIRDS

With the evolution of man the evolution of new technologies started that gave comfort to modern man for his day to day needs. This was the beginning of pollution of air, land and water; leading to disappearing biodiversity with forests and the wildlife; depletion of soil cover resulting in floods as well as droughts and then the habitat loss. The men started moving to cities increasing crowding in the cities and leaving rural areas barren. As the cities expanded requirements of men and his technologies increased leading to the increase in number of vehicles, refrigerators, air conditioners and increase in pollution in the form of CO_2 and CFC (the chlorofluorocarbons), the green house gases. The atmospheric emission of harmful heavy metals has increased along with the development of industries, energy production and traffic, adding to this pollution (Boutron and Delmas, 1980; Rühling, 1994). Kendall and Scanlon (1982) have explained the high level of lead pollution with urban pollution sources like industry and traffic emissions. All these together lead to acid rains which are considered to be responsible for changing ground level habitat and the flora and fauna supported by it. This decline in flora and fauna results in shortage of food supply and shelter around pollution point source discharge. (Freitag and Hastings, 1973; Bengtsson and Rundgren, 1984; Heliövaara and Väisänen 1990). An increase in leaching of aluminum and other metals as a result of acidification caused by acid deposition of atmospheric emissions of sulfur and nitrogen has been reported (Van Straalen and Bergema, 1995).

Rachel Carson's *Silent spring* identified the urban use of pesticides as the cause of noticeable decline in bird's singing in U.S. and also the cause of mass songbird mortalities due to primary causes of reproductive impairment that affects embryo and adults. Growing young are most sensitive to the detrimental effects of pollutants (Scheuhammer, 1987). These include reduced hatchability of eggs, failure of chicks to thrive (Wasting syndrome), Teratological effects that produce skeletal abnormalities, impaired deformities of the reproductive and also of nervous systems through mechanisms of hormonal mimicking of estrogens. The range of chemical effects on adult birds cover acute mortality, sub lethal stress, reduced fertility, suppression of egg formation, egg shell thinning and impaired incubation and chick rearing behaviour. The types of pollutants that produce these effects include organochlorine pesticides, industrial pollutants, organophosphates, petroleum hydrocarbon, heavy metals, herbicides, fungicides, *etc*. The substances like DDT, polychlorinated biphenyls (PCBs) and mixtures of organochlorines have been identified as environmental estrogens affecting populations of Gulls breeding in polluted "hot spots" in southern California, the Great Lakes, and Puget Sound. Estrogenic organochlorines represent an important class of toxicants to birds because differentiation of the avian reproductive system is estrogen dependent. -- **Environ Health Perspect** (1995)

26.2 AIR POLLUTION

Air pollution can influence reproduction in two ways. By directly influencing the physiological processes as well as genetic make up of the parents resulting in deformed calcium metabolism and deformed young and indirectly by modifying the habitat. In the later case, the habitat modification may result due to soil and plant runoff, in reduced number of nesting sites, decrease in the available nesting material or influencing the food supply. During nestling stage the young ones depend on protein rich food sources like insects and fishes while parents depend on additional food supply like molluscs rich in calcium.

The sense of olfaction (smell) is poorly developed in songbirds, the birds which occupy the major terrestrial ecosystems. These species mainly feed on invertebrates like insects and molluscs. These birds may also be affected in a polluted habitat in two ways. First, under the direct influence of air pollution they may leave the area, but if not, by the indirect pronounced influences of the pollution on the breeding activities resulting decline in their population.

The air pollution results into acid rains that influence the pH of soil. This change results in the decline in soil fauna which in turn results in short supply of these food items for higher taxa (Scheuhammer, 1991). Heavy metals accumulate in ground living, mobile, often adult prey base like ants, cockroaches and beetles than in foliage living less mobile often larval prey like caterpillars, spiders and aphids (Hunter and Johnson, 1982; Bengtsson and Rundgren, 1984; Grue *et al.*, 1986). A decline in availability of herbivorous larvae in trees close to the pollution source has been reported by several workers (Eeva *et al.*, 1997, Heliövaara and Väisänen, 1993; Koricheva and Haukioja, 1992; Ruohomäki *et al.*, 1996). Hence food limitation has been suggested to be a major factor responsible for decrease in bird population in polluted areas (Morrison, 1986; Graveland, 1990; Furness and Greenwood, 1993). House martins have been reported to avoid breeding in areas with high pollution possibly because of an association with reduction in insect food (Newman *et al.*, 1989).

Social dominance is important in some species in which dominant individuals occupy areas with better quality and force subordinate individual to breed in areas of lower quality such as areas affected by air pollution (Lundberg *et al.*,1981, Alatalo *et al.*, 1985). The other possibility is, the birds may be of same status at the beginning of the breeding but their condition deteriorates under pollution stress either directly or indirectly during the course of breeding. The healthy clutch in a polluted area does not indicate healthy population. The lower local survival rates of a species in polluted area may include higher rate of emigration from the low quality and unproductive habitat rather than by direct mortality of adult birds due to heavy metal poisoning. Emigration may occur when healthy population is breeding outside the polluted area (Eeva and Lehikoinen, 1998). Padate and Sapna (2002) have reported good diversity of birds in an ecofarm developed around an oil refinery in Central Gujarat.

26.3 CALCIUM NEEDS AND ITS MOBILIZATION IN BIRDS

The female birds have single ovary in which at a time only one egg with large amount of yolk is produced. Poultry birds lay one egg at a time at an interval of about 26 hours during light hours. This delays egg laying daily by about 2 hours and hence hens take a break per week. Whereas, wild species have their own schedules either with only one egg per day or laying at a gap of a day to few days. The clutch (the number of eggs laid in a brood) varies depending on the growth period, size of bird, type of food given, *etc*. In short, depending upon the ability to take care the clutch size is fixed and some species may lay an extra egg in anticipation of healthy food supply. If conditions are good it may survive till fledging other wise it is not fed and it dies.

The yolk is synthesized in the liver and transported via blood to the developing ovum. Three layers of blood vessels develop around ovum which helps in the deposition of yolk. Once the ovum is ovulated it has to pass through the oviduct where other layers like albumin, egg membranes and finally calcareous shell are deposited. Hence, in female birds the demand of calcium and other nutrients increases before egg laying starts. (Sturkie, 1986)

26.3.1 From where the Calcium is acquired ?

There are mainly two sources of calcium.

(i) From food via intestine

(ii) Mobilization from the bone (medullary bone).

i. *Calcium from diet*: Hens consume approximately 25% more feed on days when egg shell formation occurs than days when it does not (Johnson, 1986). If the concentration of calcium in the feed is 3.5% or higher, most of egg shell calcium is derived directly from the intestine (Hurwitz and Bar, 1969). If less than 1.95% then 30 to 40% calcium is obtained from bone and if the diet is free of calcium the skeleton is the principal source (Muller *et al.*, 1969). However, much of the shell is formed at night when no calcium is consumed and the calcium in the digestive tract is declining. During such time bone is the major source especially in the wild birds which have to spend time and energy in search of calcium rich food. During reproductive activities of birds there is increase in the intestinal absorption of calcium with increase in the Vit. D (Kenny, 1986). Vit. D is required for absorption of Calcium from intestine and retention of the same from the kidney. In laying hens the total blood calcium rises by 20-30% which helps in deposition of calcareous shell. During final 15 hours of shell formation 100-150 mg/h calcium is secreted by the shell gland. This can lead to depletion of calcium in the body within 8 – 18 minutes if calcium is not replaced via diet (Sturkie, 1986).

ii. Medullary Bone (How bone is involved?) : An increase in the weight of skeleton occurs about 10 days before a hen reaches sexual maturity. A special bone called medullary bone develops in the marrow cavity of female birds in tibia/fibula. As it develops in the marrow it is called medullary bone (Simkiss, 1967). Medullary bone grows from endothelial surface of the marrow cortex and can fill the entire marrow cavity. Simultaneously large blood supply is developed in bone which can facilitate mobilization of bone mineral during shell formation as well as its subsequent replacements. Degree of vascularisation is important in medullary bone formation (Taylor and Moore, 1954). This bone develops under synergistic action of Estrogen as well as androgens. This has been reported in Pigeons which lay a clutch of only 2 eggs at the interval of 48 hours. The shell formation is accompanied by remarkable increase in number of osteoclast, the bone resorpting cells in birds. Presence of osteoclasts indicates bone resorption. The osteoblasts – bone forming cells, dominate when egg shell is not formed. (Bloom *et al.*, 1941). The shell gland transports 2 to 2.5 g. of Calcium within 15 hours for calcification in hen which lays an egg weighing about 50 g. If a hen is laying 280 eggs per year she needs 30 times more calcium then present in her body. Poultry birds and even aviary birds (in zoos or as pets) are given extra calcium in their diet. A small stone like porous piece present with the food in caged birds is the shell of cuttle fish (*Sepia*) which is made-up of calcium carbonate and supplies additional needs of calcium to the captive laying females. Though the phenomenon of metabolism is same in the domestic hen and females of wild species of birds, the situation is different. A poultry hen produces about 260 eggs/year and receives additional calcium supplies while a wild species lays only as many eggs as she can take care of and has to spend lots of energy in search of the calcium rich food.

26.3.2 Pollution and calcium metabolism

When supply of calcium is affected, the formation of eggshell is affected, resulting in to a fragile eggshell. A fragile egg may not bear the weight of the incubating parents, may provide less calcium to the growing young and do not provide proper protection to the growing embryo (Eeva and Lehikoinen, 1995). If the breeding attempt is not successful a wild species of bird may have to delay its breeding activities for the next favourable season which can be a year later. One unsuccessful year for a species can result in to decline in the population of that species.

Direct and indirect influence of air pollution on egg formation and breeding success of adult birds include: decline in size of egg which may provide less nutrient to growing embryo and decline in thickness of shell which may lead to breakage of shell resulting in embryo death. Impaired growth can lead to deformed, weak, unhealthy young that may not survive to adulthood. Under the influence of pollution impaired calcium metabolism can result in fragile skeleton of mother as the Calcium is taken from the bone. Calcium is also required for various physiological processes in the body. The biochemical mechanisms leading to thin shell is species specific and depends on environmental conditions. Possible mechanisms for which there is experimental evidence are reduced availability of carbonate ions and premature termination of shell growth. Copper from industrial source affects thyroid and parathyroid glands which control the mobilization of calcium from medullary bone (Cooke, 1973). In urban areas birds may utilize Ca-rich items such as chicken egg shells. Calcium rich food items are absent from the most polluted soils. (Graveland *et al.*, 1994). In a study around factory emitting heavy metal pollution in air near Harjavalta in Finland, Jussila and Jormalainen (1991) reported low amount of exchangeable calcium in the humus layer and higher level of calcium in pine needle near factory complex emitting Calcium rich fertilizer dust. The difference in the calcium gradient in soil and pine needles has been reported reflect in the difference in calcium concentra-tions in the faeces of nestling with difference in foraging behaviour, *i.e.* in the species that forage in foliage during the nestling period high calcium content are found in faeces (Gibbs, 1954) then the species feeding on the flying and ground dwelling insects (Silverin and Anderson, 1984).

26.4 AIR POLLUTION AND BIRD PLUMAGE

To breed successfully in short favourable period, birds, especially the males of the species need to announce their readiness for breeding and to attract the females (Welty and Baptista, 1988). One of the methods for these is the display of the colourful plumage. Development and/or maintenance of this plumage are under hormonal control. Bright plumage signals good nutritional condition. (Hill, 1990; Palokanges *et al.*, 1994). Under heavy metal pollution, metals like aluminium, copper, zinc, cadmium and lead are deposited in the feathers of birds (Solonen, *et al.*, 1999) that fades the plumage of feathers (Eeva *et al.*, 1998). Further, metal levels in feathers reflect the level in food during the feather growth, *i.e.* during growth of young birds or during the moult period of fully grown birds. (Westermark *et al.*, 1975; Solonen and Lodenias, 1990). It has also been expected that the accum-ulation of metals increases with high trophic level of the species (Furness *et al.*, 1986; Esselink *et al.*, 1995).

In general, birds cannot synthesize the necessary carotenoid pigments themselves and obtain them from the diet (Brush, 1978). The access to the carotenoid sources may be affected by the air pollutants by reducing the amount of carotenoid– rich food items *i.e.* insects. This anthropogenically limited access to the carotenoid sources can affect the fitness of individual bird via mate selection or social dominance (Lundberg *et al.*,1981; Alatalo *et al*., 1985). Brighter the plumage healthier the pair formation and better success in breeding. In addition to mate choice, plumage colour can also influence male-male competition, social dominance in winter flocks or the ability to avoid predation via crypticism (Eeva *et al.,* 1998). Further, the deficiency of vitamin A (retinoids) is associated with a diversity of anomalies, including reproductive impairment, embryonic mortality, growth retardation and bone deformities (Peakall, 1992). Animals are unable to synthesize retinoids that are needed by the body hènce get them from the food. One molecule of carotenoid is synthesized from two molecules of retinoids. Insects, especially the caterpillars are good source of carotenoid for growing and moulting birds. A decline in the number of caterpillars around the source of air pollution (*i.e.* from a copper smelter) has been reported by several workers (Heliövaara and Väisänen, 1988; Ruohomäki *et al*., 1996; Eeva *et al*., 1997) and a smaller proportion of green larvae in diet of nestlings in the vicinity of the pollution source has also been reported correlating plumage colour intensity with the abundance of green herbivore larvae (Eeva *et al*., 1997) under air pollution gradient. The pollutants can also directly affect carotenoid metabolism and/or transportation of and block the movement of carotenoid between ingestion and its eventual deposition in the feathers (Hill *et al*., 1994).

Paler the plumage, lower the dominance of male and poorer the chances of attracting a healthy female (Eeva *et al*., 1998). This can result in failure in breeding or production of weak progeny finally influencing population of a species. In a study of feathers of pigeon Scheuhammer (1987) reported highest metal levels in urban environment. This may be due to external contamination too. The nestlings of urban birds of prey which were frequently fed with feral pigeon by their parents are also reported to have heavy load of metal pollution. Though there is a great possibility of biomagnifications of the pollutants in the birds of Prey, an alternate hypothesis suggests that such a biomagnifications is non existent (Laskowski, 1991). Hence pollution; may that be air, soil or water; can influence breeding performance and survival of birds in several ways resulting in the decline in their population which in severe cases can lead to extinction of a species from the area.

REFERENCES

Alatalo, R.V., Lundberg, A. and Ulfstrand, S. 1985. Habitat selection in the Pied Flycatchers (*Ficedula hypoleuca*). In: *Habitat selection in Birds*. Cody, M.L.(ed). Academic Press, London 59-83.

Bengtsson, G. and Rundgren, S. 1984. Ground living invertebrates in metal-polluted forest soil. *Ambio*, 13:29-33.

Bloom, M.A., Bloom, W. and McLean, F.C. 1941. The role of androgens in the production of medullary bone in pigeons by the administration of sex hormone. *Am. J. Physiol.*, 133: 216.

Boutron, C. and Delmas, R. 1980. Historical record of global atmospheric pollution revealed in polar ice sheets. *Ambio,* 9 (5):210-215.

Brush, A.H. 1978. Avian pigmentation. *Chemical Zoology* (Brush A.H. ed.). Academic Press, New York. Pp 141-164.

Cooke, A.S. 1973. Shell thinning in avian eggs by environmental pollutants. *Environ. Pollut.*, **4**: 85-152.

Eeva, T. and Lehikoinen, E. 1995. Egg shell quality, clutch size and hatching success of Great Tit (*Parus major*) and the Pied Flycatcher (*Ficedula hypoleuca*) in an air pollution gradient. *Oecologia,* 102: 312-323.

Eeva, T., Lehikoinen, E., 1998. Local survival rates of the Pied Flycatchers (*Ficedula hypoleuca*) and the Great Tits (*Parus major*) in an Air pollution gradient. *Ecoscience,* **5**(1) : 46-50.

Eeva, T., Lehikoinen, E. and Pohjalainen, T. 1997. Pollution related variation in food supply and breeding success in two hole-nesting passerines. *Ecology,* **78**:1120-1131.

Eeva, T., Lehikoinen, E. and Rönkä, M. 1998. Air pollution fades the plumage of the Great Tit. *Functional Ecology,* 12 : 607 -612.

Esselink, H., van der Geld, F.M. Jager, L.P., Posthuma-Trumpie, G.A. Zoum, P.E.F. and baars, A.J. 1995. Biomonitoring heavy metals using the Barn Owl (Teto *alba guttata*): Sources of variation especially relating to body condition *Arch Environ. Contam. Toxicol.* 28: 471-486.

Freitag, R. and Hastings, L. 1973. Kraft mill fallout and ground beetle populations. *Atmospheric Environment* 7:587-588.

Fuller, R.J., Gough, S.J. and Merchant, J.H. 1995. Bird population in new lowland woods: Landscape, Design and Management perspectives. In: *The ecology of woodland creation* (Ferris-Kaan, R., ed.). John Wiley and Sons Ltd. London.

Furness, R.W. and Greenwood, J.D.1993. *Birds as Monitors of Environmental Change*. Chapman and Hall. London.

Furness, R.W., Muirhead, S.J. and Woodburn, M. 1986. Using bird feathers to measure mercury in the environment: relationships between mercury content and molt. *Mar. Pollut. Bull.*, 17:27-30.

Gibbs, J. 1954. Feeding ecology of Tits, notes on Tree creeper and Goldcrest. *Ibis*, 96: 513-543.

Graveland, J. 1990. Effects of acid precipitation on reproduction in birds. *Experimentia*, 46: 962-970.

Graveland, J., Wal, R. van der, Balen, J. H. van, Noordwijk, A.J. van. 1994. Poor reproduction in forest passerine from decline of snail abundance on acidified soils. *Nature*, 368 : 446-448.

Grue, C.E., Hoffman, D.J., Beyer, W.M. and Franson, L.P. 1986. Lead concentration and reproductive success in European Starling *Sturnus vulgaris* nesting within highway roadside verges. *Environ. Pollu. Ser.,* 42:157-182.

Heliövaara, K. and Väisänen, R. 1988. Interactions among herbivores in three polluted pine stands. *Silva Fennica*, 22:283-292.

Heliövaara, K. and Väisänen, R. 1990. Concentration of heavy metals in the food, faeces, adults and empty cooons of *Neodiprion sertifer* (Hymenoptera : Diprionidae). *Bull. Environ. Contam. Toxicol.,* 45 : 13-18.

Heliövaara, K. and Väisänen, R. 1993. *Insects and pollution*. CRC press. Boca, Raton; Florida, USA.

Hilden, O. 1965. Habitat selection in birds: A Review. *Annales Zoologici Fennici.,* 2:52-75.

Hill, G.E. 1990. Female house finches prefer colourful males: Sexual selection for a condition dependent trait. *Anim. Behavior,* 40: 563-572.

Hill, G.E. Montgomerie, R., Inouye, C.Y. and Dale, J. 1994. Influence of dietary carotenoids on plasma and plumage colour in the house finch. Proc. *Roy. Soc. Lond. B. Biol. Sci.*, 258:47-52.

Hunter, B.A. and Johnson M.S. 1982. Food chain relationships of copper and cadmium in contaminated grassland ecosystems. *Oikos*, 38:108-117.

Hurwitz, S. and Bar, A. 1969. Intestinal calcium absorption in the laying fowl and its importance in calcium homeostasis. *Am. J. Clin. Nutr.*, 22: 391.

Immelmann, K. 1971. Ecological aspects of periodical aspects of reproduction. In: "*Avian Biology*". Vol. I. Farner, D.S. and King, J.R. eds. Academic Press, New York, London.

Johnson, A.L. 1986. Reproduction in the female. In: *Avian Physiology*. Sturkie, P.D.(ed.) Springer-Verlag, New York, Berlin, Heidelberg, Tokyo.

Jussila, I. and Jormalainen, V. 1991. Spreading of heavy metals and some other air pollutants at Pori-Harjavatta district in South western Finland *SYKE publications B*. 4:1-58.

Kendall, R.J. and Scanlon, P.F. 1982. Tissue lead concentrations and blood characteristics of Rock Doves from an urban setting in Virginia. *Arch. Environ. Contam. Toxicol.* 11:265-268.

Kenny, A.D. 1986. Parathyroid and ultimobranchial glands. In: *Avian Physiology*. Sturkie, P.D.(ed) Springer-Verlag. New York Berlin Heidelbergs Tokyo.

Koskimies, P. 1989. Birds as a tool in environmental monitoring. *Annales Zoologici Fennici*, 26: 153-166.

Laskowski, R. 1991. Are the top carnivores endangered by heavy metal biomagnifications? *Oikos*, **60**:387-390.

Louette, M., Bijnens, L., Upoki Agenong, A.D. and Fotso, R.C. 1995. The utility of birds as bioindications: Case studies in Equatorial Africa *Belgain. J. Zool.*, 125:157-165.

Lundberg, A., Alatalo, R.V., Carlson, A. and Ulfstrand, S. 1981. Biometry, Habitat Distribution and breeding success in Pied Flycatcher (*Ficedula hypoleuca*) – *Ornis Scand.*, 12: 68-79.

Morrison, M.L. 1986. Bird populations as indicators of environmental change. *Current Ornithology*, **3**: 429-451.

Muller, W.J. Brubaker, R.L. and Caplan, M.D. 1969. Egg shell formation and bone resorption in laying hens. *Fed. Proc. Fed. Am. Soc. Exp. Biol.*, **28**: 1851.

Newman, J.R.E., Novakova, E. and Mc Clave, J.T., 1989. The use of House Martin (*Delichon urbica*) as a bioindicator of air pollution. In: *Proceedings of 5th International Conference, Bioindicators Deterioresationis Regionis*. Bohae, J. and Ruzicka, V. (eds.) Institute of Landscape Ecology. CAS, Ceske Budejovice, Czech Republic.

Osenberg, C.W., Schmitt, R.J., Holbrook, K.E., Abu-Saba and Flegal A.R. 1994. Detection of environmental impacts: Natural Variability, effect size, and power analysis. *Ecological Application*, 4:16-30.

Padate, G.S. and Sapna S. 2002. Avifauna amidst Industrial Air Pollution. In: *Air Pollution Development at What cost.* Eds. Jasrai, Y.T. and Arun Arya. Daya Publishing House, New Delhi.

Palokangas, P., Korpimaki, E., Hakkarainem. H., Huhta, E., Tolonen, P. and Alatalo. R.V. 1994. Female Kestrels gain reproductive success by choosing brightly ornamental males. *Anim. Behaviour,* 47**:** 443-448.

Peakall, D. 1992. *Animal markers as pollution indicators*. Chapman and Hall. London.

Root, M. 1990. Biological monitiors of pollution. *Bioscience,* 40: 83-86.

Ruhling, A. (ed.) 1994. Atmospheric heavy metal deposition in Europe – Estimation based on moss analysis. *Nord.,* **9**: 1-53.

Ruohomäki, K., Kaitaniemi, P., Kozlov, M., Tammaru, T. and Haukioja, E. 1996. Density and Performance of *Epirrita aufumnata* (Lepidoptera: Geometridae) along three air pollution gradients in northern Europe. *J. App. Ecol.*, **33**: 773-785.

Scheuhammer, A.M. 1987. The Chronic dietary toxicity of Aluminium, Cadmium, Mercury and Lead in birds. A review. *Environ. Pollut.,* 46: 263-295.

Scheuhammer A.M. 1988. Chronic dietary toxicity of methylmercury in the Zebra finch, *Poephila guttata*. Bull. *Environ. Contam. Toxicol.***,** 40: 123-130.

Scheuhammer, A.M. 1991. Effects of acidification on the availability of toxic metals and calcium to wild birds and mammals. *Environ. Pollut.*, 71:329-375.

Silverin, B. and Anderson, G. 1984. Food Composition of Adult and Neothing Pied Flycatchers (*Ficedula hypoleuca*) during the breeding period . Vår Fågel Värld. 43:517-524.

Simkiss, K. 1967. Calcium in reproductive physiology. New York, Reinhold.

Sturkie, P.D. 1986. *Avian Physiology.* 4th edition. Springer-Verlag, New York, Berlin, Heidelberg, Tokyo.

Solonen, T. and Lodenius, M. 1990. Feathers of birds of prey as indicators of mercury contamination in Southern Finland. *Holarct. Eol.,* 13:229-237.

Solonen, T., Lodenius, M. and Tulisalo, E. 1999. Metal level of feathers in birds of various food chains in Southern Finland. *Ornis Fennica*, 76:25-32.

Taylor, T. G. and Moore, J. H. 1954. Skeletal depletion in hens laying on a low calcium diet. *Br. J. Nutr.*, 8: 112.

Van Staarlen, N.M. and Bergema, W.F. 1995. Ecological risk of increased bioavailability of metals under soil acidification. *Paedobiologia*, 39: 1-9.

Welty, J. C. and Baaptista, L. 1988. The life of Birds. 4th edition. Saunders Collage Publishing, New York, Chicago, San Francisco, Philadelphia, Montreal, Toronto, London, Sydney, Tokyo.

Westemark, T., Odsjö, T. and Johnels, A.G. 1975. Mercury content of bird feathers before and after Swedish ban on alkyl mercury in agriculture. *Ambio,* 4:87-92.

CHAPTER - 27

Corrosion and Environment: Studies with Zinc

Rakesh T. Jasrai

Department of Chemistry, Shri. R. K. Parikh Arts and Science College, Petlad – 388450 Gujarat

ABSTRACT

The corrosion of metals may be due to environmental factors or biological organisms. The oxidation or rusting of iron takes place in presence of moisture. Scientists have tried to study corrosion of metals and have correlated the chemical changes with biological effects. The corrosion product increase with the concentration of acid. The paper describes corrosion of metals in dry and wet condition. The zinc is not a corrosion resistant metal, but it is utilized as a sacrificial metal. Zinc plays an important part as protective coating over steel. The protective properties of various compounds are described.

Keywords: Corrosion, Environment, Zinc, Metal.

Corrosion is the destructive attack on a metal by chemical or electrochemical reaction with its environment. (Uhlig 1946) Corrosion has been defined in many ways, some of which are (Evan 1960)

1. Destruction of metal by chemical or electrochemical reaction with its environment.
2. Eating away of metals.
3. Destruction of materials by means of other than straight mechanical methods.
4. Extractive metallurgy in reverse.

Every year great losses have been occurring due to corrosion. Corrosion is a serious problem because it directly and definitely contributes to the depletion of our national resources.

Most environments are corrosive to greater of lesser extent *e.g.* air and moisture, fresh distilled or saline water, rural urban and industrial atmosphere, steam and other gases, mineral acids, organic acids, alkalis, soils, solvents, vegetable and petroleum oils and a variety of food products. In general inorganic materials are more corrosive than organic material.

Most metallic components eventually deteriorate in use, due to the chemical nature of process applications. Since corrosion is impractical to eliminate, the effective step is to control it rather than to preventing it. All metals and alloys are susceptible to corrosion to a greater or lesser extent. No single material is suitable for all applications. For effective control of corrosion we must have a thorough understanding of its nature and method of measurement (Ailor 1971).

Corrosion occurs in widely different forms. Classification of these forms is usually based on one of these factors.

Nature of the corrodent - Corrosion can be classified as 'wet' and 'dry'. A liquid or moisture is necessary for the former and 'dry' corrosion usually involves reaction with high temperature gases.

Appearance of the corroded metal – Corrosion is either uniform, or metal corrodes at the same rate over its entire surface or it is localized, in which case, only small areas are affected.

Later one is further classified into macroscopic or microscopic. The macroscopic form initiate at a structural defect but grow by corroding 'good' material, whereas in microscopic form considerable damage can occur before the problem becomes visible to the naked eye.

Most metals exist in their natural state as a compounds (oxide, sulfide etc.). This is their thermodynamically stable state. We dig these compounds out of the ground, extract the metals and then convert them into numerous useful articles. Unless we have substantially changed the metal (*e.g.* by alloying) the thermodynamic properties will not change, when we expose such a metal article to a natural environment. The importance of thermodynamics in the above example extends to corrosion in all media. Corrosion is possible only when the corrosion products are thermodynamically more stable than the reactants. All the thermodynamically possible reactions do not take place.

In all cases of electrochemical corrosion, the anodic reaction consist of oxidation of solid metal to its ions in the solution –

$$M \rightarrow M^{n+} + ne^{-}$$

With cathodic reactions, there are a number of possibilities (Fontana 2005).

1. Hydrogen evolution in acid solution

$$2H^+ + 2e^- \rightarrow H_2$$

2. Oxygen reduction in acid solution

$$O_2 + 4H^+ + 4e^- \rightarrow 2H_2O$$

3. Oxygen reduction in neutral and basic solutions

$$O_2 + 2H_2O + 4e^- \rightarrow 4OH^-$$

4. Metal-ion reduction

$$M^{+2} + e^- \rightarrow M^{+2}$$

5. Metal deposition

$$M^{+3} + e^- \rightarrow M^{+2}.$$

Hydrogen evolution and oxygen reduction are the most common cathodic reactions. Different cathodic reactions can occur simultaneously on a corroding metal surface. According to the law of thermodynamics if a reaction is to proceed, it must have a negative free energy change (ΔG). For electrochemical reactions,

$$\Delta G = -nFE \tag{1}$$

Thus for a reaction to be feasible, the cell potential E must be positive. When the metal is dipped in the solution of its ions, the existed potential difference E can be measured with the help of voltmeter by connecting it to an inert reference electrode, placed in the solution. Such copotential is called a 'half cell potential' because it is the potential of an electrode that can form half cell (that is the anode of a corrosion cell). Similarly a half cell potential exists for the cathode reaction.

By convention, the hydrogen reaction $2H+2e^- \rightarrow H_2$ is assigned as standard 'half cell potential' of zero volts, all other half cell potential are measured relative to the potential of hydrogen reaction. A list of such half cell potential is given in a table.

Standard Oxidation		– Reduction Potentials
Au^{+3}/Au	:	+1.498
$O_2+4H^++4e^- \rightarrow H_2O$	:	+1.229
$Ag+/Ag$	:	+0.779
Fe^{+3}/Fe^{+2}	:	+0.771
$O_2+2H_2+4e^- \rightarrow 4OH^-$	:	+0.401
Cu^{+2}/Cu	:	+0.337
$2H_2+2e \rightarrow H_2$	:	+0.000

Ni^{+2}/Ni	:	0.250
Fe^{+2}/Fe	:	0.440
Cr^{+3}/Cr	:	0.744
Zn^{+2}/Zn	:	0.763
Al^{+3}/Al	:	1.662
Na^{+}/Na	:	2.714

The cell potential E that is used in the free energy change in equation (1) is defined as the potential difference between the cathode and anode half cell. For the example of iron in acid solution.

$$E = E\ (H^+/H_e) - E\ (Fe^{+2}/Fe) = 0\text{-}(-0.440) = +\ 0.440\ V.$$

since E is positive, Δ G in equation (1) is negative and the reaction is feasible.

Sometimes the same piece of metal behaves differently due to the differences, which exist at interface on the metal surface itself. Regarding the characteristics of the metal the differences may be chemical or metallurgical in nature.

27.1 Limitation of thermodyanamic approach

When a metal M is placed in contact with a solution, it will be unstable as long as the process proceeds spontaneously

$$M \rightarrow M^{z+} + Ze^- \qquad (2)$$

If the solution does not contain species (Where the evolutions are retained by the metal), which are capable of accepting the electrons from the metal, the above reaction will soon stop as the potential difference set up between the metal and solution, opposes the further transfer of metal ions into the solution. If, however, species are available in the solution to accept the electrons, as for example

$$H^+ + e^- \rightarrow \tfrac{1}{2}\ H_2 \qquad (3)$$

$$H^+ + \tfrac{1}{4}\ O_2 + e^- \rightarrow \tfrac{1}{2}\ H_20 \qquad (4)$$

reaction can occur, it will depolarize the metal and the reaction (2) can continue to take place. Thus the metal will be unstable if one of the overall processes

$$M + ZH^+ \rightarrow M^{+z} + \tfrac{1}{2}\ ZH_2 \qquad (5)$$

$$M + ZH^+ + \tfrac{1}{4}\ ZO_2 \rightarrow M^{+Z} + \tfrac{1}{2}\ ZH_20 \qquad (6)$$

proceeds spontaneously, *i.e.* if it proceeds with a negative change of free energy, the above reaction (5) and (6) are the overall reactions for the cells.

M/M^{+z} / H^+ / H_2 (7)

M/M^{+z} / H^+, O_2 (g) / H_2O (8)

When we consider reactions on the basis of thermodynamics, the rates of the processes are neglected. Thus in actual practice, tin is found not to corrode in solution of pH 7 in contact with air. The depolarization reaction (5) and (6) are extremely slow on tin.

Let us see the thermodynamic feasibility of a certain corrosion reaction, if the electrical resistance in the solution between the anode and cathode were infinite, then the potentials of the anode and cathode would remain at their half cell level. In this condition the current would be zero (Ohm's law, I = E/R) and no corrosion would occur. If the resistance were zero, the current would be infinite and we would have instant and total corrosion. In practice, the resistance approaches zero, which results in maximum finite current.

When current flows between the anode and cathode, the potentials of both the anode and cathode changes, this change is called polarization. Polarization reduces the potential difference between anode and cathode areas as shown in Fig. 1.

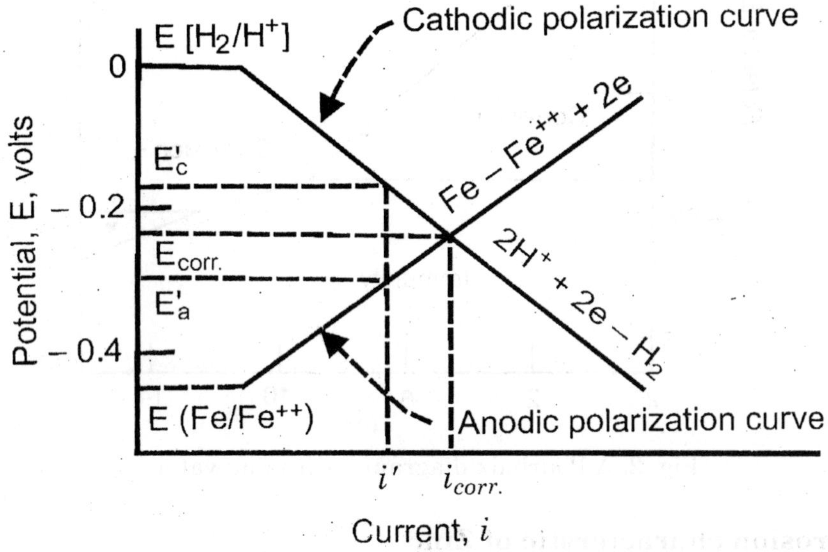

Fig. 1

The current flowing between the anode and cathode at any point is given by

$$i' = (E'c - E'a)/R$$

Once a metal has come to equilibrium in its environment and is corroding freely, it will have a corrosion potential E_{corr} and a corrosion current density between anode and cathode areas of i_{corr}

27.2 pH as a factor in corrosion

Theories of corrosion based only on the e.m.f. series, take no effect of films which may form on metals under a variety of conditions particularly in which H^+ and OH^- terms are also involved. On the basis of necessity to consider other equilibria, a very useful form of presentation of both the potential dependent and the pH dependent equilibria has been developed by Pourbaix (Evans, 1970).

A Pourbaix diagram relates the possibility of corrosion to the pH of corrodent and the potential difference between the pure metal and its ions in solution. It tells us under what pH-potential conditions we might get corrosion and where we will not get it. This diagram gives us clues as to how corrosion could be minimized or avoided.

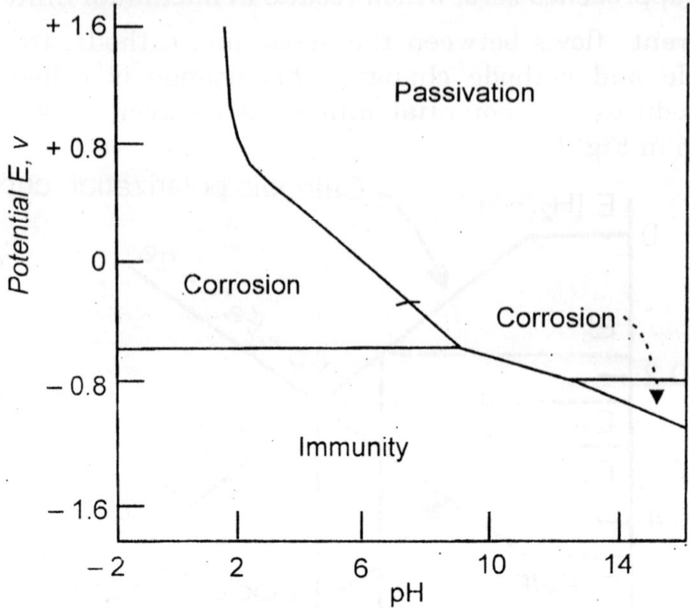

Fig. 2. A Pourbaix diagram for iron in water.

27.3 Corrosion characterstic of Zinc

Zinc is not a corrosion-resistant metal, but it is utilized as a sacrificial metal (Fontana, 2005). Zinc metal pays an important part as protective coating for steel. Both zinc and zinc alloys have excellent resistance to corrosion in the atmosphere and in most natural waters (Shreir, 1976). The Property, which gives zinc, this valuable corrosion resistance is its ability to form a protective layer consisting of zinc oxide and hydroxide, or various basic salts, depending on the nature of the environment. When the protective layers have formed and completely cover the surface of the metal, the corrosion proceeds at a greatly reduced rate. Zinc is considerable more electronegative than iron and a number of other metals with the exception of magnesium. The equilibrium potential of zinc is – 0.76 Volt and its corrosion potential in strong chloride solution is about – 0.88 Volt. Because of this and other good qualities, zinc is used extensively as a coating material for steel and some times for alloys of aluminum. Zinc coatings are cheap and more protective than other coatings and are preferred to nickel, tin, copper and lead, which tend to accelerate attack on exposed steel. Zinc and its alloys are also extensively used as galvanic anodes. The corrosion resistance of zinc is mainly determined by its insoluble corrosion products produced in contact with the corrosive environment.

The normal impurities in zinc are lead, cadmium and iron (Sunder and Boyd 1971). Zinc dissolves in liquids whose pH are below 5 and above 12.5. It has been found that some common acids attack zinc, while only the stronger solutions of most of the alkalis would cause any corrosion. Owing to the high over potential for hydrogen evolution, pure zinc is attacked only very slowly by dilute sulfuric acid and hydrochloric acid but impure zinc dissolves quite readily, evolving hydrogen. Addition of cadmium and lead reduce the solubility of technical zinc in acids because they increase the hydrogen over voltage. On the other hand, addition of iron, nickel and copper in zinc because of their low hydrogen overlarge increase corrosion of zinc in acids.

27.4 Corrosion inhibitors

The use of corrosion inhibitors has tremendously increased in the last fifty years. In view of the great ranges of substances that can act as corrosion inhibitors, it is not surprising that no single mechanism has yet been suggested to explain satisfactorily the mode of action of all inhibitors.

Putilova *et al.* (1960) classified corrosion inhibitors into the following two groups; type "A" inhibitors, which attach to the metal surface and type "B" inhibitors, which act by deactivating the corrosive environments. Inhibitors of type "A" are classified to be of three types: (I) Passivators, (ii) Immunizers and (iii) Adsorption inhibitors. Passivators are usually inorganic oxidizing substances such as chromates and molybdates, which passivate the metal and shift the corrosion potential in the noble direction. They form protective layer on the metal surface either of oxide or metallic salt. Immunizers do not so much reduce the corrosion rate as they only delay the onset of corrosion, after which the metal corrodes, as in uninhibited in solution. Adsorption inhibitors are used virtually in acidic solutions where the cathodic control is the discharge of hydrogen. Uhlig (1946) has also distinguished between passivators and adsorption inhibitors. Inhibitors of type "B" include sulphites, which deactivate the solution by combination with dissolved oxygen and therefore it retards the cathodic reaction.

27.5 Objectives of the study

Studies on metallic corrosion and its inhibition have been concentrated more on ferrous metal as compared to nonferrous metals. Detailed studies are of an academic character for zinc, which are essential in order to obtain the necessary background information, are lacking in literature for many systems. Apart from the practical importance of corrosion inhibition data in the field of corrosion prevention, it is are also useful in the interpretation of the mechanism of the corrosion inhibition.

Zinc is one of the most important nonferrous metal, which finds extensive use in metallic coatings. It is found that zinc has a good resistance to atmospheric attack, but it corrodes rapidly in polluted air, as well as in different acids. The characteristic instability of zinc in acids is clearly due to its high hydrogen voltage. The corrosion products of zinc are readily soluble both in alkalis and acids and protect the metal surface only in neutral media. Technically pure zinc energetically dissolves in dilute mineral acids. Addition of cadmium, lead or mercury reduces the solubility of zinc in acids because they increase the hydrogen over voltage on the inclusions, which acts as cathodes on the other hand, addition of iron, nickel, copper and noble metals because of their low hydrogen over voltages considerably increases corrosion of zinc. An attempt has been made to study the retardation of corrosion of zinc in hydrochloric acid by means of diverse organic substances.

Detailed investigation on the inhibition of the corrosion of zinc in hydrochloric acid, solution were undertaken with a view to study the effect of variables like inhibitor concentration, time, temperature on the rate of corrosion. The optimum conditions for satisfactory retardation of corrosion have been established.

27.6 Materials and Methods

Gravimetric method has been used to study the corrosion of zinc in hydrochloric acid. Rectangular pieces of zinc of fixed size 3 cm x 3 cm (thickness 28 S.W.G.) with a small hole of about 2 mm diameter just near the one end of the specimen for suspension was used. A specimen of the metal was first washed with distilled water for several times and dried. It was then polished using successively '0' to '0000' oakey emery papers. The final polishing was done using jwellers rough polishing, which gave a mirror like finish. The specimen was finally degreased by immersion in chloroform (AR grade). Pyrex glass hooks were used to suspend the specimens. Each specimen was suspended in 0.5N HCl at the same depth of 1.5 cm below the surface of the solution. The volume of solution for all the experiments was 230 ml. Only one specimen was suspended in each container, which was a beaker (Pyrex make) of 250 ml capacity. Experiments were performed in triplicate. In long duration tests the level of the corrosive liquid was maintained by periodic addition of distilled water. Experiments were conducted at 35 + 0.5° to 55 + 0.5°C. For the satisfactory assessment of corrosion, it is necessary to remove corrosion products from the specimen, at the same time it is essential to avoid further attack on metal during cleaning process. It was observed that for this purpose 30% nitric acid was satisfactory.

All the chemicals used in the investigation were of AR quality. Distilled water were used throughout the work. All the volumetric apparatuses like burette, pipette etc. was of standard quality and calibrated before use. The weights were taken on mettler balance, M5 type.

27.7 RESULTS AND DISCUSSION

In the course of present investigation influence of parameters like inhibitor concentration, temperature and time on the performance of inhibitors have been studied (Jasrai, 1995) The mechanism of inhibitor action have been investigated by gravimetric method. The inhibitors tested in 0.5N hydrochloric acid included vanillin, phenylthiourea and N-allyl thiourea appear to be potential candidates as inhibitor. At 35°C vanillin seems to be the best, as with the passage of time (Tables 1-4), the efficiency is around 95%. Phenyl-thiourea is in general better than N-allyl thiourea as it gives quite good inhibition in initial period, however, its efficiency falls off considerably with time (30 minutes) as compared to that of vanillin. The effect of temperature on the efficiency of inhibitors is quite interesting. Phenyl and N-allyl thiourea were satisfactory as inhibitors for 0.5N hydrochloric acid solution at 35°C, however phenylthiourea and N-allyl thiourea do not withstand the rise in temperature and at 55°C (Table 4) N-allyl thiourea does not at all function as a satisfactory inhibitor. Phenyl-thiourea also fails as an inhibitor at 55°C and this is quite so during 30 minutes time interval. Vanillin is quite efficient at 55°C during 15 minutes time interval but its efficiency drops significantly for 30 minutes duration.

Table-1 Effect of temperature on corrosion loss (mg/dm^2) of Zn in 0.5 N hydrochloric acid (values in the parentheses shows % efficiency). Duration: 5 minutes

Inhibitor & Inhibitor concentration	Weight loss in mg/dm^2 of Zn at different temperatures				
	35±0.5°C.	40±0.5°C.	45±0.5°C.	50±0.5°C.	55±0.5°C.
Nil	395	411	426	452	496
1.0% Vanillin	19 (95)	25 (94)	19 (93)	31 (93)	31 (91)
0.2% Phenyl thiourea	22 (94)	43 (90)	63 (85)	106 (77)	156 (69)
0.5% N-Allyl thiourea	56 (86)	146 (64)	241 (43)	358 (21)	465 (6)

Table 2. Effect of temperature on corrosion loss (mg/dm^2) of Zn in 0.5 N hydrochloric acid (values in the parentheses shows % efficiency). Duration: 5 minutes

Inhibitor & Inhibitor concentration	Weight loss in mg/dm^2 of Zn at different temperature				
	35±0.5°C.	40±0.5°C.	45±0.5°C.	50±0.5°C.	55±.5°C.
Nil	1087	1295	1451	1786	2257
1.0% Vanillin	26 (97)	56 (96)	61 (96)	106 (94)	154 (93)
0.2% Phenyl thiourea	165 (94)	125 (90)	181 (88)	329 (82)	595 (74)
0.5% N-Allyl thiourea	301	(72) 460 (64)	642 (56)	955 (47)	1258 (44)

Table 3. Effect of temperature on corrosion loss (mg/dm^2) of Zn in 0.5 N hydrochloric acid (values in the parentheses shows % efficiency). Duration: 5 minutes

Inhibitor & Inhibitor concentration	Weight loss in mg/dm^2 of Zn at different temperature				
	35±0.5°C.	40±0.5°C.	45±0.5°C.	50±0.5°C.	55±.5°C.
Nil	2160	2453	2688	3054	3687
1.0% Vanillin	81 (96)	99 (96)	111 (96)	294 (90)	332 (91)
0.2% Phenyl thiourea	111 (95)	236 (90)	382 (86)	758 (75)	1187 (68)
0.5% N-Allyl thiourea	481 (78)	722 (71)	1120 (58)	1187 (61)	2208 (40)

Table 4. Effect of temperature on corrosion loss (mg/dm^2) of Zn in 0.5 N hydrochloric acid (values in the parentheses shows % efficiency). Duration: 5 minutes

Inhibitor & Inhibitor concentration	Weight loss in mg/dm^2 of Zn at different temperature				
	35±0.5°C.	40±0.5°C.	45±0.5°C.	50±0.5°C.	55±0.5°C.
Nil	6046	6473	7174	8070	9126
1.0% Vanillin	151 (98)	201 (97)	408 (94)	1129 (86)	2864 (69)
0.2% Phenyl thiourea	1080 (82)	2301 (64)	3293 (54)	4717 (42)	156 (32)
0.5% N-Allyl thiourea	1831 (70)	2734 (57)	3684 (48)	4790 (41)	5983 (34)

Scientists have studied the inhibition of acid corrosion of zinc by some thiourea derivatives. The corrosion inhibition was measured using hydrogen evolution weight loss techniques. The results showed that the thiourea derivatives efficiently resist corrosion in hydrochloric acid solution.

Shanthamma and Venkatesha (2003) reported that the some aldehyde and amine group derivatives give good protection to zinc in hydrochloric acid solution. The dissolution behavior of zinc in acidic and nearly neutral medium is known to be inhibited by the nitrogen, sulfur and oxygen containing organic compounds. Such compounds contain electron donating group that decrease the corrosion rate by increasing the hydrogen over voltage on the corroding metal. Most organic inhibitors contain at least one polar group with an atom of nitrogen, sulfur and oxygen and each of them is considered a chemisorption center. The protective properties of such compounds depend on the electron densities around the chemisorption center. The higher the electron density at the center the more effective the inhibitor (Lisac *et al.*, 1994 and Salghi *et al.*, 2000).

27.8 Conclusion

The present studied inhibitors provide good corrosion inhibition for zinc in aqueous hydrochloride medium. The inhibition efficiency of the tested inhibitors follows the order

Vanillin> Phenyl-thiourea> N-allyl thiourea

These compounds are soluble in water and nontoxic, hence can be used as effective corrosion inhibitors for zinc in acidic medium.

References

Ailor, W.H. 1971. *Handbook on Corrosion Testing and Evaluation*, John Wiley and Sons, Inc., U.S.A.

Evan, U.R. 1960. *The Corrosion and Oxidation of metals*. Edward Arnold and Co.

Evan, U.R. 1970 Metallic Corrosion, Passivity and Protection.

Fontana, M.G. 2005 *Corrosion Engineering*. Tata McGraw-Hill Publishing Company Limited, Delhi.

Jasrai, R.T. 1995 "Corrosion Inhibitors" M. Phil Dissertation, Gujarat University, Ahmedabad.

Lisac, E.S., Podbrseck S and Soric T. 1994. *J. App. Elecrochem.*, 24: 779.

Putilova, I.N., Barannik V.P and Baiezin, S.A 1960. *Metallic Corrosion Inhibition* Pergmon Press, New York, R.

Salghi, Hamouti, B., Aounti, A., Berrabah M and Kertit, S. 2000. *J. Elecrochem., Soc., India*, 49(1): 40.

Shanthamma K.R. and Venkatesha T.V. 2003. *Turk. J. Chem.* 27:189-196.

Shreir, L.L. 1976. *Corrosion, Volume I*, Newness-Butterworth's.

Sunder, C.J. and Boyd W.K. 1971 *Zinc, its corrosion resistance*, Zinc Development Association.

Uhlig, H. H. 1946. *Chem. Eng. News*, 24:3154.

Uhlig, H.H. 1964. *Corrosion and Corrosion Control*, John Wiley and Sons, Inc., New York, London.

CHAPTER - 28

Air Pollution Terminology

V. S. Patel

President IAAPC, Visiting Professor,
The Maharaja Sayajirao University of Baroda, Vadodara 390 002
E mail: vpatel_usa@hotmail.com

ABSTRACT

The last few decades have witnessed a rapid industrialization resulting into severe air pollution problems. Industries operate within societies and their impact on air quality is well known. After Bhopal Gas Tragedy in 1983, 2 people died and 500 injured due to Oleum gas leak in Shri Ram Food and fertilizers in Delhi, in December 1985, 5000 injured, in Ahmedabad in 1987, 6people died and 100 injured in Ganjam district Orissa in October 1989 due to chlorine gas leak. In Dahanu, Maharashtra 100 people died and 19 injured due to chemical explosion in 1991. Similarly there are several examples which tell about chemical disasters. The vehicular pollution is yet another major cause of air pollution due to Supreme court intervention the CNG is introduced in Delhi. The efforts are on to find out alternative fuels with low pollution emission. The use of biodiesel obtained from varied sources is one, the use of hydrogen or solar energy is other alternative. Besides seed oil obtained from *Jatropha* and *Pongamia* extraction from Petrocrops may be other options. There is a rapid increase in awareness about growing hazards to our planet caused by enhanced pollution of biosphere. To control environmental pollution and lead a happy healthy life it is essential to know the terminology used in the subject. The chapter brings some of the common terms used and their explanations.

Key words: Air Pollution, Terminology

Air Pollution: Worsening of air quality resulting from unwanted chemicals or other pollutants occurring in the air.

Air Pollutants: High amounts of contaminants in the air that have harmful effects to humans, animals, vegetation, and/or materials.

Air Quality Index (AQI): A scale developed by the government to measure how much air pollution is in the air. The AQI measures five criteria pollutants – ozone, particulate matter, carbon monoxide, sulfur dioxide, and nitrogen dioxide. AQI levels range from 0 (Good air quality) to 500 (Hazardous air quality). The higher the index, the higher the level of pollutants and the greater the likelihood of health effects. The AQI also has another category – unhealthy for sensitive groups – that ranges from 101 to 150. The AQI is usually used in weather reports. When it is high, there are warnings for people not to do things like sports or hard work outside. People with asthma or other lung problems are urged to stay inside on days that the AQI is high.

Air Quality Management District (AQMD): A group of counties or parts of counties that regulate stationary, indirect, and area sources of air pollution within a region. It contains a regional air pollution control board that mostly includes elected officials from within the region.

Air Toxics: A general term referring any pollutant that is in the air and has the potential to produce harmful health effects is an air toxic.

Asthma: A chronic inflammatory ailment of the lungs characterized by wheezing, breathlessness, chest tightness, and cough.

Carbon Dioxide (CO_2): A colorless, odorless gas resulting from the incomplete combustion of fuels. CO interferes with the blood's ability to carry oxygen to the body's tissues and results in numerous harmful health effects. Over 80% of the Co emitted in urban areas is contributed by motor vehicles. CO is a criteria air pollutant.

Carcinogen: A cancer-causing substance.

Criteria Air Pollutant: Six common air pollutants that are regulated under the state and federal Clean Air Acts. They include ozone, lead, nitrogen oxide, sulfur dioxide, carbon monoxide, and particulate matter.

dB : A measure to know the intensity of sound, Decibel. The unit is named after Alexander Graham Bell.Human breathing will produce 10 dB sound, train whistle 110 db and space rocket launching 170-180 dB. A person can withstand sound of 80 dB.

dB meter: An instrument to measure noise pollution.

Emissions: Substances released into the air, usually by human activities such as burning fossil fuels.

Emission Offsets (*also known as Emissions Trading*): A regulation that allows for a new or modified stationary source of air pollution only if there is a reduction of emissions from another already existing stationary source or sources of air pollution.

Environmental Justice: The fair treatment of people of all races and incomes with respect to development, implementation, and enforcement or environmental laws, regulations, and policies. Fair treatment implies that no person or group of people should shoulder an unfair share of negative environmental, health and economic impacts results from environmental programs.

Federal Clean Air Act (FCAA):A federal law passed in 1970 and amended in 1974, 1977 and 1990, which forms the basis for the national air pollution regulation. Basic elements of the act include national ambient air quality standards for major air pollutants, mobile and stationary control measures, air toxics standards, acid rain control measures, and enforcement provisions.

Fossil Fuels: Fuels such as coal, oil, and natural gas; so-called because they are the remains of ancient plant and animal life.

Global Warming: An increase in the temperature of the Earth's troposphere. Global warming is the term most often used to refer to the warming as a result of increased emissions of greenhouse gases.

Greenhouse Effect: The warming effect of the Earth's atmosphere. Light energy from the sun, which passes through the Earth's atmosphere, is absorbed by the Earth's surface and re-radiated into the atmosphere as heat energy. The heat energy is then trapped by the atmosphere, creating a situation similar to that which occurs in a car with its windows rolled up. The release of carbon dioxide and other gases into the atmosphere may increase the greenhouse effect and contribute to global warming.

Greenhouse Gases: Atmospheric gases such as carbon dioxide, methane, chlorofluorocarbons, nitrous oxide, ozone and water vapour that slow the passage of re-radiated heat through the Earth's atmosphere.

Hazardous Air Pollutant (HAP): An air pollutant listed under section 112 (b) of the Federal Clean Air Act as very hazardous to health. U.S. EPA identifies emission sources of hazardous air pollutants, and emission standards are set accordingly.

Haze (Hazy): A phenomenon that results in reduced visibility due to the scattering of light caused by aerosols. Haze is caused in large part by human-made air pollutants.

Indirect Source: Any facility, building, structure, or installation, that causes increased mobile source pollution, such as car traffic. Examples of indirect sources include employment sites, shopping centers, sports facilities, housing developments, airports, commercial and industrial development, and parking lots and garages.

Indoor Air Pollution: Air pollutants that occur inside buildings or other enclosed spaces; some examples of indoor air pollutants are nitrogen oxides, tobacco smoke, asbestos, formaldehyde, carbon monoxide.

Indoor Bio-pollution: different bio-particles like fungal spores, pollen grains cause damage to organic objects present in rooms. Bio-pollutants can also cause asthma, allergy, corneal infections and skin diseases in children and human beings.

Major Source: A stationary facility that emits a regulated pollutant in an amount exceeding the threshold level depending on the location of the facility and attainment with regard to air quality status.

National Ambient Air Quality Standards (NAAQS): Standards established by the United States EPA that apply for outdoor air throughout the country. There are two types of NAAQS. Primary standards set limits to protect public health and secondary standards set limits to protect public welfare.

National Sources: Non-human made emission sources, including biological and geological sources, wildfires, and windblown dust.

New Source Performance Standards (NSPS): Uniform national EPA are emission standards that limit the amount of pollution allowed from new sources or from modified existing sources.

New Source Review (NSR): A Clean Air Act requirement that State Implementation Plans must include a permit review, which applies to the attainment areas, to ensure acceptable levels of national ambient air quality standards. The two major requirements of NSR are Best Available Control Technology and Emission Offsets.

No attainment Area: A geographic area identified by the U.S. EPA and/or CARB as not meeting either NAAQS or CAAQS standards for a given pollutant.

Office of Environmental Health Hazard Assessment (OEHHA): A department within the California Environmental Protection Agency that is responsible for rating chemicals for harmful health impacts and establishing safe exposure levels. OEHHA also assists in performing health risk assessments ad developing risk assessment procedures for air quality management purposes.

Ozone: A harmful form of oxygen that is produced when sunlight creates a reaction between nitrogen oxide and hydrocarbons. Net to the earth's surface, ozone is a major component of smog. Ozone exists in the upper atmosphere ozone layer (stratospheric ozone) as well as at the Earth's surface in the troposphere (ozone). Ozone in the troposphere cases numerous harmful health effects and is a criteria air pollutant.

Ozone Depletion: The reduction in the stratospheric ozone layer. Stratospheric ozone shields the Earth from ultraviolet radiation. The breakdown of certain chlorine and/or bromine-containing compounds that destroy ozone molecules in the stratosphere can cause a reduction in the ozone layer.

Ozone Precursors: Chemicals such as non-methane hydrocarbons and oxides of nitrogen, occurring either naturally or as a result of human activities, which contribute to the formation of ozone, a major component of smog.

Particulate Matter (PM): Any solid material, except pure water, that is suspended or carried in the air. The size of particulate matter can vary from coarse, wind-blown dust particles to fine particle combustion products.

Permit: Written authorization from a government agency (e.g., an air quality management district) that allows for the construction and/or operation of emissions generating facility or its equipment within certain specified limits.

PM2.5: Includes tiny particles with an aerodynamic diameters less than or equal to a nominal 2.5 microns. This fraction of particulate matter penetrates most deeply into the lungs.

PM10 (Particulate Matter): A criteria air pollutant consisting of small particles less than 1/7 the width of a single human hair. Their small size allows them to make their way to the air sacs deep within the lungs where they cause harmful health effects. PM10 also causes visibility reduction.

Pollution Prevention: The use of materials, processes, or practices to reduce, minimize, or eliminate the creation of pollutants or wastes.

Public Workshop: A workshop held by a public agency for the purpose of informing the public and getting its input on the development of a regulatory action, or a development that may add to the air pollution levels.

Regional Haze: The haze produced by many different sources and activities that release fine particles across a large area.

Smog: A combination of smoke and other particulates, ozone, hydrocarbons, nitrogen oxides, and other chemically reactive compounds. Under certain weather conditions they may produce a murky brown haze that causes harmful health effects. The primary source of smog in California is motor vehicles.

Smoke: A form of air pollution made up mostly of particulate matter. Other components of smoke include gaseous air pollutants such as hydrocarbons, oxides of nitrogen, and carbon monoxide.

Soot: Very fine carbon particles that have a black appearance when released into the air.

Source: Any place or object from which air pollutants are released. Sources that the fixed in space are stationary sources and sources that move are mobile sources.

State Implementation Plan (SIP): Non-mobile source such as power plants, refineries, and manufacturing facilities, which release air pollutants.

Sulfur Dioxide (SO2): A strong smelling, colorless gas that is formed by the burning of fossil fuels. Power plants, which may use coal or oil high in sulphur content, can be major sources of SO2. SO2 and other sulphur oxides contribute to the problem of acid deposition. SO2 is a criteria air pollutant.

Toxic Air Contaminant (TAC): An air pollutant that may cause or contribute to an increase in deaths or in serious illness, or it may pose a present or potential hazard to human health.

Toxic Hot Spot: A location where emissions from specific sources may expose people to higher risks of harmful health effects – including but not only cancer.

United States Environmental Protection Agency (U.S. EPA): The federal agency charged with setting policy and guidelines, and carrying out legal mandates for the protection of national interests in environmental resources.

Variance: Permission granted for a limited time (under stated conditions) for a person or company to operate outside the air pollution limits stated in a regulation.

Visibility: The ability to see and identify objects at different distances. Visibility reduction from air pollution is often due to the presence of sulphur and nitrogen oxides, as well as particulate mater.

Volatile Organic Compounds (VOCs): Carbon-containing compounds that evaporate into the air. VOCs contribute to the making of smog and may be toxic. VOCs often have an odor, and some examples include gasoline, alcohol, and the solvents used in paints.

CHAPTER - 28

Facts: It makes you think

1. One tone of recycled steel save the energy equivalent of 3.6 barrels of oil and 1.49 tones of iron ore over the production of new steel.
2. Only 1% of the earth's water is available for drinking. 2% is currently frozen.
3. Humans require about 2.8 liters of water a day.
1. Preventing 1000 kg of paper waste will ultimately save between 16 and 18 mature trees.
5. About 44% of junk mail is never opened or read.
6. The term EL Nino literally means "the little boy". The term was originally used by fishermen along the coast of Ecuador and Peru to indicate a period of warm ocean water that occurred around Christmas, lasting for a few months thereafter.
2. Cooking with wood, dung and coal in traditional stoves each year kill 1.5 million people, who breathe in the poisonous fumes.
8. Swedish scientists have found that pine and spruce produce 1000-2000 particles of monoterpenes per cubic centimeter of air which help to cool the planet by bouncing energy from the sun back into space.
9. Americans throw away enough aluminum every three months to rebuild their entire commercial air fleet, and enough office paper each year to build a 4 meter high wall of paper from New York to Los Angeles (about 4000 km.).
10. Fiber optics, made of glass, will eventually replace copper cables throughout the world.

3. Home owners use up to ten times more toxic chemicals per acre than farmers.
4. Energy saved from one recycled aluminum can will operate a TV set for 3 hours.
5. Recycling a glass bottle saves enough energy to light a 100 watt bulb for 4 hours.
6. The world ocean with a coastline of 312,000 km and a volume of 137 x 106 km^3 is the largest ecosystem on earth.
7. A fungus that colonizes the plant, lives within the tissues, for all or part of there life cycle and cause no apparent infections is called in endophytic fungus. *Pestalotiopsis microspora* produced anticancer drug Taxol in Yew tree *Taxus wallichiana*.
8. The average 5- minutes shower takes between 56-94 liters of water.
9. Every ton of recycled paper saves 380 gallons of oil .
10. A PC left on overnight uses the same energy as laser printing 800 sheets of A4 paper.
11. Glass produced form recycled glass instead of raw materials reduces related air pollution by 20% and water pollution by 50%.
12. Honey contains only slight traces of residues from industrial emissions, car exhaust and agricultural chemicals because bees act as a biological filter : They die if they come in contact with toxins and thus do not bring pollutants into the hive.
13. Lichenometry is a technique used to find relative or absolute date of rock surface exposure by the use of lichens made up of algae and fungi.
14. Paper and card make up approximately one third of the waste produced by Britain's homes.
15. On average, every family in the UK consumes around 500 glass bottles and jars a year.
16. If all the aluminum cans in the UK were recycled there would be 12 million fewer dustbins every year.
17. It takes approximately 17 trees to make one tonne of paper.
18. Aluminium cans have a value worth 6 – 20 times that of any other used packaging material.
19. Five two – litres recycled PET bottles produce enough filling for ski jacket.
20. Producing glass from raw materials requires 30 % more energy than producing it from crushed, used glass.

21. It takes 39,000 gallons of water to manufacture a new car, including tyres.
22. "A nation that destroys its soils destroys itself. Forests are the lungs of our world, purifying the air and giving fresh strength to our people." Attributed to Franklin Roosevelt.
23. Recycling steel and tin cans saves 60-75% of the energy used to produce them from raw materials.
24. Enough sunlight falls on the Earth's surface every hour to meet world energy demand for an entire year.
25. The UK target is to deliver 10% of its electricity form renewable sources by 2010 and 15% by 2020.
26. The energy needed to make batteries is 50 times than the energy they give out.
27. Water is the most commonly used renewable energy resource, providing enough power to meet the needs of 28.3 million consumers.
28. Silicon from just one tonn of sand, used in photovoltaic cells, could produce as much as electricity as burning 500,000 tonnes of coal.

REFERENCES

Jones D. 2007 Factoids Did you know. TPI News 55: 14-15.

SCIENTOON

OZONE LAYER

Ozone layer is thinning over Australia, Russia, America, Canada, Europe and major part of Latin America.

UV radiation coming to earth due to ozone layer damage, may cause skin cancer, eye cataract and reduces our immunity.

"Sir! He says he is ready to go even to hell, but says he will never go on space mission to earth."

SCIENTOON

1 Crore 50 lakh acres of land is converted into **DESERT** every year in the world.
Main reason for this, is deforestation and global warming. It has caused 10 crore people to become homeless so far and also a very big damage to **BIODIVERSITY**.

I still very well remember that we used to go in our car from India gate to Rashtrapati Bhawan and the place around was full with greenery"

SCIENTOON

In our nations capital Delhi, more than **700 people die every month due to heavy air pollution**. Even children are suffering from various respiratory problems including Asthma. Culprit perhaps are around 30 lakhs vehicles which are running in Delhi, everyday.

"The first photo belongs to my great grand father who died at the age of 95. Second is dadaji who died due to plague at 85 and third one is of my father who left this world at the age of 35 only. Why? Because he lived in Delhi".

वृक्ष : हमारी अनमोल नियामत

पीढ़ियों दर पीढ़ियों से
हमारी अमानत रहे हैं ये पेड़
वृक्ष हमें भोजन ही नहीं देते
ये वातावरण के सजग प्रहरी भी है।

मुस्कराते, खिलखिलाते
ये हमें जीवन की राह दिखाते हैं,
ये आपसी प्यार को तो बाँटते ही हैं।

लेकिन हमारी तरह
इनमें ईर्ष्या, द्वेष नहीं होते
इनमें जाति, धर्म नहीं और
न इनमें हैं भाषा के झगड़े
आपसी राजनैतिक दाँवपेंच
इनके खून में नहीं
गर्मी सर्दी हर मौसम में
प्रसन्नता से खडे रहने वाले
इन वृक्षों से
काश हम सीख पातें
बाधाओं से जूझना
प्रदूषण हटाना
मिलजुल कर रहना और
सदा एक–दूसरे के काम आना।